GRAPHS,
DYNAMIC PROGRAMMING,
AND
FINITE GAMES

This is Volume 36 in
MATHEMATICS IN SCIENCE AND ENGINEERING
A series of monographs and textbooks
Edited by RICHARD BELLMAN, *University of Southern California*

The complete listing of books in this series is available from the Publisher
upon request.

Originally published in the French language under the title
"Méthodes et Modèles de la Recherche Opérationnelle,"
Tome II in 1964 by Dunod, Paris.

GRAPHS,
DYNAMIC PROGRAMMING,
AND
FINITE GAMES

by A. Kaufmann

Professor at the E.N.S. des Mines de Paris
and at the Polytechnic Institute of Grenoble
Scientific Adviser to the Compagnie des Machines Bull

Translated by Henry C. Sneyd

1967

ACADEMIC PRESS　　　　New York　San Francisco　London
A Subsidiary of Harcourt Brace Jovanovich, Publishers

ACADEMIC PRESS, INC.
111 Fifth Avenue, New York, New York 10003

United Kingdom Edition published by
ACADEMIC PRESS, INC. (LONDON) LTD.
24/28 Oval Road, London NW1

LIBRARY OF CONGRESS CATALOG CARD NUMBER: 66-29729

PRINTED IN THE UNITED STATES OF AMERICA
80 81 82   9 8 7 6 5

To my wife, Yvette, who has always forgiven my passion for mathematics and has, indeed, never failed to encourage me in my work.

FOREWORD

The last two decades have seen an awesome development of mathematical theories focusing on decision-making and control theory. What is so remarkable about these new theories is their unifying effects in mathematics itself. Whereas many other subdisciplines have tended to become highly specialized, generating arcane languages with a narrow, intense focus of interest, control theory and decision-making have grown wider and wider, requiring all of the resources of modern mathematics, analysis, algebra, and topology for their effective treatment. Furthermore, the requirement of obtaining numerical answers to numerical questions demands a study of numerical analysis and analog and digital computers which, in turn, brings in questions of logic and the theory of algorithms. The study of adaptive control processes forces one into contact with psychology, neurophysiology, and so on, and so on.

Much of this interdisciplinary effort has been going on at the upper levels of the mathematical world. Without good books written by mathematicians in the forefront of research who are also masters of exposition, there is no efficient way of training the next generation to work in these important areas or of informing the current generation as to what is going on and encouraging them to go into these promising fields now. Fortunately, in my friend and esteemed colleague, Arnold Kaufmann, we have exactly the desired combination. It therefore gives me particular pleasure to welcome his book into this series and to recommend it highly for those who wish a graceful introduction to three important areas of contemporary research and application.

RICHARD BELLMAN

University of Southern California
Los Angeles

ix

FOREWORD TO THE FRENCH EDITION

I have sometimes wondered what would have happened if modern science, instead of turning from mathematics in the direction of mechanics, astronomy, physics, and chemistry, and focusing the whole of its effort on the study of matter, had concentrated instead on the study of the human mind. Our knowledge of psychology would probably bear much the same relation to our existing psychology as modern physics bears to the physics of Aristotle.

The place of mathematics in the human sciences is already an important one and before long it will become predominant. Into such diverse fields as psychology, economics, semantics, and philology, mathematics brings clarity and precision of method. Operations research will be the future science of action.

HENRY BERGSON

PREFACE TO THE FRENCH EDITION

The encouraging reception accorded to the first volume of this work,[1] which was translated into several languages and published in various countries, has led us to adopt the same method of presentation in the present volume. The first part of the book is again devoted to simple and concrete examples, for which only an elementary knowledge of mathematics is needed, and in the second part these examples, together with further concepts, are developed in a fuller and more theoretical manner requiring a considerably higher level of mathematics. Even in the second part, however, lack of space has permitted us to give the proofs of only the more important theorems and for the other proofs, the reader is referred to specialized works given in the bibliography.

Linear programs, the phenomena of expectation, problems of stocks and the wear and operation of machinery were the subject of the first volume. We turn in the present volume to the theory of graphs and its applications, to dynamic programming, and to the theory of games of strategy. In the third volume[2] we shall deal with combinatorial problems, linear programs with parameters, programs with integral values, quadratic programs, and applications of Boolean mathematics.

The concepts which are treated in this volume are of interest at the present time to engineers, management personnel, organizing experts, economists, and specialized accountants: to anyone, in fact, who has to prepare for or make decisions in industry. Students at higher technical colleges and those trainees who are studying different scientific disciplines now have a number of concepts connected with operations research included in their courses, and they, too, should find this work useful to them.

[1] A. Kaufmann, "Méthodes et Modèles de la Recherche Opérationnelle," Tome I (in French). Dunod, Paris, 1962. English language edition published by Prentice-Hall, Englewood Cliffs, New Jersey in 1963 under the title "Methods and Models of Operations Research."

[2] To be published in French by Dunod, Paris.

At this point we must recall our purpose of writing works in applied mathematics which would be suitable for people at all levels of scientific knowledge, so that the reader, whatever his level might be, could gain insight into the subject or increase his knowledge of it without being confused by too complex a treatment. Among all scientific workers, from the technician to the advanced analyst, there should be a suitable language of communication; this has always been our belief, and the encouragement which we have received for this concept makes us hold it more firmly than ever.

As a result of our efforts, the material in the first volume and the present work has already been taught in several advanced colleges of engineering in France and elsewhere. It has been a source of profound satisfaction to us to discover the great interest which the material has aroused—not only have a number of students decided to adopt operations research as their career, but several of them have already published material of high quality on the subject. The cooperation of those taking part in the instructional courses which we were instrumental in establishing in France and other countries made it possible for us to introduce improved methods for teaching it.

In our opinion, perhaps the greatest satisfaction for a teacher is to discover that aspects of the subject which he had not had the opportunity to treat himself have been developed by some of his own students. Scientific progress can take place only in an atmosphere of generosity and friendship, and we have always believed that the large number of readers who have written to us have added a valuable contribution to our knowledge.

I wish also to emphasize how agreeable it is for an author to receive the full cooperation and constant encouragement of his publisher and to take this opportunity of extending my gratitude to Monsieur Georges Dunod and his staff.

Professor André A. Brunet, who decided to include the first volume of this work in the series of which he is the editor, has honored me by the inclusion of the present volume as well. He has also greatly assisted me in improving certain parts of the book by suggesting ways whereby the explanations could be clarified and made more precise.

My thanks are also due to my friends in the Compagnie des Machines Bull: MM. Faure, Cullmann, Malgrange and Pertuiset, who have helped me with many constructive suggestions. Nor do I forget Monsieur R. Cruon who helped me over a number of awkward passages and who was kind enough to read over the manuscript. Finally, my daughter, Anne, recently graduated in mathematical science, helped me both in the final editing and in reading the proofs.

A. KAUFMANN

CONTENTS

Part I. Methods and Models

Chapter I. Graphs

Chapter II. Dynamic Programming

Chapter III. The Theory of Games of Strategy

Part II. *Mathematical Developments*

Chapter IV.   *The Principal Properties of Graphs*

Chapter V.   *Mathematical Properties of Dynamic Programming*

Chapter VI. *Mathematical Properties of Games of Strategy*

Bibliography

PART I

METHODS AND MODELS

---

1. Introduction

An entirely new theory, that of the diagrammatic "graph,"[1] can be of great assistance in dealing with those combinatorial problems which occur in various economic, sociological, or technological fields. Indeed, the realization of the value of this "theory of graphs" has given it a place of great importance in education. It is, perhaps, that aspect of the theory of sets which can produce the most fruitful results, not only for the "pure" mathematician, the engineer, and the organizer, but also for the biologist, the psychologist, the sociologist, and many others.

To deal exhaustively with the theory of graphs would entail explaining a great variety of concepts and theorems, some of them of an extremely complex nature. To do this would not be appropriate in a work with the scope of the present one. We shall limit ourselves, therefore, to an explanation of some of the simplest concepts, showing how they may be usefully employed, and setting out some of the more interesting methods which can be used in applying them.

2. The Use of Points and Arcs to Represent Structures

A. DEFINITION

Let us consider a set of points, which may be finite or infinite in number, provided they are denumerable, such as points $A, B, ..., F, G$ in Fig. 2.1. In the diagram, these points, which may also be referred to as *vertices*, are connected by oriented lines that we shall call *arcs*. Thus A is connected directly to B by arc a, to D by arc e, and to E by arc d. Vertex D is joined to vertex E by arc f, to vertex F by arc i, to vertex G

---

[1] According to the particular meaning given to this word in the theory of sets dating from the works of König.

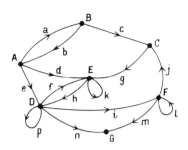

FIG. 2.1.

by arc *n* and, in addition, to itself by arc *p*. The diagram in question constitutes a *graph*. Graphs can be used to represent structures of the most diverse nature, such as

(1)   A network of roads or streets (Fig. 2.2).
(2)   An electrical circuit (Fig. 2.3).
(3)   A group of people of which a psychosociological study of communication is being undertaken (Fig. 2.4).
(4)   The circulation of documents within an administrative system (Figs. 2.4 and 2.7).
(5)   The growth of populations in demographic phenomena (Fig. 2.6).
(6)   The family relationships of a group of people (Fig. 2.8).
(7)   The rules of certain games such as chess or checkers.
(8)   The ranking of participants in a tournament.
(9)   The operations of assembling and dismantling a technological system, etc.

In considering the above concepts, the graph in each case must not be confused with the concept associated with it; it is merely the structure, in which the use of *vertices* and *arcs* provides a useful representation of certain properties, which is of interest to us.

B. IDEA OF MAPPING A SET INTO ITSELF

It is important to understand how a mathematician defines a graph, since it is due to his structures and abstract deductions that these graphs are of such value to those who are concerned with the application of mathematics to the physical or human sciences.

In the first place, the mathematician defines a collection of objects, which may be of any kind, but must be separate and distinct, as a *set*; for example, a set of seven objects: (A, B, C, D, E, F, G). To

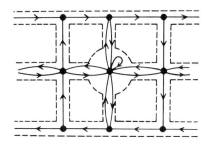

FIG. 2.2. Urban traffic system.

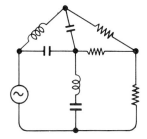

FIG. 2.3. Electric circuit.

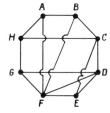

FIG. 2.4. Communications among
a social group.

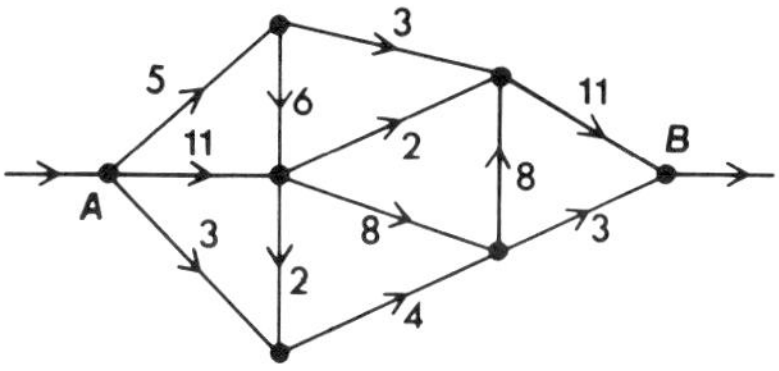

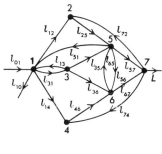

FIG. 2.5. Flow in a network.

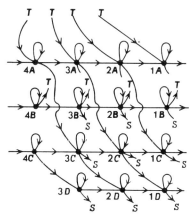

FIG. 2.6. Demographic change in
a population of students.

FIG. 2.7. Circulation of files in
a business office.

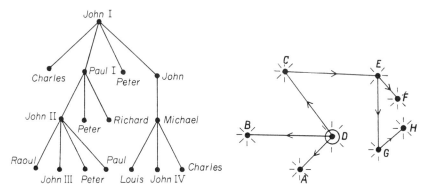

FIG. 2.8.   Genealogical tree.   FIG. 2.9.   Television network.

each of the objects of this set he attaches a correspondence with 0, 1, 2, or several objects of the same set. Representing the law which determines this correspondence as Γ, he defines a graph as *a set and its law of correspondence*.

Let us suppose that, in the set of seven objects, the law of correspondence is the following one:

$$\Gamma(A) = \{B, D, E\}, \qquad \Gamma(B) = \{A, C\}, \qquad \Gamma(C) = \{E\},$$
$$\Gamma(D) = \{D, E, F, G\}, \qquad \Gamma(E) = \{D, E\}, \qquad (2.1)$$
$$\Gamma(F) = \{C, F, G\}, \qquad \Gamma(G) = \varnothing,$$

where $\varnothing$ is a symbol meaning "null set" (it is preferable not to use the symbol 0, since it is a question of objects, rather than of numbers); thus, the set $(A, B, ..., G)$ and the correspondences given in Eq. (2.1) constitute a graph, which can be shown by Fig. 2.1. To provide a better understanding of the meaning of the correspondences defined by Eq. (2.1), these have been represented by Figs. 2.10–2.16.

Whenever a structure reveals a set of separate objects and a law of correspondence between these objects, there will be a "graph." It is to be understood that such a concept is fairly common and may be encountered in a great diversity of structures.

To designate a graph it is therefore necessary to define both the set of vertices, which will be termed **X**, and the correspondences between these vertices (mathematicians describe this structure as a "mapping[1]

---

[1] In many works the word "mapping" or "correspondence" is synonymous with "function," where these functions are univocal (possess one-to-one correspondence). The word is employed here in the more general sense used by Berge [H2].

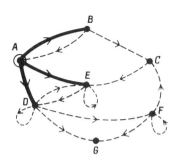

FIG. 2.10. $\Gamma(A) = \{B, E, D\}$.

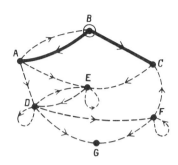

FIG. 2.11. $\Gamma(B) = \{A, C\}$.

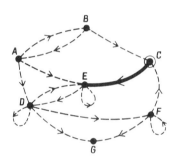

FIG. 2.12. $\Gamma(C) = \{E\}$.

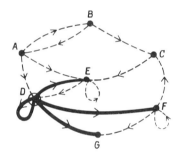

FIG. 2.13. $\Gamma(D) = \{D, E, F, G\}$.

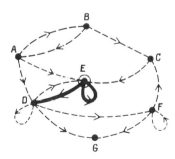

FIG. 2.14. $\Gamma(E) = \{D, E\}$.

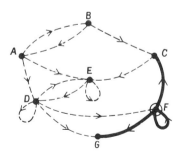

FIG. 2.15. $\Gamma(F) = \{C, F, G\}$.

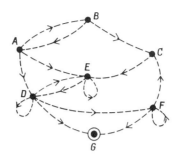

FIG. 2.16. $\Gamma(G) = \varnothing$.

of **X** into **X**"), which will be represented by Γ. Thus a graph may be designated by the symbol $G = (\mathbf{X}, \Gamma)$. One may also refer to a graph as the set **X** of vertices and the set **U** of arcs, which amounts to the same thing. In the latter case one writes $G = (\mathbf{X}, \mathbf{U})$. For the reader who requires greater exactitude, these definitions and symbols will be dealt with in the second part of this work with much more precision.

3. Principal Concepts Used in the Theory of Graphs

Before explaining various interesting applications of this theory we will introduce some important concepts. To make our explanations sufficiently clear it will first be necessary to define a number of technological terms.

A. Oriented Concepts

Path. This is a series of adjacent arcs giving access from one vertex to another. Thus the arcs (i, j, g) in the graph of Fig. 2.1 form a path connecting point D with point E. The path may also be referred to by the vertices that it contains: for example, D, F, C, E.

Circuit. This is a path in which the initial and final vertices coincide. The arcs (i, j, g, h) form a circuit (Fig. 2.1) which may be referred to as (D, F, C, E, D).

Length of a path or circuit. This is the number of arcs contained in the path or circuit.

Loop. A loop is a circuit of length 1. *Example*: loop k in Fig. 2.1.

Symmetrical graph. If a vertex X is joined to a vertex Y, then Y should be joined to X. If this property is valid for all the vertices between which a correspondence exists, the graph is "symmetrical."

Antisymmetrical graph. If a vertex X is connected with a vertex Y, then Y must not be joined to X. If this property can be verified for all the vertices between which a correspondence exists, the graph is "antisymmetrical."

Strongly connected graph. Whatever the position of the vertices X and Y $(X \neq Y)$, which are being considered, a path exists from X to Y. Thus the graph of Fig. 3.1 is strongly connected but that of Fig. 2.1

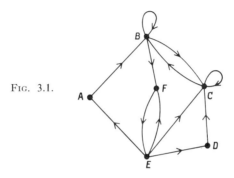

FIG. 3.1.

is not. Point G is not joined to the other points and one cannot go from C to A or B, etc.

These concepts are very important, as we shall see later on.

B. NONORIENTED CONCEPTS

Link. It can be said that a link exists between two vertices X and Y if there is an arc from X to Y and/or from Y to X. This is the case, for example, in Fig. 3.1, for the pairs (A,E) (F,E), etc., but not for the pair (A,F).

Chain. This is a series of consecutive links. For example, in Fig. 3.1 (A, E, F, B, C) is a chain.

Cycle. This is a closed chain. For example, in Fig. 3.1 (B, C, E, F, B) forms a cycle.

Connected graph. Whatever the position of the vertices X and Y which are being considered, there is a chain between X and Y. Thus the graph in Fig. 2.1 is not strongly connected, but it is connected.

Complete graph.   Whatever the position of the vertices *X* and *Y* which are being considered, there is a link between *X* and *Y*; see, for example, Fig. 3.2. It should be noted, however, that we are speaking of a link between *X* and *Y*, and not of an arc.

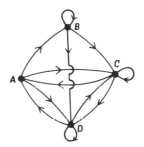

FIG. 3.2.   Complete graph.

C. PARTIAL GRAPH AND SUBGRAPH

Partial graph.   If one or more arcs are omitted from a graph, a partial graph is formed from the reference graph (Figs. 3.3a and 3.3b).

The map of all the roads in a country, including the major and secondary ones, forms a graph. It becomes a partial graph only if the major routes are shown on it.

Subgraph.   If, in a graph, one or more vertices are omitted, together with the arcs to and from these points, the remaining portion of the graph is a subgraph of the reference graph (Figs. 3.3a and 3.3c). In addition, there may be partial subgraphs.

The definitions which we have given above may not always seem sufficiently precise, but it has been our aim to avoid undue complications, and to defer a greater exactitude until the theory of sets is explained in the second part of this work.

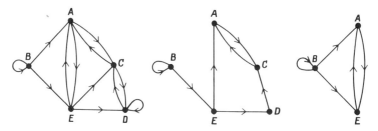

FIG. 3.3.   (a) Reference graph;   (b) partial graph of *G*;   (c) subgraph of *G*.

Any instances dealing with the theory of graphs which occur in the first part will lead to the introduction of new concepts. At the present time we do not wish to overburden either the memory or the patience of our readers.

4. Scheduling and Sequencing Problems

We intend to show, by means of some examples, in what way the theory of graphs can help in the solution of various problems of scheduling and sequencing.[1] Let us recall that scheduling and sequencing consist of placing in a certain chronological order operations which are not, as a rule, separate and which involve the consideration of expenditures and assets, of technical performances, and of time. We do not propose at this stage to undertake a comparative study of all the known methods of scheduling and sequencing, but to show how the theory of graphs may be effectively used to solve various numerical problems that arise in applying these methods. Let us not forget that the purpose of this book is to show how mathematics may be employed in the study of the phenomena of organization.

A. FINDING THE CRITICAL PATH

The construction of a house is, in our opinion, an excellent example of a problem of scheduling, in which one is concerned with the intervals of time separating the different operations which comprise the project.

In building a house a certain number of fundamental operations have to be considered, such as

(a) Purchase of the site.
(b) Obtaining a building permit.
(c) Construction of the foundations.
(d) Erection of the main body of the house.
(e) Roofing.
(f) Painting and decorating.
(g) Locks.
(h) Final completion of the work.

---

[1] The single French word *ordonnancement* is applied to problems which are described in English by two different terms: *scheduling*, when the problem is concerned with the instant at which a certain event takes place; and *sequencing*, when various operations are to succeed one another in a certain order. The first example is a problem of scheduling, the second one, of sequencing.

Some basic operations can only be carried on if certain others have been completed, but various operations may be carried on simultaneously. Let us suppose that each operation has to be completed in its entirety. Let us also take for granted that the interval which elapses between the commencement of one operation and that of the next has been calculated and is known. To simplify the problem, let us further suppose that the construction of the house does not involve more than twelve basic operations, and that P_1, P_2,..., P_{12} represent the different stages to be considered.

The graph for this problem (Fig. 4.1) has been drawn in the following manner. Let P_{12} represent the final operation (the completion of the

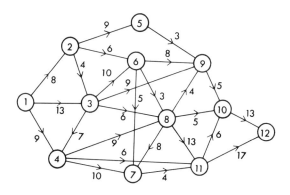

FIG. 4.1.

work) and P_1 the initial operation. Between the start of P_1 and that of P_2 there is an interval of eight weeks; we therefore draw an arc (1,2) and assign the value of 8 to this arc. From the start of P_1 to that of P_3 there is a lapse of thirteen weeks, and a value of 13 is assigned to the arc (1,3). Between the commencement of P_2 and that of P_3 the interval is four weeks and the arc (2,3) is given a value of 4; and so forth. Wherever the interval has been calculated between two clearly defined operations the value of this interval is written beside the corresponding arc.

It is of great importance for what follows to note that a graph drawn in this way must not contain any circuit. If it did the problem would be meaningless, for there cannot be any delay between the start of an operation and itself.

Let us now determine the total delay in execution, or, in other words, the interval of time, from the commencement until the termination of the work, which is minimal for the completion of the required operations.

This total delay from P_1 to P_{12} should be greater than or equal to the sum of the delays taken from "the least favorable" or critical path. To evaluate this critical path the following procedure will be followed. Starting with P_1, to which the value 0 will be assigned, the arcs arriving at each vertex will be evaluated. For each of these arcs the sum will be found of the delay carried by the arc and of that given to its vertex of origin. The results will be compared, and the highest value will be chosen and transferred to the vertex under consideration. This procedure can be explained very simply from our example (Fig. 4.2).

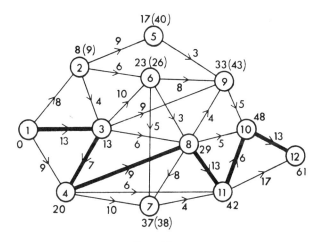

FIG. 4.2.

At P_2, there is only the arc of arrival (1,2) and, as the value of P_1 is 0, P_2 is given the value $0 + 8 = 8$. Two arcs arrive at P_3 [(2,3) and (1,3)] and the two sums, $8 + 4 = 12$, and $0 + 13 = 13$, are compared, and the value 13 is transferred to P_3, which indicates the delay required to reach P_3 must be at least equal to 13. Two arcs [(3,4) and (1,4)] are drawn to P_4, and their values, $13 + 7 = 20$ and $0 + 9 = 9$, are compared, the larger number 20 being transferred to P_4. To P_5, the value 17 is assigned. At P_6, two arcs [(2,6) and (3,6)] arrive, so that $8 + 6 = 14$ is compared with $13 + 10 = 23$, and 23 is transferred to P_6. Three arcs [(6,8), (3,8), and (4,8)] arrive at P_8, and the three sums $23 + 3 = 26$, $13 + 6 = 19$, $20 + 9 = 29$ are compared, and the largest, 29, is transferred to P_8.

This procedure is followed until the last vertex P_{12}, to which the value 61 is given. This number represents the delay in reaching P_{12} if the certainty exists that all the operations will be completed within the delays which separate them from each other. The path corresponding to 61, which is easily obtained step by step by returning towards P_1, constitutes the *critical path*. Any delay between the operations of which it is composed, cannot fail to disrupt the unfolding of subsequent operations, and the operations belonging to this path are *critical operations*, the fulfillment of which must be supervised with great care.

It is interesting to find out in the case of noncritical operations the maximal delay which may be allowed without increasing the total delay. To do this, let us first of all consider P_7, which is separated from P_{11} by a delay of 4, so that P_7 may begin on a date between 37 and 38 without causing any disruption. P_9 is separated from P_{10} by a delay of 5, and hence P_9 may commence at any given date between 33 and 43 without disturbing the planning. P_6 is separated from P_9 by a delay of 8, from P_8 by a delay of 3, and from P_7 by a delay of 5; hence, the values $43 - 8 = 35$, $29 - 3 = 26$, $38 - 5 = 33$ are compared, and it is found that the operation P_6 cannot extend beyond date 26 (the smallest of the three), without interfering with the project. The numbers that we have just calculated have been placed in parentheses on Fig. 4.2 to distinguish them from those entered on it when the critical path was evaluated. At P_5 the number 40 will be written. For P_2, $40 - 9 = 31$ will be compared with $26 - 6 = 20$ and $13 - 4 = 9$, and 9 will be entered beside it.

To complete the problem, the dates for the start of the different operations may be given as follows:

$$
\begin{aligned}
P_1 &: \quad 0, \\
P_2 &: \quad \text{between 8 and 9,} \\
P_3 &: \quad 13, \\
P_4 &: \quad 20, \\
P_5 &: \quad \text{between 17 and 40,} \\
P_6 &: \quad \text{between 23 and 26,} \\
P_7 &: \quad \text{between 37 and 38,} \\
P_8 &: \quad 29, \\
P_9 &: \quad \text{between 33 and 43,} \\
P_{10} &: \quad 48, \\
P_{11} &: \quad 42, \\
P_{12} &: \quad 61,
\end{aligned}
\tag{4.1}
$$

where these numbers represent weeks.

Let us now make certain observations:

(1)   There may be several critical paths.

(2)   The method employed consists of first calculating a maximal

path, that is, a path connecting the vertex of origin with the last vertex. This is done by ensuring that the sum of the values assigned to the arcs of which the path is composed is maximal, and by then calculating for each vertex the maximal path until the last vertex is reached.

(3)   There are a number of methods for obtaining either a maximal or minimal optimal path. Some of these are given in Section 36, but it will be observed that when dealing with graphs that do not include a circuit, the calculation involved is very simple.

(4)   If two operations occur simultaneously, they are considered to form one.

(5)   If there are several separate vertices of origin, the problem can easily be reduced to one of a single vertex of origin by the addition of suitable supplementary arcs.

(6)   It is necessary, before making a practical drawing of a graph which is to represent a sequence of operations, that the operations should be scheduled in such a manner that, when considered in reverse order, no operation must appear on the list until all its successors have done so (Fig. 4.3). If this requirement is not met, the problem is insoluble and it then becomes necessary to discover the error of evaluation which resulted in the introduction of a circuit.

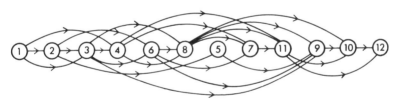

FIG. 4.3.

We have assumed that the values carried by the arcs of the graph in Fig. 4.1—values consisting of the data—were precisely known, which is scarcely realistic. As a rule, such data are far from exact. Consequently, the process which has been described has to be modified when the data consist of uncertain values arrived at by subjective methods, and it now seems an appropriate time to give a brief description of the PERT method (Program Evaluation Research Task).[1]

---

[1] Method developed by the Special Project Office, U.S. Navy, Washington, D.C. [H42]. In the PERT method, the vertices of the graph represent events, and the arcs, the operations which separate these events. In considering the example shown in Fig. 4.1, we have taken the vertices as representing operations and the arcs as showing constraints of time. This form of representation is more often used in France, where it was introduced by Roy [H56], who has termed it "the method of potentials."

This method consists of evaluating the delays and costs which intervene in a program of research or expansion or, for that matter, in one of investment or production, for which purpose a graph, such as that in Fig. 4.1, is drawn. A team of specialists evaluate the uncertain delays and the costs. In the case of every basic delay, the most pessimistic value, the most optimistic value, and the most probable value (the mode or value of the variable which has the greatest probability) are evaluated. Using these three values, and employing formulas that provide a suitable approximation, a mean value $\bar{t}_{ij}$ and a variance σ^2_{Tij} are obtained for each delay between the operations P_i and P_j. The mean values $\bar{t}_{ij}$ are transferred to the arcs of the graph. The critical path, or paths, are then determined, as well as the delays that are considered acceptable for the noncritical points. With each value thus obtained (which is transferred to a vertex of the graph) we associate its variance, the latter being easily calculated by adding the variances in the optimal path under consideration (it is assumed that the delays are uncertain variables distributed in accordance with the distribution law, β, an assumption which makes it possible to arrive at a solution, but which is difficult to justify theoretically, in view of the manner in which the estimates are arrived at).

The knowledge of the mean values and the variances enables us to decide with a greater degree of assurance that the total mean delay found from the critical path will be observed, since a more complete assessment has been provided by the law of probability of the total delay. Once these results have been obtained, and the total cost of the operations has been calculated, the graph is studied with the object of determining whether a modification of some of the operations might make it possible to draw a new graph giving a lower total delay, or, if certain variants of the method are used, a lower total cost.

The PERT method is a heuristic one[1] in which the theory of graphs, the theory of probabilities, and processes of estimation and readjustment are combined to solve a problem of a practical nature.

In our first example, only an elementary knowledge of the theory of graphs was needed, but as the scope of the PERT method or of similar methods is extended, we shall encounter various questions of greater complexity for which the theory of graphs will frequently provide an answer.

At this point we shall discuss an entirely different type of problem of sequencing.

---

[1] The term "heuristic" is often used for a method that is the result of invention. In the *Regulae and directionem ingenii* in his *Ars Inveniendi*, Descartes defines heuristic as the set of rules of inventive thought. A number of writers apply a somewhat different meaning to the word.

B. FINDING AN OPTIMAL PERMUTATION. OPTIMAL HAMILTONIAN PATH

We are concerned with a production line using a basic material which can be diversified in a hundred different ways by altering the operations performed in the chain of production. To change from one product to another in accordance with the orders that are received, alterations of varying cost have to be made to the production line. Thus in changing from a product P_i to another P_j, there is a launching cost or transition cost c_{ij}. Every week the factory receives a tabulation of the orders, and it is necessary to discover the sequence in which the production line will produce the various articles with a minimal launching cost.

So as to make it easier to understand the method which will be used to solve this practical problem, we will reduce its scope and limit the number of products to be considered to 9:

$$P_1, P_2, ..., P_9,$$

which must be produced in the order which will ensure that the total cost of launching them is minimal. Expressed differently, this means that the permutation or permutations of the nine products giving a minimal cost has to be found. For this purpose, let us assume that the matrix shown in Table 4.1 provides the costs associated with each

TABLE 4.1

	(1)	(2)	(3)	(4)	(5)	(6)	(7)	(8)	(9)
(1)	0	8	9	6	8	3	2	4	6
(2)	5	0	3	3	6	4	4	8	3
(3)	3	4	0	4	3	7	2	7	6
(4)	7	8	4	0	8	4	4	3	8
(5)	4	5	5	7	0	3	2	8	7
(6)	11	5	9	5	4	0	3	2	12
(7)	10	7	9	8	6	4	0	4	5
(8)	5	9	8	2	11	1	3	0	2
(9)	7	6	11	4	12	6	8	9	0

product, so that if, for example, one passes from P_3 to P_7 the cost is $c_{37} = 2$.

If each product is represented by a vertex and each change of product by an arc, a graph can be drawn which has the property of containing an arc joining every vertex P_i with every vertex P_j, and which is therefore a complete and symmetrical graph. Our aim is to find a path which passes once and only once through each vertex, and which includes all of the vertices, such a path being termed "a Hamiltonian path." In the case of a complete and symmetrical graph containing nine vertices (a generalization for n vertices can be made), there are[1]

$$9! = 1 \cdot 2 \cdot 3 \cdot 4 \cdot 5 \cdot 6 \cdot 7 \cdot 8 \cdot 9 = 362{,}880 \qquad (4.2)$$

Hamiltonian paths or 362,880 permutations. If the graph is not complete and/or is not symmetrical, the number of such paths is reduced, and sometimes substantially so, in certain cases.

The reader has probably already become acquainted with a variation of the present problem in Volume I, Section 19 (the problem of the traveling salesman).[2] There is no theoretically precise solution of this problem except by using the method of linear programming with integral values (the method of Gomory,[3] for example), and this method frequently leads to extensive calculations out of proportion to problems such as the one we are now considering. Accordingly, we intend to find, in a suitable subset of solutions, an optimal solution for this subset. It will not necessarily be the best solution of the given problem, but it can be accepted as a practical answer to a question of industrial sequencing such as we are studying.

Let us now consider the table of costs (Table 4.1), with which we shall connect Table 4.2, assuming

$$r_{ij} = 1 \quad \text{if} \quad c_{ij} < c_{ji}, \quad i \neq j; \quad r_{ij} = 0 \quad \text{if} \quad c_{ij} > c_{ji}, \quad i \neq j$$
$$r_{ij} = 1 \quad \text{if} \quad i = j \qquad (4.3)$$

and, if $c_{ij} = c_{ji}$ we take $r_{ij} = 1, r_{ji} = 0$, or the opposite by arbitrary choice.

Otherwise stated, the cost of $P_i \rightarrow P_j$ is compared with that of $P_j \rightarrow P_i$, and the numeral 1 is placed in Table 4.2 in the square (i, j),

---

[1] In Volume I, page 31, we explained the meaning of $N!$ (factorial N). The number of permutations of N separate objects is equal to $N!$

[2] *Note to Reader*: Throughout the present work, Volume I refers to A. Kaufmann, *Methods and Models of Operations Research*. Prentice-Hall, Englewood Cliffs, New Jersey, 1963.

[3] This will be studied in Volume III (to be published by Dunod, Paris).

TABLE 4.2

	(1)	(2)	(3)	(4)	(5)	(6)	(7)	(8)	(9)
(1)	1	0	0	1	0	1	1	1	1
(2)	1	1	1	1	0	1	1	1	1
(3)	1	0	1	1	1	1	1	1	1
(4)	0	0	0	1	0	1	1	0	0
(5)	1	1	0	1	1	1	1	1	1
(6)	0	0	0	0	0	1	1	0	0
(7)	0	0	0	0	0	0	1	0	1
(8)	0	0	0	1	0	1	1	1	1
(9)	0	0	0	1	0	1	0	0	1

provided the cost $P_i \rightarrow P_j$ is less than the cost $P_j \rightarrow P_i$, and 0 is placed in the symmetrical square (j, i). If the costs $P_i \rightarrow P_j$ and $P_j \rightarrow P_i$ are equal, an arbitrary decision is made to place 1 in square (i, j) or (j, i) and 0 in the symmetrical square. Along the main diagonal, that is to say, in squares (i, i), the numeral 1 will be entered everywhere.

If we represent Table 4.2 by a graph in which an arc is drawn in the direction $P_i \rightarrow P_j$, if $r_{ij} = 1$, except where $i = j$, the graph thus drawn is a complete and antisymmetric one (Fig. 4.4), which will only contain a small number of Hamiltonian paths to be compared. An important theorem (see Part II, Section 39, page 285) will prove that any complete graph must contain at least one Hamiltonian path. The lowest costs correspond to this antisymmetric graph when one compares a cost and its symmetrical. Thus, the solution or solutions obtained from the antisymmetric graph will be useful, but not necessarily optimal. They will, nevertheless, as we shall discover, prove more than sufficiently accurate.

Let us now show how to enumerate all the Hamiltonian paths in a graph. A number of methods are used for this purpose, and we shall describe several of them in Section 39 of Part II. In selecting one of these methods we are less concerned whether it is the best or the simplest for the present problem, than with the interest of its content, which will

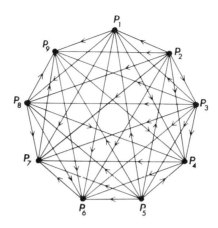

FIG. 4.4.

enable us to introduce a number of fresh concepts. We shall employ an algorithm[1] for which we are indebted to Foulkes [H30], and which is more fully explained in Section 39.

From Table 4.2 we can tell whether there is a path of length 1 between P_i toward P_j. How can we obtain a table which will enable us to discover whether a path of a length less than or equal to 2 exists from P_i toward P_j, and, after that, whether there is a path less than or equal to 4, 8,..., etc? Various methods are available for answering this question, but the one we shall use is that introduced by Foulkes.

We shall now consider the graph of Fig. 4.4, and in order to discover whether a path of length 2, for example, exists between P_5 and P_3 we shall examine the chains of length 2 having their extremities at P_5 and P_3, and passing through one of the vertices P_1, P_2, P_4, P_6, P_7, P_8, P_9. If such a chain includes a path of length 2 between P_5 and P_3 it can be assumed that at least one path of this length exists between P_5 and P_3. (P_5, P_2, P_3) is discovered, which is the only path of length 2, and by repeating the process which we have just carried out, for each vertex of the graph, a new table can be drawn up with 9×9 squares, from which it will be possible to discover whether or not there is a path of length 2. However, instead of using Fig. 4.4, let us employ matrix or Table 4.2, which is its equivalent.

---

[1] Let us recall that in the terminology of mathematics, an algorithm is defined as "a process of calculation." The word is an altered transcription of the name of the Arab mathematician Al Khovaresmi whose work was translated into Latin during the 12th century. The mutilation of his name is the result of Greek influence.

Let us carry out a boolean matrix multiplication of the matrix (Table 4.2) by itself. As matricial calculation[1] and boolean[2] calculation are, we imagine, equally new to the reader, we will explain how an operation of this kind is carried out, and what it corresponds to in the graph under consideration.

Let us first take line (5) and column (3) of the matrix (Table 4.2) and carry out the "multiplication" of line (5) by column (3) in the following manner:

$$
\begin{aligned}
&= (1)(0) + (1)(1) + (0)(1) \\
&\quad + (1)(0) + (1)(0) + (1)(0) + (1)(0) \\
&\quad + (1)(0) + (1)(0) = 1
\end{aligned}
$$

(4.4)

which proves the existence of a path of length 2. If the sum of the products had proved to be 0 we should have concluded that there was no path of this length. If it had been 2 or 3 or 4,..., it would have shown that 2 or 3 or 4,..., paths of length 2 existed. If 1 had been in square (5,3) of line 5, the presence of 1 in square (3,3), which is on the main diagonal, would in any case have shown that there was a path of length 1 from P_5 to P_3, for this is the purpose of writing the numeral 1 along this diagonal.

Instead of following the usual rules of addition:

$$0 + 0 = 0, \quad 0 + 1 = 1, \quad 1 + 0 = 1, \quad 1 + 1 = 2, \quad (4.5)$$

we will now affirm that we are no longer concerned with numbers but with the propositions:

(1)   A path exists of length less than or equal to 2;

(2)   Such a path does not exist.

---

[1] See, for example, M. Denis-Papin and A. Kaufmann, *Cours de calcul matriciel appliqué*, Albin Michel, Paris.

[2] See M. Denis-Papin, R. Faure, and A. Kaufmann, *Cours de calcul booléien appliqué*, Albin Michel, Paris, 1963.

Then, using the sign $\dot{+}$ instead of the normal sign for addition, we can affirm the following laws:

$$0 \dot{+} 0 = 0, \quad 0 \dot{+} 1 = 1, \quad 1 \dot{+} 0 = 1, \quad 1 \dot{+} 1 = 1. \quad (4.6)$$

Hence, by modifying the result of each operation such as (4.4) in the following manner:

$$N > 0 \quad \text{is replaced by 1,}$$
$$N = 0 \quad \text{is not modified,}$$

we obtain a matrix such as that shown in Table 4.3.

TABLE 4.3

	(1)	(2)	(3)	(4)	(5)	(6)	(7)	(8)	(9)
(1)	1	0	0	1	0	1	1	1	1
(2)	1	1	1	1	①	1	1	1	1
(3)	1	①	1	1	1	1	1	1	1
(4)	0	0	0	1	0	1	1	0	①
(5)	1	1	①	1	1	1	1	1	1
(6)	0	0	0	①	0	1	1	0	①
(7)	0	0	0	①	0	①	1	0	1
(8)	0	0	0	1	0	1	1	1	1
(9)	0	0	0	1	0	1	①	0	1

This table shows us whether a path of length less than or equal to 2 exists between one vertex and another.

Next, in the matrix of Table 4.3, we shall circle every numeral 1 which did not appear in the original matrix, a process whereby we shall discover the vertices between which there is a path of length 2, but not of length 1.

Using the matrix (Table 4.3), let us try to obtain a new table showing the vertices which are connected by a path of length[1] less than or equal

---

[1] It would be possible first to find the paths of length less than or equal to 3 by "multiplying" the matrix by itself and by 3 (Table 4.2), or to obtain the paths of length less than or equal to 4 without this preliminary step.

to 4. For this purpose we shall repeat for Table 4.3 the same procedure we used for Table 4.2. The multiplication of lines by columns, in conjunction with formulas (4.6) enable us to obtain Table 4.4 on which a new 1 appears (6th line, 4th column), which was absent in Table 4.3. By adopting the same procedure for Table 4.4, we shall discover the vertices which are connected by a path of length less than or equal to 8, and it will be observed that this table is the same as Table 4.4.

We shall end our calculations at this stage. If we had carried them further we should have circled each new 1 and have continued the "multiplications" of the matrices by themselves until we had obtained a new matrix that did not include any 1 which had not appeared in the previous matrix.

Studying Table 4.4 to find a line which contains only the numeral 1, we discover that there are three such lines [(2), (3), and (5)]. This indicates that the operations P_2, P_3, and P_5 must precede all others, taking into account the fact that columns (2), (3), and (5) only contain 0, except on the main diagonal.[1] We say that operations P_2, P_3, and P_5 form class 1, which takes precedence over all the others.

By eliminating the lines and columns (2), (3), and (5), Table 4.5 is obtained. In this new table we shall look for a line containing only the numeral 1, and find that there is one, namely line (1). As the corre-

TABLE 4.4   TABLE 4.5

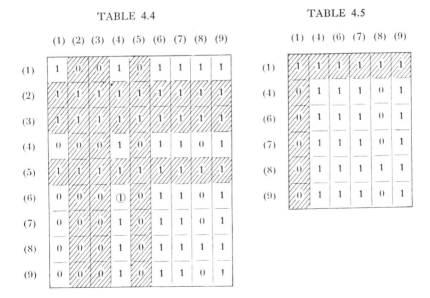

TABLE 4.4

	(1)	(2)	(3)	(4)	(5)	(6)	(7)	(8)	(9)
(1)	1	0	0	1	0	1	1	1	1
(2)	1	1	1	1	1	1	1	1	1
(3)	1	1	1	1	1	1	1	1	1
(4)	0	0	0	1	0	1	1	0	1
(5)	1	1	1	1	1	1	1	1	1
(6)	0	0	0	①	0	1	1	0	1
(7)	0	0	0	1	0	1	1	0	1
(8)	0	0	0	1	0	1	1	1	1
(9)	0	0	0	1	0	1	1	0	1

TABLE 4.5

	(1)	(4)	(6)	(7)	(8)	(9)
(1)	1	1	1	1	1	1
(4)	0	1	1	1	0	1
(6)	0	1	1	1	0	1
(7)	0	1	1	1	0	1
(8)	0	1	1	1	1	1
(9)	0	1	1	1	0	1

[1] The full explanation given in Part II, Section 39 will serve to clarify these different properties.

sponding column contains only 0, except on the main diagonal, it follows that operation P_1 precedes all the remaining ones and will form Class II.

By eliminating line (7) and column (1), Table 4.6 is obtained; in this table there is one line (8) that contains only the numeral 1, while the corresponding column contains nothing but 0, except on the main diagonal. Operation P_8 will comprise Class III.

By eliminating line and column 8, we shall be left with Table 4.7 in which 0 does not appear, and operations P_4, P_6, P_7, and P_9 will therefore form Class IV.

TABLE 4.6   TABLE 4.7

	(4)	(6)	(7)	(8)	(9)
(4)	1	1	1	0	1
(6)	1	1	1	0	1
(7)	1	1	1	0	1
(8)	1	1	1	1	1
(9)	1	1	1	0	1

	(4)	(6)	(7)	(9)
(4)	1	1	1	1
(6)	1	1	1	1
(7)	1	1	1	1
(9)	1	1	1	1

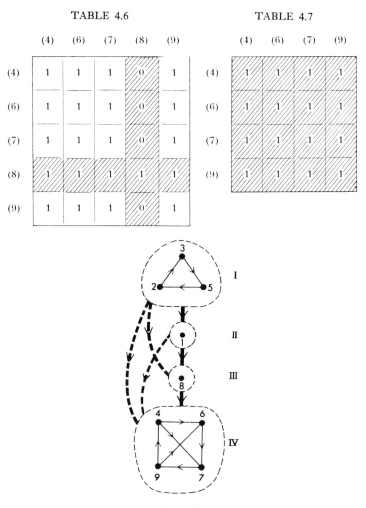

FIG. 4.5.

Each class forms a strongly connected subgraph in which a path leads from each vertex to every other vertex in it. Also, since the graph is complete, every vertex (that is to say every operation) of Class I is connected with every vertex of Classes II, III, and IV. Again, every vertex of Class II is joined to the vertices of Classes III and IV, and each vertex of Class III is connected with every vertex of Class IV.

On Fig. 4.5 the graph is shown split up into classes. Any reader with even an elementary knowledge of modern mathematics will realize that we are dealing with equivalent classes related to the property. A path exists from each point to every other one of the same class (strong connection); in addition, there is a strict order among the classes. These aspects of mathematics are explained in Sections 31 and 39 of Part II.

In order to enumerate all the Hamiltonian paths of the graph of Fig. 4.4, all that is now required is to find those that are present in each class or subgraph, and then link all these paths to one another.

Class I :   (2, 3, 5), (3, 5, 2), (5, 2, 3)
  II :   (1)
  III :   (8)
  IV :   (4, 6, 7, 9), (4, 7, 9, 6), (6, 7, 9, 4), (7, 9, 4, 6), (9, 4, 6, 7).

From the above we find the 15 Hamiltonian paths of the complete antisymmetric graph of Fig. 4.4, of which the respective costs are as follows:

(2, 3, 5, 1, 8, 4, 6, 7, 9)	cost	28
(2, 3, 5, 1, 8, 4, 7, 9, 6)	cost	31
(2, 3, 5, 1, 8, 6, 7, 9, 4)	cost	27 ←
(2, 3, 5, 1, 8, 7, 9, 4, 6)	cost	30
(2, 3, 5, 1, 8, 9, 4, 6, 7)	cost	27 →
(3, 5, 2, 1, 8, 4, 6, 7, 9)	cost	31
(3, 5, 2, 1, 8, 4, 7, 9, 6)	cost	34
(3, 5, 2, 1, 8, 6, 7, 9, 4)	cost	30
(3, 5, 2, 1, 8, 7, 9, 4, 6)	cost	33
(3, 5, 2, 1, 8, 9, 4, 6, 7)	cost	30
(5, 2, 3, 1, 8, 4, 6, 7, 9)	cost	29
(5, 2, 3, 1, 8, 4, 7, 9, 6)	cost	32
(5, 2, 3, 1, 8, 6, 7, 9, 4)	cost	28
(5, 2, 3, 1, 8, 7, 9, 4, 6)	cost	31
(5, 2, 3, 1, 8, 9, 4, 6, 7)	cost	28

It is found, accordingly, that among this subset of Hamiltonian paths the two best are (2, 3, 5, 1, 8, 6, 7, 9, 4) and (2, 3, 5, 1, 8, 9, 4, 6, 7), each of which yields 27.

Instead of seeking the best Hamiltonian path—in other words the best permutation from $9! = 362,880$—the calculation is confined to a subset of 15, but there is no certainty that the absolute best has been discovered. Indeed, it is possible to find a path with a lower total cost, but for practical purposes the result obtained by this method should be sufficiently accurate. In the above subset of 15, the cost varies from 27 to 34.

It is useful to compare this result with the one which would be obtained by using the highest symmetrical costs in Table 4.1 instead of the lowest. If this is done, another antisymmetrical graph is obtained which is the *complement* of the first. The splitting up into classes now gives Fig. 4.6. There are again 15 Hamiltonian paths (the same ones as before, reversed in direction), and in this subset the highest is 58 and the lowest 42. In accepting a path with a cost of 27, it is obvious that a very satisfactory result has been obtained.

A semianalytical method of this kind provides a provisional process of a very simple kind for sequencing technological operations, and for the solution of similar problems, until a more precise analytical method is discovered that can be applied to problems of a highly combinatorial nature.

Various improvements can be introduced in the present method; for

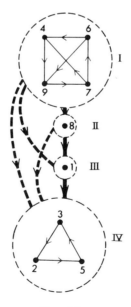

Fig. 4.6.

instance, by eliminating certain costs which are either symmetrically high or symmetrically low in relation to the others. In some problems of this nature, various constraints, which may be linear or not, are sometimes introduced; in such cases it is advisable to find which subset of the subset which has been obtained satisfies these constraints. This subset may be null, and it then becomes necessary to go back to the data in the table of costs and make the requisite readjustments; the particular process needed to make them depends on the nature of the constraints, and it is not always easy to discover it.

It should be realized that a combinatorial problem of production involving 100 products has 100! theoretical solutions, and that 100! is a number consisting of 158 digits. By contrast, the number of solutions in the antisymmetrical graph formed from the least symmetrical costs will be infinitely smaller (only a few hundred or thousand), so that the value of an approximation method using favorable subsets will be fully appreciated.

Alternative Method. In the case of a complete symmetrical graph, the most favorable Hamiltonian path can be found in the following way. Let us go back to Table 4.1 of the costs, and making an arbitrary choice of vertex (1) as our point of origin, cross to the vertex with the least transitional value, ignoring the zeros of the diagonal, which is (7). Let us eliminate line and column (1) and enter the cost of the transition, which is 2, in Table 4.8. In line (7) we look for the least cost, which is 4, and corresponds to $(7) \to (6)$ or $(7) \to (8)$. We make an arbitrary choice of $(7) \to (6)$ and enter 4 in the table, next eliminating line and column (7). The process is continued as follows:

$$(6) \to (8): \ 2; \quad (8) \to (4): \ 2; \quad (4) \to (3): \ 4; \quad (3) \to (5): \ 3; \quad (5) \to (2): \ 5;$$

$$(2) \to (9): \ 3.$$

If the present method gives a better result than the previous one, this is due to the nature of the example. We now give a list of some of the Hamiltonian paths found by this method.

$$
\begin{array}{ll}
(1, 7, 6, 8, 4, 3, 5, 2, 9): & 25 \\
(2, 3, 7, 6, 8, 4, 1, 9, 5): & 38 \\
(3, 7, 6, 8, 4, 1, 9, 2, 5): & 35 \\
(4, 8, 6, 7, 9, 2, 3, 5, 1): & 28 \\
(5, 7, 6, 8, 4, 3, 1, 2, 9): & 28 \\
(6, 8, 4, 3, 7, 9, 2, 1, 5): & 34 \\
(7, 6, 8, 4, 3, 1, 9, 2, 5): & 33 \\
(8, 6, 7, 9, 4, 3, 1, 5, 2): & 33 \\
(9, 4, 8, 7, 6, 5, 2, 3, 1): & 29.
\end{array}
\qquad (4.7)
$$

TABLE 4.8

	c_{ij}
$(1) \rightarrow (7)$	2
$(7) \rightarrow (6)$	4
$(6) \rightarrow (8)$	2
$(8) \rightarrow (4)$	2
$(4) \rightarrow (3)$	4
$(3) \rightarrow (5)$	3
$(5) \rightarrow (2)$	5
$(2) \rightarrow (9)$	3
Total:	25

If an approximation method of this kind is used, it is advisable to examine all the Hamiltonian paths which can be obtained by it. If a number of costs in the table are equal, this may necessitate finding a much greater number of paths than n, where n represents the number of vertices. In (4.7), for example, only a few of these paths were given; but, starting from each vertex of origin, it is possible to obtain more than one Hamiltonian path to certain vertices by using the rule of the arc of least cost.

Other semianalytical methods of a similar kind can be devised, and if they are used in succession, the best Hamiltonian path provided by any of them can be chosen.

Note. In the problem of allocation, for which an exact solution can be found by the method invented in Hungary (see Section 38,D; see also Vol. I, Section 18)[1], it is not possible to obtain a lower bound for the minimum in the problem of determining a minimal Hamiltonian path, for problems of allocation lead to independent circuits or to a single circuit, and not to Hamiltonian paths; that is to say, to permutations.

[1] See footnote 2, p. 18 regarding Vol. I.

5. Tree of a Graph

A. SEARCH FOR AN OPTIMAL DISTRIBUTION NETWORK

It is planned to construct a network of television relays for a country or province. In this country there are seven cities or highly populated regions between which communication must be established. Every possible connection is shown in Fig. 5.1, and although the main transmitter is assumed to be located at A, it could, as we shall see, equally well be situated at any of the other points. The graph of Fig. 5.1 is clearly not oriented, and it is also obvious that a network of relays cannot includes cycles.

Let us assume that a cost can be assigned to each link of the graph, the costs being those of the various possible connections, and let Table 5.1

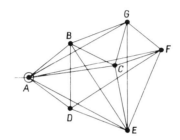

FIG. 5.1.

TABLE 5.1

		A	B	C	D	E	F	G
		1	2	3	4	5	6	7
A	1	0	8	8	3 I	11	10	9
B	2	8	0	4 II	7 V	12	∞	4 III
C	3	8	4	0	∞	9	7	6
D	4	3	′7	∞	0	13	9	∞
E	5	11	12	9	13	0	9	8 VI
F	6	10	∞	7	9	9	0	5 IV
G	7	9	4	6	∞	8	5	0

represent the cost of each hertzian connection. Thus the cost of A toward E (or E toward A)[1] is equal to 11. In each case where a connection is impossible the cost has been considered as infinity (∞). Which network must be chosen to ensure a minimal total cost?

To find it, let us introduce the concept of a *tree*.

B. TREE

A strongly connected graph which does not contain *any cycle* is a tree.

We say that a tree belongs to a given graph if it constitutes a partial graph which does not form any cycle.

It is evident that the number of links in a strongly connected graph is one less than the number of vertices in the graph. Figure 5.2 shows an example of a tree belonging to the graph of Fig. 5.1.

Among all the possible trees, which one or ones are optimal and give a minimal aggregate cost? A very simple algorithm of Kruskal's [H39] enables us to discover the optimal solution or solutions:

"Among the links which are not yet a part of the tree, choose the link which has the least value and which does not form a cycle with those already selected."

Let us make use of this algorithm (see Proof, Section 44, Part II) in the given example. In Table 5.1 we look for the link with the least value. By representing the value of link (i, j) as c_{ij}, we find $c_{14} = 3$ as the minimal value in the table, and we shall therefore start along this link, which will be marked I on Fig. 5.3. We next have a choice between $c_{23} = c_{27} = 4$, in other words between links BC or BG; if we mark BC first, and then BG, we shall not form a cycle. We next mark FG, but not

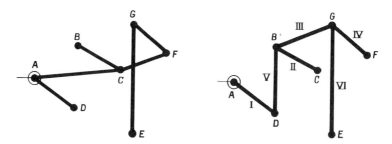

FIG. 5.2.   Example of a tree.   FIG. 5.3.   Optimal tree.

[1] A similar problem could be envisaged with asymmetrical costs due to the particular type of retransmission relays.

CG, for that would form a cycle. We pass on to BD, which will be marked, ignore CF (a cycle), and mark EG. Neither AB nor AC can be chosen, for they would form cycles. Hence, the optimal tree is that shown in Fig. 5.3 which gives a total cost of

$$3 + 4 + 4 + 5 + 7 + 8 = 31. \tag{5.1}$$

In this particular example there is only one solution, but if, for instance, it had been found that $c_{56} = c_{57} = 8$, there would be two optimal solutions.[1]

The search for a minimal tree causes difficulty in many problems, and we shall give as an example of a detailed study of this problem, a lecture given by Sollin[2] in the course of a seminar organized by Berge at the Henri Poincaré Institute in 1961.

C. Problem of the Minimal Tree and of Supplying a Network of Customers

General Remarks.   As soon as a certain level of industrial centralization is reached, whatever the type of business, correlated problems of distribution inevitably arise, and a compromise has to be made between centralization (which reduces the cost of production) and decentralization (which reduces that of transport and distribution). Since this search for a compromise has become one of the main subjects of research, its study has, not surprisingly, led to an examination of particular problems of distribution such as the supplying of a network of customers by means of "channels" or "pipes."

The problem to be solved, the solution of which only represents a suboptimum in a complete study of production and distribution, is the following: "Given customers and a factory, how are they to be interconnected in such a way that the expenses will be minimal?"

To simplify the problem, it will be agreed that the cost used to define the criterion may be expressed as the product of a distance and a fixed price, the latter being decided by the sales of the product to be channeled.

The Distributive Tree.   It is found in practice that the majority of the customers to be supplied by "channeling" can be connected by channels

---

[1] It can be proved that if all the costs are separate there is a single solution (see Reference [H2], p. 151).

[2] Sollin, who is an engineer with *Air Liquide*, has kindly given us permission to reproduce the text of his lecture, and we take this opportunity to thank him, as well as compliment him for his remarkable study. The text has been slightly altered in order to conform with the presentation of other concepts.

in which the cost per mile is constant. This results from the fact that relatively low consumption leads to small diameters for the "pipes" for which the cost of the installing is a predominant factor. Hence, it is possible to approach the problem of piped distribution by a first stage in which the problem to be answered is as follows: "Given customers and a factory, how can they be interconnected so that the network will have a minimal length?"

It is assumed that the corresponding cost will be obtained by a simple process of multiplication.

Since all the customers can be supplied by the same network, it is obvious that the retention of a single diameter does not necessitate provision to be made for the linkage of certain groups of customers, and that in a general sense, the network will form a tree.

Since every network of pipes must be connected to ensure distribution, it is bound to contain at least one partial graph which is a tree; the problem is to discover the minimal one.

Kruskal's algorithm could be used for this purpose, but rigorous as its proof is, this algorithm contains a serious practical drawback due to the continual necessity of finding "the shortest link." In cases where a large number of points have to be connected, a great many links exist, of which only one (or at most a very few) is the shortest link. For this reason we have used another algorithm[1] which, like Kruskal's, is valid whether or not the values assigned to the links represent distances on a map.

Algorithm to Be Used. "We shall proceed in stages by joining any vertex to the one nearest to it,[2] and by so doing form subtrees (in other words trees of the subgraphs of the given graph, none of these subgraphs having a vertex in common with any other). We shall then treat these subgraphs as vertices, repeating the algorithm until the subtree becomes a tree of the given graph."

The operation of joining any vertex or subgraph to the one nearest to it is a simple one involving only a few comparisons, the number of such comparisons being in practice independent of the number of vertices to be connected. In addition, this algorithm possesses the practical advantage that it can begin with any customer and can be worked out by different technicians, each dealing with a separate region.

Let us now take an example to see how this algorithm can be used: the one shown in Fig. 5.1 and Table 5.1 and reproduced in Fig. 5.4.

_____

[1] This will be referred to as *Sollin's algorithm.*

[2] "Nearest" in this sense means a vertex connected to the vertex considered by a link bearing the lowest value.

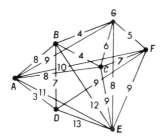

FIG. 5.4.

Commencing arbitrarily at B, we find that the "nearest" vertex is C or G; by choosing C we obtain the partial subtree BC. We now take any point other than B or C; let it be A, the nearest vertex to which is D. Choosing a further point other than A, B, C, or D, we select G, which has F as its nearest vertex. Finally, we take the last remaining point E, to which the nearest vertex is G. We have thus formed three subtrees $\mathbf{X}_1$, $\mathbf{X}_2$ and $\mathbf{X}_3$.

Only a brief examination is needed to discover that the shortest link between $\mathbf{X}_1$ and $\mathbf{X}_2$ is BD, which is worth 7. Between $\mathbf{X}_1$ and $\mathbf{X}_3$ the shortest distance is either AG or DF, each of which is worth 9. (Fig. 5.5). This shows that $\mathbf{X}_2$ is nearer to $\mathbf{X}_1$ than $\mathbf{X}_3$ (Fig. 5.6), and we can now form the subtree $\mathbf{X}_4$ (Fig. 5.7).[1] Between $\mathbf{X}_4$ and $\mathbf{X}_3$ the shortest link is BG which is worth 4, and we can now form $\mathbf{X}_5$, which is the optimal tree. In this example there is only one solution, but other problems might yield several solutions.

The proof of this algorithm has been left until Section 44 of Part II, and we will now proceed to show how Sollin develops the problem of supplying a network of customers (Fig. 5.8).

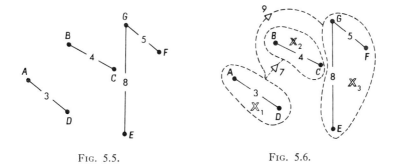

FIG. 5.5.   FIG. 5.6.

[1] This example was not included in Sollin's original lecture and has been added to enable the reader to compare this method with Kruskal's.

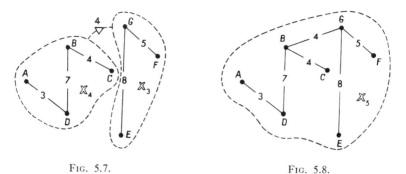

FIG. 5.7.   FIG. 5.8.

Metric Graph.[1]   In the problem to be considered, it is assumed that the values carried by the arcs are proportional to actual distances on a map. When a graph is drawn on a map with a certain scale (provided the links are straight segments connecting the points specified on the map, and the links are given values proportional to the lengths of the segments connecting the points with one another), a *metric* graph or *nontopological* graph is produced, of which Fig. 5.10 is an example. A graph of this nature is not, in the strict sense, a graph as defined by König or Berge, on account of the supplementary property required. Nevertheless, it is customary to accept the term "metric graph" or "nontopological graph." Figure 5.9 shows a topological graph, or more strictly the model of such a graph, in which we are not concerned with distances, but only with the links connecting the vertices. The values associated with the links (or eventually by the arcs) do not correspond to the metric length of the links, which might equally have been represented by some form of curve.

Indeed, a metric graph is drawn according to Euclid's system of orthogonal axes.[2] A vertex is a point on the plane; an arc becomes a

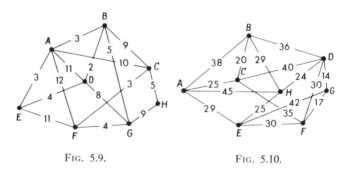

FIG. 5.9.   FIG. 5.10.

[1] Author's addition to Sollin's text.

[2] These are the systems of rectangular axes customarily represented as xOy which are studied in elementary mathematics.

directed segment, a link a nondirected segment. It is assumed that the scale of the two coordinate axes is the same, so that the idea of an angle acquires meaning, which it obviously does not possess in a graph drawn in accordance with the theory of sets; that is, in the sense given to it by König and Berge. Sollin's method is obviously equally suitable for a nontopological graph drawn on a map.

Final Optimization in the Case of a Metric Graph.   If it is decided that a network of pipes must be a tree, there is no *a priori* way of knowing the number of vertices required in an actual case. It may easily happen that the addition of a vertex, which represents a fictitious cutomer, will reduce the value of the sum of the distances on the map. If, for example, it is desired to connect three points to the vertex of an equilateral triangle, the optimization of the metric graph constituting this triangle gives a length equal to two sides, whereas the sum of the altitude and one side gives 1.86 sides (Fig. 5.11). In this case the "improved" number of vertices is increased by a unity corresponding to the foot H of the altitude.

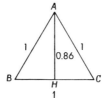

Fig. 5.11. $AB + BC = BC + CA = CA + AB = 2;$ $AH + BC = 1.86.$

By allowing ourselves the latitude of creating as many vertices as we wish, it is possible to make a notable improvement of the minimal tree derived from the geographical position of the customers. As we shall discover, this optimization may amount to as much as 15% of the value of the network which has been optimized, and it leads in a natural way to the addition of vertices, that is to say new locations for customers. Paradoxical as it is, these new customers result in a reduction of the total cost of channelling, so that it is almost possible to speak of "marginal" customers.

The Case of Three Points.   The optimization with which we shall now deal leads to very simple rules. Apart, however, from the paradoxical nature of some of the results obtained by it, the study of these rules is somewhat complex when a general case involving n points is considered. It is for this reason that we shall begin by considering the case of three points, before proceeding to a generalization which will not, in any case, receive a very strict proof.

Let there be three points A, B, C to be joined, such that the minimal tree is formed of the two shortest links. Let us state, with the required proof to support the statement, that the fact we are not drawing the channel corresponding to the longest side AB, means that the angle[1] between the two channels of the minimal tree is always the largest, and therefore can never be less than 60°.

Various Important Properties. The following interesting property is one which can be proved: "The number of supplementary vertices which will reduce the length of a minimal tree with n vertices cannot exceed $n - 2$."

Let us suppose that two vertices Y_1 and Y_2 are added to a tree containing 3 vertices A, B, and C; the number of links is then equal to $3 + 2 - 1 = 4$ in accordance with property (3) which was given at the beginning of the present section. If the two vertices Y_1 and Y_2 are "hanging" vertices (the extremity of one link not being connected with any other), the lengths of the links increase the length of the tree. If either of the points Y_1 or Y_2 is a hanging vertex it must be rejected for the same reason. If neither of the vertices is hanging it necessarily follows, since the tree has four links, that a chain exists of the form Y_1Y_2X or Y_2Y_1X, where X may be either A, B, or C. If these vertices are aligned, this means that one of the two Y's is not to be considered. If these vertices are not aligned, the broken line coinciding with the chain is longer than the straight line coinciding with link Y_1X or Y_2X, which equally leads to the rejection of one of the Y's.

Since the number of links increases in accordance with the number of vertices of the graph, this proof remains valid regardless of how many vertices the original minimal tree may contain.

Another very important property can now be introduced: "The point from which the sum of the distances to the three vertices of a triangle is minimal is the one at which the three sides of the triangle form equal angles[2] of 120°."

---

[1] We must repeat that we are now dealing with metric graphs. The possibility of representing lengths results in the possibility of defining angles. In reproducing Sollin's study, it seemed to us that, apart from the considerable pedagogic interest it would have for our readers, it would also reveal the similarities which exist between Euclidean geometry and topology. We should remember that numerous properties introduced in the theory of graphs are also considered in topology.

[2] The proof is due to de Comberousse, and was carried a further stage by Polya (see *Les mathématiques et le raisonnement plausible*, Gauthier-Villars, Paris, 1958). This theorem has had an important place in the following problems in physics: the study of the trajectory of a luminous ray reflected by a cylindrical mirror; the study of ellipses; and the study of the equilibrium of three converging forces.

The construction required for this point is obtained by the intersection of two arcs with inscribed angles of 120° drawn on two sides of the triangle. The following cases can arise:

(1) Each of the three angles of the triangle is less than 120°. The required point exists, and the minimal tree will always be improved by the addition of this supplementary vertex (Fig. 5.12).

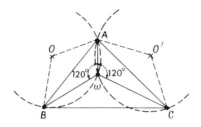

FIG. 5.12.

(2) One angle of the triangle is equal to 120°. The required arcs constructed on the two shorter sides which include this angle are tangent at the vertex of the angle of 120°. Since this angle is the largest angle of the triangle, the minimal tree cannot be improved.

(3) One angle of the triangle is greater than 120°. In this case the three required arcs do not possess a common point of intersection; the intersection of the two arcs constructed on the shorter sides coincides with the vertex of the angle which is greater than 120°, and which marks the optimal limit of the required point. In this case, also, the minimal tree cannot be improved.

Equilateral Triangle (Length of Side = a). The minimal tree with three vertices gives a length equal to $2a$. The minimal tree with four vertices, of which one is the center of the circumscribed circle, gives $a\sqrt{3} = 1.73a$. The minimal tree with four vertices, one of which is the foot of an altitude, gives $(1 + \sqrt{3}/2)a = 1.86a$.

Isosceles Triangle. If we vary the angle α opposite the smallest side belonging to the minimal tree we produce a variation between the minimal tree with three vertices and the one with four vertices, which (expressed as a percentage of the minimal length, and with $\delta(\alpha)$ representing the percentage variation) is as follows:

$\delta(5°) = 1.2;$ $\delta(10°) = 2.4;$ $\delta(15°) = 3.6;$ $\delta(20°) = 4.8;$ $\delta(30°) = 7.3$
$\delta(40°) = 9.9;$ $\delta(45°) = 11.1;$ $\delta(50°) = 12.6;$ $\delta(60°) = 15.5$

(case of the equilateral triangle).

Generalization with More than 3 Vertices. Let us consider an important property: "A supplementary vertex must have three and only three oriented links."

It is obvious that a supplementary vertex must have three oriented links; otherwise, this vertex would require the replacement of one original rectilinear link by two others forming a broken line. If a supplementary vertex W produces more than three oriented links, there are at least two pairs of links which include an angle of less than 120°, and each of this pair of links can therefore be optimized by creating a supplementary vertex.

The number of supplementary vertices after optimization is therefore equal to 3. If n represents the number of vertices X of the original three improved by the first supplementary vertex W, the number of links which include, pair by pair, an angle of less than 120° is equal to or greater than $n - 2$, and cannot be less than 2. Hence, the total number of supplementary summits is equal to $n - 2 + 1 = n - 1$, which contradicts what we proved earlier, namely that the number of supplementary vertices which makes it possible to reduce the length of an original tree with n vertices, by modifying it, can never exceed $n - 2$.

If we apply this "rule of the angles of 120°" to the case of a tree with four vertices (Fig. 5.13) $ABCD$, we obtain the following properties:

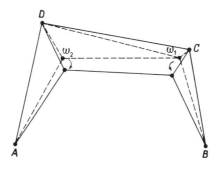

FIG. 5.13.

(1) The number of supplementary vertices cannot exceed 2.

(2) If we improve the minimal partial tree BCD, we find that

$$B\omega_1 + C\omega_1 + D\omega_1 < BC + DC.$$

If we then improve the partial minimal tree AD, we find a point ω^2 such that $A\omega_2 + D\omega_2 + \omega_1\omega_2 < AD + D\omega_1$.

But it is clear that the angle $\omega_2\omega_1 B$ is less than $120°$, which means that the tree $\omega_2\omega_1 B$ may again be improved, which will produce a fresh displacement of ω_1, and will therefore modify the angle $\omega_1\omega_2 A$ which, in turn, will result in a new improvement, and so on.

Since the initial tree $ABCD$ has all its adjacent links inclined at less than $120°$, we can easily verify the appearance of two supplementary vertices with three links issuing from them. And, as the limit, we shall obtain a tree with the same aspect as the one produced by the first improvement, but in which *all the adjacent links include angles of 120°*.

It is possible to reproduce on transparent paper a sheaf of optimal links for quadrilaterals (Fig. 5.14), thus avoiding successive approxima-

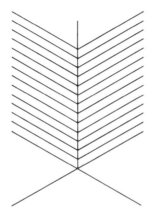

FIG. 5.14. Sheaf of "fasces" of optimal links for reproduction on transparent paper.

tions and enabling us, once the positions of all points ω_1 and ω_2 have been fixed, to trace the three which optimizes the original tree of any quadrilateral.

Whatever the number of points, the algorithm must lead to trees whose adjacent links include angles of the greatest possible magnitude and which are at least *greater than* $60°$, since it has been explained above that if channelling does not correspond to the longest side, this is equivalent to saying that the angle between two channels of the three is always the largest, and is always greater than or equal to $60°$. Hence, the size of the angles between adjacent links of polygons which are optimized will depend on the number of sides as in Table 5.2.

The value of the final optimization decreases when the number of vertices of a polygon exceeds k. Nevertheless, the investment corresponding to the construction of a network of pipes is of such importance that

TABLE 5.2

Number of links, k	Sum of angles, θ	Angle, θ/k
3	180	60
4	360	90
5	540	108
6	720	120
7	900	128
8	1080	135

the absolute value of any saving is never negligible, especially since such economy may be obtained in a very simple manner and with little expense, as we shall now see.

Final Optimization by Means of Mechanical Process. If we retain the hypothesis of simplification introduced at the beginning of this section, namely the assumption of a constant unit of price for the pipes of the whole system, it is possible to improve all the triangles and quadrilaterals by applying the method which we have just discussed in the case of metric graphs. For this purpose, we make use of a transparency with the diagram of Fig. 5.14; but in the case of a pentagon, this work may be both lengthy and awkward.

In addition, the elimination or addition of a fresh point of consumption or production often modifies a whole part of the network. It is therefore useful, in order to make a study of successive approximations possible, to be able to increase or decrease the number of vertices of the system at will, and thereby obtain the corresponding optimal network with the least possible delay.

For this purpose, the "rule of the angles of 120°" has a well-known mechanical counterpart, which is that if three convergent forces are equal and in a state of equilibrium, they form angles of 120° between them. The proof of this is self-evident and consists of showing that the sum of two of the forces balances the third. If we make the axes of these forces coincide with the links of the trees which are to be improved, we can "automatically" obtain the optimal network.

The mechanical process will be carried out in the following way. A geographical map, showing the locality of points of consumption or of production which are to be connected by channels, is fixed to a rigid frame. Holes are bored at each point marked on the map, so that threads attached to small, equal weights may slide across it without appreciable friction. To the free ends of the threads are attached split rings, which are used to make the connections corresponding to the minimal tree found from the algorithm of optimization which has already been given, by joining them successively to each other.

When the minimal tree obtained for the points originally chosen has thus been fitted into position, only a slight impulse need be given to the assembled system (friction having either been eliminated or rendered uniform), for the threads to trace the shape of the improved network.

This mechanical simulator can finally be made to allow for obstacles which cannot be crossed, or for enforced changes of direction. All that need be done is to ensure that a vertex always remains connected to its nearest neighbor at the correct distance (either rectilinear or in a curve formed by the thread), and to place the obstacles on the map so that the thread is forced to slide around them, measuring the new lengths of the arcs when applying the algorithm.

Case of a Variable Unit of Price for the Network of Pipes. The hypothesis of simplification explained earlier, namely that the unit of price for the channels is the same in all the links of the network, is valid for the great majority of customers, but not for all of them. Admittedly, it would be possible to devote a preliminary stage to the study of the more important customers, replace it by a second network with a smaller diameter of piping, then by a third, and so on; but this process does not easily lead to the optimum, and it is liable to prove a lengthy one.

If, however, we examine more closely the mechanical simulator (Fig. 5.15), we see that "the theorem of virtual work," so well known in mechanics (if applied to the system which can be changed into equilibrium) can be expressed, for example (friction being ignored), in the case of three points:

$$F_A \cdot d\omega_A + F_B \cdot d\omega_B + F_C \cdot d\omega_C = 0, \tag{5.2}$$

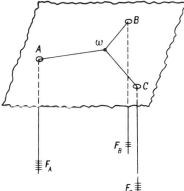

FIG. 5.15.

where F_A, F_B, F_C are the forces acting at A, B, C and $d\omega A$, $d\omega B$, $d\omega C$ are the differences in length of the channels corresponding to the vertical displacements.

If we now write the equation of price for that part of the network and take P_A, P_B, P_C for the price of the unit of length in the different sections of the piping, we have

$$P_A\omega_A + P_B\omega_B + P_C\omega_C . \tag{5.3}$$

Assuming that in the general case the supplementary vertices are subject to the same rules of denumeration, and that there is only one optimum solution, we can state that, as regards the optimum, a slight variation of position of point ω does not produce any change in the overall price for this section of the network, so that

$$P_A \cdot d\omega_A + P_B \cdot d\omega_B + P_C \cdot d\omega_C = 0. \tag{5.4}$$

By comparing (5.2) and (5.4), it can be stated that

$$F_A = kP_A , \qquad F_B = kP_B , \qquad F_C = kP_C . \tag{5.5}$$

In other words, all that has preceded this can be carried to completion by balancing the forces of the mechanical simulator proportionally to the units of cost of the pipes carrying them.

Naturally, the cost of the piping is determined in a preliminary stage, as a function of the relative importance of the customer and of the possible interconnections.

The algorithm based on the theory of graphs enables the optimum to be found, whether the graph is a metric one or not. The second stage, which involves the addition of fictitious customers, does not depend on the theory of graphs, but on Euclidean geometry and mechanics. For this second stage it is also possible to envisage a repetitive process using an electronic computer.[1]

6. Search for an Optimal Flow in a Network. The Ford–Fulkerson Algorithm

A. TRANSPORT NETWORK

Different ports of shipment A_1, A_2,..., A_m, respectively handle tonnages x_1, x_2,..., x_m of a certain product which other ports, B_1, B_2,..., B_n, require in quantities represented by y_1, y_2,..., y_n.

[1] Author's observation.

Each overseas connection between a port A_i and a port B_j has a limited capacity c_{ij}. The problem is to organize the shipments so as to fulfill all the orders in the best possible way.

For this purpose, a transport network is constructed by connecting a point of entry A to the m vertices A_i by arcs with a capacity x_i, and the n vertices B_j to an exit B by arcs with a capacity y_j. By letting ζ_j represent the total tonnage reaching a port B_j, and η_i the shipments from the ports A_i, where r_{ij} stands for the tonnage carried from A_i to B_j, the total tonnage will be maximal if

$$[\text{MAX}]R = \sum_{i=1}^{m} \sum_{j=1}^{n} r_{ij}, \tag{6.1}$$

$$0 \leqslant r_{ij} \leqslant c_{ij}, \qquad i = 1, 2, ..., m; \qquad j = 1, 2, ..., n, \tag{6.2}$$

$$\sum_{i=1}^{m} r_{ij} = \xi_j \leqslant y_j, \qquad j = 1, 2, ..., n, \tag{6.3}$$

$$\sum_{j=1}^{n} r_{ij} = \eta_i \leqslant x_i, \qquad i = 1, 2, ..., m. \tag{6.4}$$

It should be added that certain capacities c_{ij} may be null on the hypothesis that there is no overseas connection in their case.

In Fig. 6.1 we have shown this problem by means of a graph in which we have placed a common point of origin A and a common point of destination B, with arcs of which the respective capacities are the x_i's and y_j's.

Care must be taken not to confuse the present problem, in which the maximum flow of goods to be obtained with a limited capacity, is being

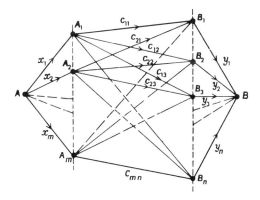

FIG. 6.1.

sought, with the one treated in Vol. I (Section 17),[1] where the aim was to minimize the over-all cost of transportation.

The problem defined in (6.1)–(6.4) could be solved by linear programming, but there is a much neater way of obtaining a solution based on the theory of graphs. We shall explain it further on; but first of all let us show, from another example, how useful this type of model may be.

B. TRAFFIC SYSTEM

An oriented system of traffic such as that of Fig. 6.2 is to be considered. At instants $0, 1, 2, 3, ..., \theta - 2, \theta - 1$, the road junctions A, B, C receive streams $a_0, a_1, a_2, ...$; $b_0, b_1, b_2, ...$; $c_0, c_1, c_2, ...$; $c_{\theta-2}, c_{\theta-1}$. Vehicles may stop at A, B, C, or D.

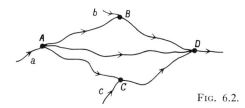

FIG. 6.2.

The capacity of roads AB, AC, BD, and CD are known, and also the capacity for stationary vehicles at A, B, C, and D, capacities which vary according to the weather. The average time required to traverse arcs AB, AC, etc., is known (see Fig. 6.3):

$$t_{AB} = 1, \qquad t_{AC} = 1, \qquad t_{AD} = 2, \qquad t_{BD} = 1, \qquad t_{CD} = 1.$$

The aim is to determine the number of vehicles which can enter each road or must be checked at the junctions between instants 0 and 1, 1 and 2,..., $\theta - 1$ and θ, so that the total flow at junction D may be maximal from 0 to θ.

To work out this problem of traffic control, all that need be done is to treat it in terms of the flow in a network. To do this, a graph will be drawn in the following manner. First, we define the vertices:

$$A_0, \quad A_1, \quad A_2, ..., A_{\theta-1}, \quad A_\theta; \quad B_0, \quad B_1, ...; \quad C_0, \quad C_1, ...; \quad D_0, \quad D_1, ..., D_\theta,$$

symbols which indicate the location at any given instant.

We begin the graph by connecting A_0 to A_1 by an arc, then A_1 to A_2,..., $D_{\theta-1}$ to D_θ, in order to indicate the different capacities for

---

[1] See p. 18, footnote 2.

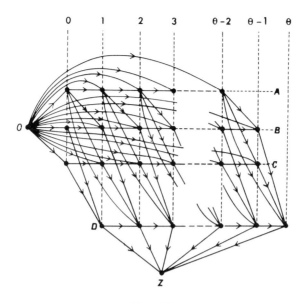

FIG. 6.3.

stationary vehicles between the designated instants, some of which may be null and others infinite. Next, we shall connect the junctions, taking account of the existing roads and the times required to traverse them. Thus, A_0 will be connected with B_1, C_1 with D_2, C_3 with D_4, $C_{\theta-1}$ with D_θ,..., etc. A supplementary point 0 will be added, which will be joined to all the points A, B, C, distinguished from each other by their subscripts, and the corresponding arcs will be assigned capacities of a_0, a_1,... ; b_0, b_1,... ; c_0, c_1,..., $c_{\theta-1}$. A second supplementary point Z will be added, and to it all D points with their various subscripts will be connected, the arcs in this case being assigned an infinite capacity. In this way, the problem becomes one of finding a maximal flow in a network, where the vertices are situated both in space and in time.

C. Flow in a Network

The two examples which we have discussed are special cases of a more general problem which may be entitled "problem of the flow in a transport network," and which can be expressed in Fig. 6.4.

Given a connected graph with two vertices X_0 and X_n called respectively "entry" and "exit," all arcs with a vertex at X_0 are pointed toward the interior of the graph, and all those with a vertex at X_n are pointed toward the exterior of the graph (Fig. 6.4). Each arc is assigned a non-

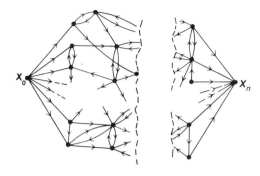

FIG. 6.4.

negative number called its "capacity." A certain quantity of material
will enter at X_0 and the same quantity will leave by X_n . The amount of
material which enters an arc must be less than or equal to its capacity.

The problem, with which we are confronted, is to find the quantity of
material which must enter the different arcs so that the total flow crossing
the transportation network will be maximal, with the restriction that the
partial flow reaching each vertex must be equal to the partial flow
leaving it (conservation of flow).

D. THE FORD–FULKERSON ALGORITHM [H26, H27]

This algorithm makes it possible to calculate the maximal flow
through a network, and its development will be based on the example
shown in Fig. 6.5, in which the numbers placed beside the arrows on the

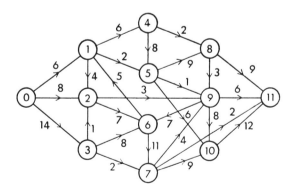

FIG. 6.5.

arcs represent their respective capacities. In this figure the point of entry is at 0 and the exit at 11.

(1) FINDING THE REQUIRED FLOW.   At 0, a given flow which is expressed in integers,[1] enters the network and is passed on from vertex to vertex in such a way that the property of conservation of the flow is ensured at each vertex. Thus at vertex 1 in Fig. 6.6:

$$6 + 1 = 3 + 1 + 3.$$

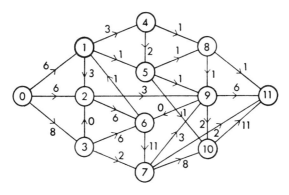

FIG. 6.6.

If the flow is too great for the given capacities, it is reduced to the point where they can absorb it, and it is by this procedure that Fig. 6.6 is obtained.

(2) FINDING A COMPLETE FLOW.   A flow is complete if each path leading from the point of entry to that of departure includes at least one *saturated* arc.

To obtain a complete flow it is only necessary to consider the graph restricted to the nonsaturated arcs. If the flow is not complete a path exists leading from the entry to the exit. The flow in this path is increased by units to the point at which there is at least one saturated arc. The process is then repeated until every path contains at least one saturated arc.

In Fig. 6.7 the saturated arcs are first marked by heavy lines, and a path from (0) to (11) is then sought which does not include a saturated arc. There is one (0, 2, 6, 1, 4, 8, 11), and the flow passing through each

---

[1] If the capacities are not given as integers it is a simple matter to transform them into whole numbers by an appropriate multiplication.

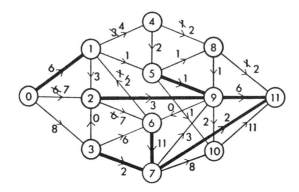

FIG. 6.7.

arc of this path in increased by unity, which results in the saturation of
arcs (2,6) and (4,8).

We now proceed to Fig. 6.8 where we find another path leading from
the entry to the exit which does not pass through a saturated arc: this is
path (0, 3, 6, 1, 5, 8, 11). The flow in the arcs of this path is increased by
unity, resulting in the saturation of (1,5), and leading to Fig. 6.9. There
is still a path (0, 3, 6, 1, 4, 5, 8, 11); we increase this by 1, thus saturating
arc (3,6) and giving us Fig. 6.10, in which it can be easily seen that it is
no longer possible to find a path from the entry to the exit which does
not include a saturated arc. In this way we have now found a complete
flow.

Given a complete flow mark the vertices in the following manner:

(a)   Mark the entry with a [+];

(b)   After marking a vertex X_i, mark every X_j, which has not been

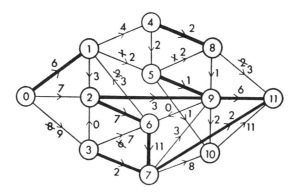

FIG. 6.8.

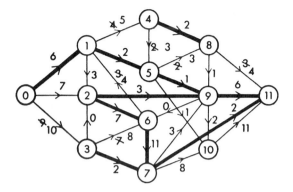

FIG. 6.9.

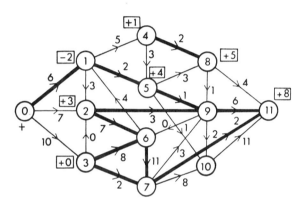

FIG. 6.10.

marked, and where there is a *nonsaturated arc* (X_i , X_j) with $[+X_i]$; mark every vertex X_j, which has not been marked, and where there is an arc (X_j , X_i) traversed by a flow which is not null, with $[-X_i]$.

(c)   If this process leads to the exit being marked, the flow is not maximal. In that case, one examines a chain of points marked (+ or −), leading from the entry to the exit and the sequence of vertices which it contains. If an arc in this sequence is oriented in the order of the sequence the flow in this arc is increased by unity; if it is oriented in the inverse direction unity is subtracted from the flow in this arc.

In the example given, the process is as follows (Fig. 6.10): (0) is marked with [+]. We mark (3) with [+0] and (2) with [+3]; (1) with [−2]; (4) with [+1]; (5) with [+4]; (8) with [+5]; and (11) with [+8]. Since the exit has been marked, the flow is not optimal. We must therefore add 1 to the flow in the arcs pointing in the direction of the path, and subtract 1 from the others in the path. Hence (0,2) will change from 7 to 8; (2,1) from 3 to 2; (1,4) from 5 to 6; (4,5) from 3 to 4; (5,8) from 3 to 4; and (8,11) from 4 to 5.[1]

A fresh arc (1,4) is saturated, and we turn to Fig. 6.11. We mark (0) with [+]; (3) with [+0]; (2) with [+3]; (1) with [−2]; and (6) with [−1]; but we cannot proceed further, since arc (6,9) is pointed in the opposite direction to the path and has a null flow, and hence cannot be diminished by unity.

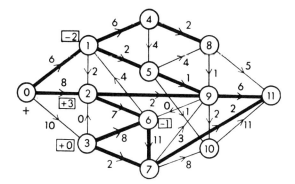

FIG. 6.11.

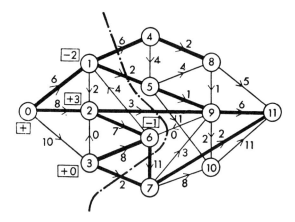

FIG. 6.12.

[1] We might also consider the chain passing through X_3 which would lead to another solution.

It is easy to see that the exit cannot any longer be marked, and a maximal aggregate flow has therefore been found. To verify this, let us make what is termed a *cut* (Fig. 6.12) by drawing a line separating the marked from the unmarked vertices. This cut is formed by saturated arcs or a nonsaturated one, but with a null flow oriented from the non-marked vertices in the direction of those that are marked. Hence, it is realized that the flow can no longer be increased. The cut gives the value of the maximal flow: $6 + 2 + 3 + 0 + 11 + 2 = 24$.

Finding a Minimal Flow. In certain problems the minimal flow[1] may be required. The Ford–Fulkerson algorithm, if suitably altered (see Part II, Section 37), enables this to be found.

E. Problem of the Maximal Traffic of a Railroad

One of the most remarkable practical applications of the theory of the flow in a transportation system has been given by Matthys and Ricard [H44] who have studied the maximal expense of transporting freight by railroad between two areas of a section of noninterchange double-track railroad which is subject to passenger priorities. It is as the result of this study that the expense has been appreciably reduced on the Paris–Lyons line, on which there are a number of passenger trains with a priority, and over which it is necessary to maintain an important freight service. The method employed consisted of solving a problem of flow in a traffic system, of which the vertices are coordinates expressing *time-distance* (as in Fig. 6.3), but in which the arcs (pointed towards positive times and corresponding to a particular station or siding), have a given capacity g_i, whereas the vertical arcs (representing a particular instant, but leading in the direction of the positive distances), have a capacity equal to 1. It is to be noted that a path of this system, that is to say, an unbroken succession of consecutive arcs, leading from the point of origin (A) to the terminal point (Z), reproduce the timetable of a freight train in such a way that it is compatible with the priorities. The vertical arcs correspond to the passage of the train from one station or siding to the next one, and the horizontal arcs to the periods when the train is stationary. The problem therefore becomes that of finding the maximal number of paths leading from A to Z, which must also be compatible with one another. With an increase in the number of trains and stations, it became necessary to use an electric computer, on which the general algorithm was programmed.

---

[1] It is clearly a case of finding a quantity of matter greater than or equal to the capacity of each arc.

7. Application of the Theory of Graphs to Psychosociology

A. SYSTEMS OF ASSOCIATION

Psychologists and sociologists were among the first to make use of the theory of graphs, and in order to explain the form which their interest took, we shall first define some concepts frequently used in this sphere. These concepts will be considered in the case of nonoriented graphs.[1]

B. OTHER NONORIENTED CONCEPTS

ARTICULATED POINT. This is a vertex P which separates the vertices of the graph into two subsets with vertices $\mathbf{E}_1$ and $\mathbf{E}_2$, having only P in common and such that any chain between a point of $\mathbf{E}_1$ and a point of $\mathbf{E}_2$ must pass through P.

Example (Fig. 7.1). Vertices C and D are articulated points.

NONARTICULATED OR CLUSTERED GRAPH. This is a connected graph which does not possess an articulated point.

Example. When considered separately, the subgraphs *DEFGHI* and *ABC* in Fig. 7.1 form clusters.

DISTANCES. The distance between two vertices of a connected graph is the length of the shortest chain, measured by the number of arcs in it, which may not be unique. This distance is referred to as $\lambda(X_1, X_2)$, namely "the distance from X_1 to X_2."

Example (Fig. 7.1).

$$\lambda(A, F) = 4, \qquad \lambda(E, I) = 2.$$

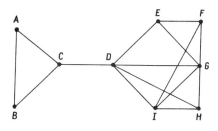

FIG. 7.1.

---

[1] These concepts can equally be extended to oriented graphs. See, for example: **Part II,** Sections 40, 41, and 46.

DIAMETER. This is the greatest distance which can be found in the graph.[1] To find the diameter, the table of distances (Table 7.1) is drawn up and the maximum in each line is selected. The *maximum maximorum* is then found, and this constitutes the diameter. Thus, the diameter of the graph in Fig. 7.1 obtained from Table 7.1 is equal to 4.

TABLE 7.1

MATRIX OF DISTANCE

	A	B	C	D	E	F	G	H	I	MAX
A	0	1	1	2	3	4	3	3	3	4
B	1	0	1	2	3	4	3	3	3	4
C	1	1	0	1	2	3	2	2	2	3
D	2	2	1	0	1	2	1	1	1	2
E	3	3	2	1	0	1	1	2	2	3
F	4	4	3	2	1	0	1	2	1	4
G	3	3	2	1	1	1	0	1	1	3
H	3	3	2	1	2	2	1	0	1	3
I	3	3	2	1	2	1	1	1	0	3

SEPARATION OF A VERTEX.[2] This is the maximum of the distance of this vertex to all other vertices, the separation of vertex X being represented by $s(X)$.

Example (Fig. 7.1 and Table 7.1). $s(A) = 4$, $s(D) = 2$, $s(I) = 3$, which are the highest values in the MAX column in Table 7.1.

The greatest separation gives the diameter.

CENTRAL POINT OR CENTER. This is the point in a connected graph which has the minimal separation, and it need not always be a single point.

[1] See Section 41 for a more precise and restrictive definition of this concept.

[2] It is also referred to as the *number associated with the vertex*.

Example (Fig. 7.1 and Table 7.1). *D* is the center of the graph and is found by taking the smallest number in the MAX column of the matrix (Table 7.1).

RADIUS. This is the separation of the center (or centers) of the graph.

Example (Fig. 7.1 and Table 7.1). The radius of the graph is equal to 2.

POINT ON THE PERIPHERY. This is the point with the maximal separation.

Example (Fig. 7.1 and Table 7.1). Points *A*, *B*, and *F* are points on the periphery.

C. APPLICATION TO PSYCHOSOCIOLOGY

Various mathematicians and psychologists, in particular, Harary [H5] and Leavitt [H40], have studied the structure of systems of association between individuals, and have sought different characteristics and numerical values in these systems which are of special significance for a psychologist. These are chiefly the presence of articulated points, the centrality or remoteness of the members of a group possessing possibilities of association. To understand this better, let us consider four types of associative systems, each consisting of five people, shown in Fig. 7.2 with the different tables of distance. In the case of each of these graphs, we have the diagram shown by Fig. 7.2 and Table 7.2.

Leavitt has introduced the concept of "an index of centrality" β_i for each vertex X_i, such that:

$$\beta_i = \frac{\sum_i \sum_j \lambda(X_i, X_j)}{\sum_j \lambda(X_i, X_j)}, \tag{7.1}$$

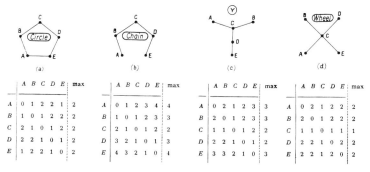

	A	B	C	D	E	max			A	B	C	D	E	max			A	B	C	D	E	max			A	B	C	D	E	max
A	0	1	2	2	1	2		A	0	1	2	3	4	4		A	0	2	1	2	3	3		A	0	2	1	2	2	2
B	1	0	1	2	2	2		B	1	0	1	2	3	3		B	2	0	1	2	3	3		B	2	0	1	2	2	2
C	2	1	0	1	2	2		C	2	1	0	1	2	2		C	1	1	0	1	2	2		C	1	1	0	1	1	1
D	2	2	1	0	1	2		D	3	2	1	0	1	3		D	2	2	1	0	1	2		D	2	2	1	0	2	2
E	1	2	2	1	0	2		E	4	3	2	1	0	4		E	3	3	2	1	0	3		E	2	2	1	2	0	2

FIG. 7.2.

TABLE 7.2

	Circle	Chain	Y	Wheel
Articulated point	none	B, C, and D	C and D	C
Center	none (or all)	C	C and D	C
Points on the periphery	none (or all)	A and E	A, B, and E	A, B, D, E

which is an index obtained by dividing the sum of the elements in the table of distances by the sum of the distances of X_i to the other vertices (the sum of the elements in line i of the table of distances). He has also introduced a "relative index of peripherism" γ_i for each vertex X_i, which is the difference between the value β_k of the center X_k of the graph and the value β_i of vertex X_i.

For the graphs of Fig. 7.2 we obtain Table 7.3.

TABLE 7.3

	Circle		Chain		Y		Wheel	
	β_i	γ_i	β_i	γ_i	β_i	γ_i	β_i	γ_i
A	5	0	4	2.66	4.5	2.7	4.57	3.43
B	5	0	5.71	0.95	4.5	2.7	4.57	3.43
C	5	0	6.66	0	7.2	0	8	0
D	5	0	5.71	0.95	6	1.2	4.57	3.43
E	5	0	4	2.66	4	3.2	4.57	3.43

From these results, it is possible to make some interesting deductions about the systems of association under consideration. The arrangement in the form of a *wheel* is the most centralized, and in the form of a *circle* the least centralized. In the wheel, the elimination of C ends all connection; in the circle the elimination of a potential association, for instance of A with E, changes the circle into a *chain* and increases the centrality of C. Eliminating an element such as A also transforms the circle into a chain, but with a proportional increase of the centrality of C and D.

With such simple graphs as those of Fig. 7.2, the deductions which can be made with the help of the theory are almost self-evident, but this is not so if the system of association is a much more complex one. To find the articulated points and the different indices in such cases it becomes necessary to make use of matricial calculation. For this purpose, Harary [H5] has introduced numerous properties, and the references in [H7] and [H35] can also be profitably studied.

Using the theory of graphs, psychologists are able to decide to what extent the role of a particular person is crucial (a result which occurs if he happens to occupy a point of articulation with a high degree of centrality); what the effect of a new association or of the elimination of others will be; and what conditions of redundancy[1] may be produced by certain cycles, etc.

D. MINIMAL ARTICULATED SET

This is a set[2] consisting of the least number of vertices whose removal from a system of communication will divide it into two separate, non-connected systems. If this set is reduced to a single vertex, it is an articulated point.

An example given by Berge [H2] shows how useful the discovery of a minimal articulated set can prove.

A study has been made of the connections (roads, railroads, etc.) between two regions with the intention of destroying the centers of

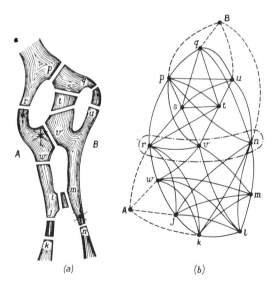

(a)   (b)

FIG. 7.3.

---

[1] According to mathematical definition, a *redundancy* is a repetition within a structure.
[2] This set may not necessarily be unique.

communication (bridges, tunnels, construction works, etc.), so as to prevent communication between the two regions. For instance, in the map of Fig. 7.3, what is the minimal number of bridges which must be destroyed to end communication between regions A and B on different sides of the river?

To discover this, we convert the map into a graph representing the connections between the various bridges. Thus, the possibility of going from bridge r to bridge s is shown by a symmetrical arc between r and s, and so on for s to u, etc., but not in the case of r to m, or u to v, etc. The two regions A and B are treated as vertices which must be connected by links representing the bridges which have an approach in one or other of the regions. The problem has thus become one of finding from the graph of Fig. 7.3b a minimal set of bridges whose elimination will separate the graph into two disjoined subgraphs, one of which contains A and the other B. Several methods exist for effecting this optimal separation (see Part II, Section 46, and reference H2), and it is found that the destruction of r, v, and n will separate regions A and B, between which there will no longer be any means of communication.

Similar problems may arise in connection with such questions as finding the sensitive parts of an organization, management, a delivery network, telecommunication connections, etc.

E. Multigraphs

A multigraph is a generalization of the concept of a nonoriented graph. Instead of connecting any two vertices by zero or by a link, the vertices are now connected in pairs by more than one link. If the greatest number of links between any pair of vertices is p, the graph is referred to as a p-graph.[1]

Figure 7.4 shows an example of a 3-graph and represents a gathering of four people who are able to converse with each other in different languages. Other examples can be thought of: for instance, in chemistry, ethylene can be represented by a 2-graph, and acetylene by a 3-graph, and several such examples are given in Fig. 7.5. Using the theory of graphs as their basis, specialists in organic chemistry can make some very interesting deductions. [2]

---

[1] Some authors define the idea of a multigraph or p-graph for oriented concepts. We prefer, as will be seen in Part II, Section 34, to introduce a different terminology (p-colored and p-applied graphs) for this concept.

[2] See, for instance, the thesis "Fundamental numbers of the graphs and matrices of incidence of organic chemical compounds" by E. N. Albino of the University of Buenos Aires, 1962. Although this application of the theory of graphs does not deal with problems of organization, it seemed to us sufficiently interesting to be cited.

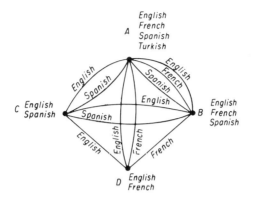

FIG. 7.4.

$$H - C \equiv C - H$$

Acetylene (3-graph)
(a)

Benzene (2-graph)
(b)

Ethane (1-graph)
(c)

Anhydrous ethane
(2-graph)
(d)

FIG. 7.5.

8. Automatic Textual Emendation.[1]
Search for Lost Manuscripts

A. HISTORICAL BACKGROUND

If the reader should be surprised at finding an example so alien to the sphere of economics, he may be reminded that the field of operations research covers far more than economics or its application to military affairs. The present example is one which is of interest to the philologist and, in a wider sense, to those engaged in documentary or historical research.

Some twelve years age, the monks of the Abbey of Solesmes undertook, under the supervision of Dom Froger, the compilation of an emended edition of the Roman Catholic Gradual, the collection of chants which may be used during Mass. To restore the original liturgical words and even the original music from the copies of them which existed, they began by applying the ordinary methods of textual emendation; and as these methods still contained obscure and controversial details, they were obliged to study and evaluate these methods as well.

In the course of this work, Dom Froger became aware that some of the processes employed had a completely logical form, and that it might be possible to feed them to an electronic computer, using the theory of graphs to define and handle the structures which they had encountered. An exposition of these new methods was made public at the Congress of Lexicography at Besançon in June, 1961. It bore the imprint of the Order of St. Benoit, known as the "Order of the Four Fathers," which was already well known for its earlier studies.

It soon became clear that this method of programming, which was henceforward available to interested philologists, could be used in a much broader sphere. In fact, from the time that the older methods of textual emendation were first programmed[2] on a machine, they have been made so much more profound and so generalized that it is now possible to deal successfully with most imaginable cases of textual uncertainty. In particular, cases which were notorious for their "contamination" at several stages have been satisfactorily solved.

B. GENEALOGY OF DOCUMENTS

The new theory will be shown by means of an example from the

---

[1] This section, prepared in collaboration with P. Poré, is taken from the article "Critique textuelle automatique" by J. Poyen and P. Poré (Engineer, SEMA and Technical Advisor, Compagnie des Machines BULL, respectively). We thank the authors for their kind permission to present their work.

[2] The work was carried out on GAMMAET, manufactured by the *Compagnie des Machines BULL*.

study undertaken by the Order of the Four Fathers and made public at the Congress in 1961.

Let us consider an original manuscript ω which is now missing and has to be traced and restored. Three copies of this manuscript are known to have been made and will be referred to as C, H_a, and N. Later, three copies of H_a were also made, and will be known as H, S, and T. The same capital letter H is used for two manuscripts, H_a being the original text in which certain passages were erased by an emendator and were replaced by others. There are therefore parts of two separate versions on the same parchment, and we are now dealing with "palimpsests."

In any case, whether there is a palimpsest or not, we can represent the sequence of different versions by a genealogical tree, an oriented figure which philologists have termed a "stem." Mathematicians will recognize that it can be shown as a "graph" in which the vertices are the manuscripts, and the arcs the filiations.

A stem (graph) is characterized by two basic properties:

(1) The *linking*, which shows that we pass from one manuscript to another through a series of intermediate ones without being concerned with the direction.

Example (Fig. 8.1). (S, H_a, ω, N) comprise a *chain*, and also (ω, H_a) and (T, H_a, H);

(2) The *orientation*, which assumes the choice of a point of origin and a sense of direction: from father towards son.

These two properties may be disjoint. The linking presupposes a structure which is independent of the orientation and can be found separately; the same linking would be found with a different point of origin (see, for example, Figs. 8.1 and 8.2). Hence, it can be said that genealogy is "absolute" or "true" if the point of origin is the same as the historical ancestor; it will be "relative in relation to C" if the origin represents any of the other manuscripts, such as C.

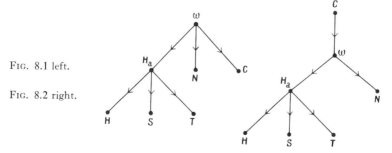

FIG. 8.1 left.

FIG. 8.2 right.

C. ERRORS

In certain parts of the text known as "points of variance," the manuscript H_a, which was copied on ω contains errors due either to carelessness or to a deliberate, but unsuccessful attempt to correct the model. These errors comprise a class which we shall call $F(H_a)$, and have been transmitted to manuscripts H, S, and T, which are derived from H_a, with the result that they are common to these four manuscripts. Hence, $F(H_a)$ may represent either:

(a) The "reponsible" manuscript H_a, in which the errors $F(H_a)$ first appeared:

$$F(H_a) \rightarrow H_a.$$

(b) The faulty "group" of manuscripts which have the errors $F(H_a)$ in common, a group which will be referred to as G_{H_a}, with the result that

$$F(H_a) \rightarrow G_{H_a} = (H_a, H, S, T).$$

The error is said to be "absolute" or "true" if the manuscript of reference is the original, true ω; it is "relative in relation to C" or "variant in relation to C" if the manuscript of reference C is one of the other manuscripts.

In order to illustrate our particular example, we have given a list of variant points, each represented by the number of the line in the manuscript of reference C where it appears. In each case the alternative words are given, and opposite them the list of manuscripts in which either appears.

Example. In line 69 of Table 8.1, manuscript C has the word "quoque," which is reproduced in manuscript N, but is omitted from manuscripts H, H_a, S, and T.

On line 69, the manuscript of reference has the word "quoque"; manuscripts H, H_a, S, T, from which it is missing, commit a "relative error" in relation to C.

The set of variants $F(H_a)$ is made up of the places of variance underlined in Table 8.1, so that

$$F(H_a) = \{35, 40, 62, 69\}.$$

D. GRAPH OF THE FAULTY GROUPS AND GRAPH OF THE RESPONSIBLES

In our example (Fig. 8.2) a vertex represents a manuscript and an arc an affiliation. Each manuscript, such as H_a, is responsible for a class

TABLE 8.1

34	SEDENTIBUS	C, H, H_a, S, T
	RESIDENTIBUS	N
35	NOSTRUM	C, N
	( )[a]	H, H_a, S, T
36	DISTRIBUERET	C
	TRIBUERET	H, H_a, N, S, T
38	DE PRIMA CONJONCTIONE COENOBITARUM	C
	( )	H, H_a, N, S, T
40	DIXIT	C, N
	( )	H, H_a, S, T
41	MISERICORDIA DOMINI	C, H, H_a, S, T
	DOMINI MISERICORDIA	N
45	HABITARE FRATRES	$C. H, H_a, S, T$
	FRATRES HABITARE	N
49	QUO	C
	( )	H, H_a, N, S, T
52	JAM NUNC	C, H, H_a, N, S
	NUNC JAM	T
54	PROSEQUAMUR	C, H, H_a, S, T
	PROSEQUAR	N
57	UNANIMITAS IPSA	C, H, H_a, N, S
	IPSA UNANIMITAS	T
62	PRAESSE	C, N
	ESSE	H, H_a, S, T
69	QUOQUE	C, N
	( )	H, H_a, S, T
73	DOMINO	C, H, H_a, S, T
	DOMINI	N
106	PRIMATUM	C, H, H_a, N, S
	PRINCIPATUM	T
117	HUJUSCEMODI	C, H_a, N, S, T
	HUJUSMODI	H
154	QUIS	C
	EJUS	H, H_a, N, S, T
155	OBLATIO SUSCIPIATUR	C, H, H_a, N, T
	SUSCIPIATUR	S
296	DEBET	C, H, H_a, S, T
	DEBEAT	N
304	PSALLENDI ORDINEM	C
	PSALLENDUM	H, H_a, N, S, T
317	AMPUTANDA	C
	AMPUTANDAE	H, H_a, N, S, T
335	LATEBRIS	C, H, H_a, S, T
	ILLECEBRIS	N

[a] The parentheses indicate the word in question has been omitted.

of errors $F(H_a)$ transmitted to its descendants. Thus, from the original graph (G_e) (Fig. 8.3) a new graph (R_e) (Fig. 8.4) can be drawn of "the responsibles" in which the vertices are still manuscripts, but in which the arcs now represent the transmission of errors. In our example, (G_e) and (R_e) are identical, but this is not so in the more frequent cases where there has been contamination. If each vertex X in (R_e) is now made to represent the faulty group which have the errors of X in common (that is to say, of X itself and its descendants), the "graph of the faulty group" or the "graph of the errors" (Fig. 8.5) is obtained. In this graph each group contains its descendants.[1]

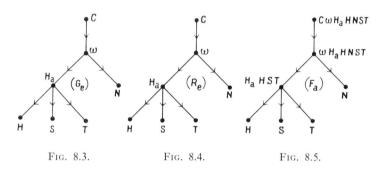

FIG. 8.3.   FIG. 8.4.   FIG. 8.5.

E. INVERSE RESTORATION OF THE GENEALOGICAL GRAPH GIVEN THE CLASSES OF ERRORS

We have explained this system from the didactic standpoint, which consists of first describing structures and then drawing deductions from them. In this case the structure is as follows: manuscripts follow one another in accordance with a genealogy in which each "father" transmits to his "son" the faults which he has inherited, as well as his own. If contamination occurs, the errors of each parent are divided among several classes: those which he transmits to his descendants and those which are corrected through the contributions of the other parents. Hence, we are dealing with a genealogy containing heredity and systematic mutation at each individual level.

Starting from these premises, we have deduced the form of the graph of responsibles, and then of the graph of errors.

Hence, our reasoning has followed the deductive channel:

$$\text{Genealogy} \to \text{Error} \to \text{Graph } (R_e) \to \text{Graph } (F_a).$$

---

[1] In (F_a) the arc represents a strictly inclusive relationship in the mathematical sense of the concept.

In practice, the problem which has to be solved is the inverse of the didactic exposition that has been given. Instead of describing causes and deducing their consequences, we examine the consequences in order to discover their causes. It is the inductive path which we now have to follow.

COLLATION OF MANUSCRIPTS. The longest manuscript C is chosen as that of reference. Every manuscript x is then compared with C, and all the differences are noted in the form of omissions or additions. In point of fact, every variation may be included under these two headings, of which we give an example (variant point 106 in Table 8.1):

word in C :   PRIMATUM,

and in x :   PRINCIPATUM.

This may be written as: omission of PRIMATUM, addition of PRINCIPATUM.

These comparisons will finally emerge as a text x which will include all the observations which have been noted, and also in the form of perforated cards bearing the numbers of the variant places and the group of manuscripts in which the variants appear. The philologist is then asked to eliminate minor variations, such as differences of spelling, which are not truly characteristic of the errors.

We are eventually left with a list of variant places (Table 8.2), the

TABLE 8.2

34	N
35	H, H_a , S, T
36	H, H_a , N, S, T
38	H, H_a , N, S, T
40	H, H_a , S, T
41	N
45	N
49	H, H_a , N, S, T
52	T
54	N
57	T
62	H, H_a , S, T
69	H, H_a , S, T
73	N
106	T

Observations $\rightarrow (F_a) \rightarrow (R_e) \rightarrow$ Genealogy and text of lost manuscripts.

group associated with each place consisting of those manuscripts which disagree with the manuscript of reference C, and which we have hitherto referred to as "faulty with regard to C."

The detailed restoration will be given further on, but at this stage we wish to explain the basic principle, for it includes a preliminary concept of importance—that of the lost manuscript.

Let us suppose, therefore, that the list of the classes of errors $F(x)$ is known, as well as the associated groups G_x. The graph (F_a) is drawn by plotting groups G_x, the largest including the smaller ones immediately beneath them. With (F_a) as the vertex of each group G_x, the manuscripts in the smaller groups below the vertices are cut off and the graph (R_e) is obtained. Each vertex will now either have the name of a manuscript or be blank.

An unnamed vertex of this kind may represent a lost manuscript, a fact which would explain the errors common to two lines of descent: a mutual "ancestor," which is unknown, has transmitted its errors to each of the lines descended from it. Such is the case with manuscript ω in our example.

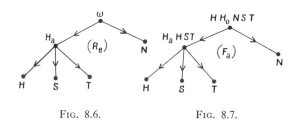

FIG. 8.6.   FIG. 8.7.

F. PRACTICAL ASPECTS OF THE PROBLEM

The ultimate aim is to reconstruct the original manuscript from the copies of it which remain. To achieve this, the work is divided into two main stages: the search for the sequence of concepts, then the quest for the true orientation.

The first step itself can also be divided into several phases:

(1)   Collation of the manuscripts to obtain the list of errors.

(2)   Regrouping the basic errors into classes.

(3)   Groups G_x are ordered by inclusion: graph $(F_a) \rightarrow$ graph (R_e).

(4)   Graph (R_e) is interpreted and the genealogy is deduced from it.

The first three steps are carried out with the assistance of an electronic computer, the results being interpreted by a philologist, who is alone responsible for the final stage.

TABLE 8.3

117	H
154	H, H_a, N, S, T
155	S
296	N
304	H, H_a, N, S, T
317	H, H_a, N, S, T
335	N

Regrouping the Errors in Classes

The errors disclosed by the same group of manuscripts do not have to be distinguished from one another. Their lot, in fact, can be regarded as a common one: they were created by the same responsible manuscript which transmitted them to its descendants. All these "equivalent" errors will accordingly be rearranged in one class, sometimes called a *constellation*, which will be defined by

(1)   Its number of precedence, F_x.
(2)   The group of manuscripts containing the errors, G_x.
(3)   Their number, ϖ_x.
(4)   The list of basic errors which have been regrouped.

This last quantity defines the "representativeness" of the class, and will be called the "weight." Each class distinguishes a named manuscript, whether the latter is in existence or not. In our example, we obtain Table 8.4.

TABLE 8.4

x	F_x	G_x	ϖ_x	Basic places of variance[a]
1	F_1	N	7	34, 41, 45, 54, 73, 296, 335
2	F_2	H, H_a, S, T	4	35, 40, 62, 69
3	F_3	H, H_a, N, S, T	6	36, 38, 49, 154, 304, 317
4	F_4	T	3	52, 57, 106
5	F_5	H	1	117
6	F_6	S	1	155

[a] See Table 8.1.

G. DETERMINING THE GRAPH OF ERRORS AND RESPONSIBLES

We now know the various classes F_x with their weight ϖ_x as well as the groups of manuscripts G_x, and will next ordinate the groups of constellations, each group being placed at a lower level than the one which includes it. To do this we shall proceed by stages.

First of all we shall define the levels on the graph of the chain

$$p = 0, 1, 2,\ldots,$$

by using a vertical scale (Fig. 8.8). These levels are numbered from the top downwards, and define the "depth," the highest level containing the largest groups.

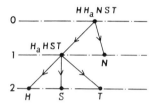

FIG. 8.8.

The method of calculation is as follows. Given two groups G_x and G_y, we take the elements common to them,[1] which will be termed $G_x \cap G_y$, and called the "intersection of G_x and G_y."

Examples.

G_x	G_y	$G_x \cap G_y$	
$\{H, H_a, S, T\}$	$\{H, H_a, N, S, T\}$	$\{H, H_a, S, T\}$	(8.1)
$\{S\}$	$\{T\}$	$\varnothing$	

(The sign $\varnothing$ is that of an empty set in the theory of sets; it shows that G_x and G_y do not possess any common element).

If $G_x \cap G_y$ differs both from G_x and G_y, these two groups cannot be ordinated in relation to each other, since nothing is known as to their respective levels. Hence, when we consult Table 8.4, we find that we cannot ordinate G_3 and G_4, G_1 and G_4, etc.

If, on the other hand, $G_x \cap G_y$ is equal to one of the two, for instance G_x, then G_x is included in G_y (that is to say, G_y contains G_x), and we

---

[1] It is a question of the *intersection* of the sets formed by G_x and G_y, which is customarily written $G_x \cap G_y$. For instance, if G_x represents the set $\{a, b, l, m, s\}$, and G_y represents the set $\{b, c, k, l, r, s, t\}$, then $G_x \cap G_y$ contains the elements $\{b, l, s\}$ common to G_x and G_y.

can state that G_x is "at a greater depth" than G_y. The same deduction can be made, for example, in the case of G_2 and G_3.

Hence, if p_x is the depth of G_x, and p_y that of G_y, then p_y must necessarily be equal to $p_x + 1$. If this were not the case, $p_x + 1$ would be taken as the new value of p_y.

Everywhere, at first, p is equal to 0, but little by little the depths increase, for each pair (x, y) is considered in sequence. Besides, this part of the calculation is repeated until the values which have been found no longer change.

For our example, this stage of the calculations shows the condition shown in Table 8.5.

TABLE 8.5

F_x	p_x	G_x
F_1	1	N
F_2	1	H, H_a, S, T
F_3	0	H, H_a, N, S, T
F_4	2	T
F_5	2	H
F_6	2	S

The next stage consists of preordinating the elements by classifying them according to their increasing depth. We have not yet obtained the graph, but we are progressing towards it, the larger groups now being arranged above, and the smaller ones below them (Table 8.6).

TABLE 8.6

F_x	p_x	G_x
F_3	0	H, H_a, N, S, T
F_1	1	N
F_2	1	H, H_a, S, T
F_4	2	T
F_5	2	H
F_6	2	S

We still have to define how the groups are linked, and for this purpose we can draw an arc (x, y) which goes from x towards its direct descendants. In this way we shall obtain the required chain (Fig. 8.9).

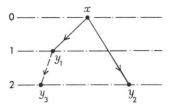

FIG. 8.9.

We shall next consider in turn all the elements y lower than x, starting with the nearest level. Given a pair (x, y), we shall take the common element, so that $G_z = G_x \cap G_y$. If y is to be a descendant of x, G_z must equal G_x, as we explained earlier. If this is not the case, it proves that no anterior relationship can be established between x and y. If it is the case, it shows we are concerned with a descendant, but this descendant must still be in direct descent. To verify this, we assume that a list of the direct descendants, of whom we already know, exists. If our group G_y were included in one of these groups ($G_y \cap G_r = G_y$), we could deduce that y was, in fact, descended from x, but through r as an intermediary, so that it is not in direct line. On the other hand, if there is no G_r which includes G_y, then G_y is truly a direct descendant and will appear on the list of x's descendants; and we shall obviously include G_y, which has thus been selected, among the direct descendants of G_x.

In practice, the tabulation for the computer sets out the results in the form of Table 8.7.

TABLE 8.7

F_x	Direct descendants
F_3	$F_1 F_2$
F_1	$\varnothing$
F_2	$F_4 F_5 F_6$
F_4	$\varnothing$
F_5	$\varnothing$
F_6	$\varnothing$

We have now collected all the information needed to enable us to draw graph (F_a): we know at which levels the elements should be placed, and we know how the arcs should be drawn (Fig. 8.10).

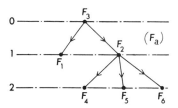

FIG. 8.10.

Let us now turn from the graph of the faulty groups (F_a) to that of the responsibles (R_e). Which manuscript, for example, is reponsible for F_2? To find it we remove from group $G_2 = \{H, H_a, S, T\}$ all the manuscripts which appear in the groups directly descended from it; in other words, all the elements, whether they are common or not.[1] Thus,

$$G_4 = \{T\}, \qquad G_5 = \{H\}, \qquad G_6 = \{S\},$$

then

$$G_2 - G_4 \cup G_5 \cup G_6 = \{H, H_a, S, T\} - \{H, S, T\} = \{H_a\}. \qquad (8.3)$$

This calculation was made at the same time that the direct descendants were determined, and we obtain the full list of the reponsible manuscripts (Table 8.8).

The manuscript responsible for F_3 is not named, thus indicating a "lost manuscript" which we will call ω. Let us now draw the graph of the responsibles (R_e), replacing each manuscript F_x in (F_a) by its responsible manuscript.

H. DETERMINING THE GENEALOGICAL GRAPH (G_e)

From the graph of the responsibles (R_e) the genealogical graph (G_e) can be obtained merely by adding C as the ancestor of ω (Fig. 8.12).

[1] It is a question of the *reunion* or *union* in the sense of the theory of sets. The union of G_x and G_y is written as $G_x \cup G_y$. If, for example,

$$G_x = \{a, b, l, m, s\} \qquad \text{and} \qquad G_y = \{b, c, k, l, r, s, t\},$$

then

$$G_x \cup G_y = \{a, b, c, k, l, m, r, s, t\}.$$

Owing to the simplicity of the graph in the present example, G_4, G_5, and G_6 are *disjoint*, that is to say, they have a *null intersection* without any common element, but this is a special case.

TABLE 8.8

F_x	Responsible manuscript
F_3	∅
F_1	N
F_2	H_a
F_4	T
F_5	H
F_6	S

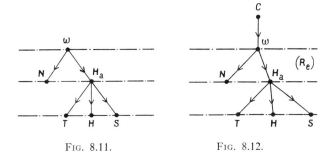

FIG. 8.11.   FIG. 8.12.

It is obvious that the example which has been chosen to illustrate the present paragraph is a very simple one, the reason being that the genealogy is "pure" or "normal," which means that a manuscript has only one ancestor. Hence, the genealogical graph in this case is an "arborescence"[1] in the sense attached to it in the theory of graphs.

Such problems are usually more complex, for when (R_e) and (G_e) are separate there is "contamination," a situation which we shall examine later on.

I. TEXT OF A LOST MANUSCRIPT

We have discovered the relative genealogy in relation to C, but we still have to find the true orientation; in other words, the original ancestor. There are other questions, too, for which we require answers; in particular, the text of each manuscript, essential in the case of a lost manuscript which we are unable to examine. For example, what is the text of ω? It is F_3. By going back to the list of variant places,

---

[1] See Part II, Section 44, p. 304.

we find that at numbers 36, 38, 49, 154, 304, 317, ω will have a text "antagonistic" to C, that is to say,

TRIBUERET   and not   DISTRIBUERET

Omission of: DE PRIMA CONJUNCTIONE COENOBITARUM

QUAE   and not   QUO

EJUS   and not   QUIS

PSALLENDUM   and not   PSALLENDI ORDINEM

AMPUTANDAE   and not   AMPUTANDA

In the remainder of the text ω will be the same as C. How, in general, can we decide the text of a manuscript, such as H_a? It contains the errors which have been perpetrated on it, and also those transmitted by its ancestors, in this case F_2 and F_3, which are the set of variant places:

F_2	35			40		62	69			
F_3		36	38		49				154	304	317

We must verify, from the complete list of variant places, that, for those given above, H_a is included in the group antagonistic to C, and that everywhere else it is in the same group as C.

CORRECT ORIENTATION OF THE GENEALOGICAL TREE. If a link (x, y) connects two manuscripts x and y, which is the ancestor of the other? This question means that we must decide which is the true arc: (x, y) or (y, x)? Two methods, taken from traditional textual emendation, enable us to answer the question.

(1) *The Conjectural Method.* Reliance is placed on the style of the author, and where the texts differ, the philologist may decide that the form found in y "can only be that of the author," in which case y is the ancestor of x and the true orientation is (y, x).

(2) *The Method of External Evidence.* Observations made with regard to the subject matter of the manuscript, its format, the condition of the ink, and external annotations, such as signatures and dates, enable a philologist who knows the historical background to decide that one manuscript is older than another and therefore its ancestor.

It is not necessary to orient all the arcs to discover the orientation of the system; a few arcs will suffice, in conjunction with some study of the figure. The origin of the orientation shows which is the original manuscript.

In our example, H_a is the historical ancestor of H: in fact, since it is a palimpsest, both are written on the same parchment. It can easily be seen that H_a is written underneath H, and the direction is from H_a towards H, which gives us an oriented arc. It can be admitted, nevertheless, that ω is the true ancestor: in point of fact, C, H, and N are manuscripts written at the same period, but discovered in widely separated places, so that there is little likelihood that one of them served as a model for the other two. With greater probability, all three were copied from the same model, and were then dispersed to the four corners of Europe. To verify this hypothesis, let us discover whether ω was the ancestor of C or the reverse. To do so, let us examine the differences between the two texts which we noted earlier:

ω	C
TRIBUERET	DISTRIBUERET
omission of	DE PRIMA, etc.
PSALLENDUM	PSALLENDI ORDINEM

From the above list it will be noticed that ω is more concise than C, so that a philologist would gladly decide that ω has a style of severe grandeur, whereas the variants in C are the products of a verbose mind, and that the correct orientation is therefore from ω towards C.

Thus, the last word rests with traditional methods: it is the philologist who in the end determines the historical archetype.

J. CONTAMINATION

The example which we have studied depended on normal genealogy: a manuscript has a single ancestor, all of whose errors it inherits, as well as adding some of its own. In this case we were able to pass directly from the graph of the responsibles (R_e) to that of the manuscripts (G_e).

In practice this is, unfortunately, not always the case owing to the phenomena of contamination, and when this is present (G_e) is not identical with (R_e). Contamination occurs when a manuscript has several ancestors: it inherits some of the errors of each, and corrects others by comparing the two models.

To explain this problem, let us assume the existence of a manuscript c which has two ancestors a and b (Fig. 8.13a). Contamination occurs, and c takes a part A' of the errors contained in a and a part B' of those in b. The whole process takes place as if a fictitious manuscript a' exists which is responsible for the errors A', some of which are transmitted to c and the remainder to a itself (Fig. 8.13b). The same process is true not only for b, but for all their ancestors as far back as the point

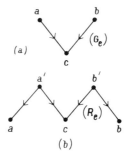

FIG. 8.13.

where the two lines of contamination are united at a common ancestor d.

For this reason the two graphs (G_e) and (R_e) are different. Every manuscript n in one of the lines of contamination divides into itself and another fictitious n', which is its ancestor. Since the parentage of these manuscripts is duplicated in the fictitious ones, we obtain graphs of the type shown in Fig. 8.14.

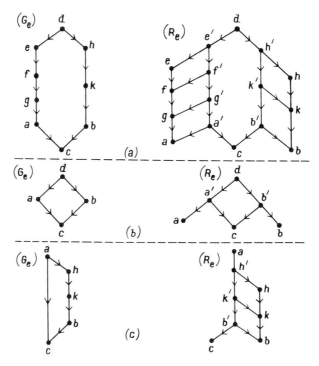

FIG. 8.14.

K. GENERALIZING THE APPLICATIONS

In the example just studied we were dealing with a monastic text. A single language, Latin, was involved, and careful and conscientious copyists preserved the text with such fidelity that even after the lapse of several centuries only a handful of words had variants.

The texts of lay manuscripts are seldom preserved with such care. Several languages are often used, as in the *Chanson de Roland*, and the handing down of manuscripts is subject to greater hazards. Indeed, in their earliest days, some of the originals were not even written, but were transmitted by word of mouth.

The quest for genealogies of this kind can still be carried on by the methods explained in this section. The only difference is that the comparison of two versions is carried out, not on the basis of words, but rather on that of phonemes, details of the setting, actions in the scenes which are described (the dying Roland turns either in the direction of France or of Spain), particular epithets (Christians, Frenchmen), turns of phrase (the Precious One of the Emir, the Silken One of Charlemagne), even on important happenings (in some texts, Roland, the lone survivor on the field of battle, sounds his horn to summon Charlemagne; in others, it is sounded to rouse his companions who had hidden in the woods at the beginning of the battle).

Even under these conditions, the methods which we have used, as well as programming on a computer, are still as useful and generally more effective than philological studies alone.

It should be noted that the philologist is always in charge of the proceedings, but his work is scientifically prepared, directed, furnished with details and brought to fruition by a scientific method, which is the outcome of the theory of graphs, and by the use of an electronic computer.

CHAPTER II

DYNAMIC PROGRAMMING

9. Introduction

One of the most efficacious methods of discovering the optimum of a criterion function connected with an economic phenomenon was formulated some ten years ago by the American mathematician Richard Bellman.[1] This method is of particular interest because of its application to the field of economics, and it can be regarded as equally useful for research and evaluations in physics or pure mathematics. The keystone of Bellman's method is a certain "principle of optimality" which we prefer to qualify by the term "theorem," it being understood that in mathematics one deals with axioms and theorems rather than with principles. This "principle" of optimality is so simple that it appears almost self-evident once it has been clearly comprehended; yet its importance and the efficacy of the methods of multistage optimization to which it has given rise have to be acknowledged in proportion to one's realization that the basic nature of numerous economic problems is of a sequential character.

The method will be introduced to the reader by means of numerous examples and by the frequent use of graphs whenever the nature of the problem implies the existence of discrete variables.[2] Moreover, where continuous variables occur, difficulties of mathematical treatment may arise, creating developments which we have deliberately eliminated. Nevertheless, we hope that this simplified presentation, which we have endeavored to render as interesting as possible, will persuade the reader to make the requisite effort to study the second portion of this work and, later on, specialized books by means of which he will obtain a mastery of the subject.

[1] As early as 1946, Massé had made use of a similar method which did not at the time receive the attention it deserved (P. Massé, *Le choix des investissements*. Dunod, Paris, 1959).

[2] It must be remembered that a variable is *discrete* when it is defined only for certain numerical values which must not be infinitely close to each other.

10. Bellman's Theorem of Optimality

A. STUDY OF AN EXAMPLE

Dynamic programming is a method for the optimization or mathematical representation of systems in which one works by *stages* or *sequences*. The basis of this method is *the theorem of optimality* which can be proved for very numerous cases, and which, in its most general form, may be enunciated as *the principle of optimality*.

To introduce this theorem we shall make use of a very simple multistage problem. Figure 10.1 represents a decision process.

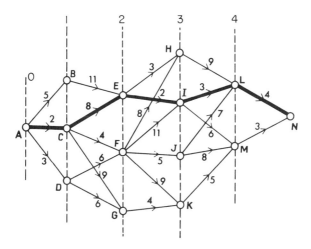

FIG. 10.1.   Minimal route.

In Stage 0, one can, with A as the starting point, choose to go to B, C, or D, which are situated in Stage 1; in the same way, in Stage 1 one can decide to go to either E, F, or G, provided corresponding arcs of the graph are available, and so on through Stages 3 and 4, in order to arrive finally at N. To each possible decision a *cost* is attached: for example, to go from C to F costs 4; from J to M costs 8, etc.

Any route going from A to N will be called a *policy*. Thus $(ADFILN)$ is a policy. A continuous section of route will be termed a *subpolicy*. Examples: (DFK), $(CEIM)$, (HL), and $(CGKMN)$ are subpolicies.

In enumerating all the possible policies one perceives that the policy $(ACEILN)$ is the one which has a *minimal total cost*. This cost is equal to 19. If one plans to minimize the costs, one will therefore choose this policy, which will be said to be *optimal*. Let us now consider all the partial routes, or subpolicies, contained in (*ACEILN*). These are (*ACEIL*), (*CEILN*), (*ACEI*), (*CEIL*), (*EILN*), (*ACE*), (*CEI*), (*EIL*), (*ILN*), (*AC*), (*CE*), (*EI*), (*IL*), and (*LN*). If one examines each of these subpolicies one can verify that it is optimal from the point at which its partial route begins to the point at which it ends; for instance, from *C* to *L* the optimal path is (*CEIL*); from *A* to *I*, the optimal path is (*ACEI*); from *E* to *L* it is (*EIL*) (see Fig. 10.2).

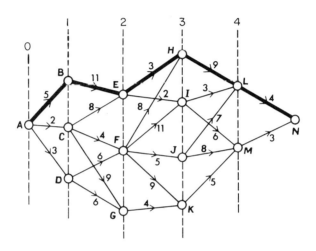

FIG. 10.2.   Maximal path.

Let us suppose that, on the other hand, one wishes to discover a route of *maximal cost*; by enumeration one will ascertain that the optimal policy is that provided by the path (*ABEHLN*), for which the cost is equal to 32; one can verify that all the subpolicies contained in this optimal policy are optimal from the point where their partial route begins to that at which it ends.

Hence, one has established the following fact: every optimal policy contains only optimal subpolicies; this property does not result from a special case for it can be expressed in general terms and enunciated as a theorem.[1]

B. THEOREM OF OPTIMALITY (DISCRETE AND DETERMINISTIC SYSTEMS)

Let us assume a system capable of a change of state at each stage *k* as the result of a decision where the possible states in each stage *k*

[1] One can extend this theorem without any difficulty to cases where the variables concerned *lie in continuous intervals* (cf. Part II, Section 50).

$(k = 0, 1, 2,..., N)$ are finite or infinite in number. Provided they are denumerable, the term *policy* is applied to a certain sequence of decisions from $k = 0$ to $k = N$, and the term *subpolicy* to a sequence of connected decisions which form part of a policy. Then if one selects a criterion function related to these changes of state and decides to optimize this function, the following theorem holds true:

An Optimal Policy Must Contain Only Optimal Subpolicies

Proof.[1]   Let us consider a subpolicy extracted from an optimal policy. If such subpolicy were not optimal there would exist a better one which, if added to the remaining portion of the policy under consideration, would improve the latter, a deduction contrary to the hypothesis.

Later on this theorem will be extended to those cases where the states to be considered are defined by continuous variables.

Bellman [I1] has enunciated the above theorem in the form of a general principle:

"A policy is optimal if, at a stated period,[2] whatever the preceding decisions may have been, the decisions still to be taken constitute an optimal policy when the result of the previous decisions is included."

11. First Example of Multistage Optimization

A. CONSTRUCTION OF A HIGHWAY

A highway is to be built between Aville and Feville (Fig. 11.1) which will bypass various towns and will be composed of five sections. For each of these sections or loops the cost of various alternatives has been studied and evaluated, this cost taking into account such items as the road building itself, necessary construction works, compulsory purchases, social cost,[3] etc. All of this is eventually expressed in the form of a graph, represented by Fig. 11.2, in which a cost is assigned to each point of intersection. Supposing, for the purpose of our calculations, that one leaves from point A in the direction of N, one will pass each of the points of intersection in the same direction; but one might equally have started from N in the direction of A, crossing them in the opposite direction. Our aim is to find *the path of minimal cost* between A and N.

---

[1] Indeed, a more precise enunciation and proof can be given if based on the concepts of modern algebra (see our reference I18). But this theorem and its proof are really so simple that the theorem may be regarded as a truism.

[2] Referred to earlier as a *stage*.

[3] The social cost includes, for example, the differences in value attributed to the various activities in the regions traversed.

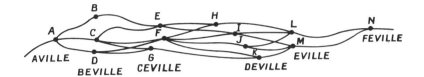

FIG. 11.1.

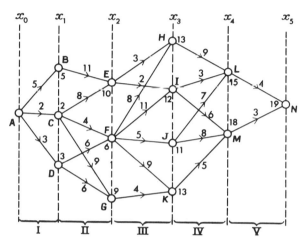

FIG. 11.2.

If we assign the symbols x_0, x_1, x_2, x_3, x_4, and x_5 to the decision variables connected with each of the sections these variables will not acquire numerical values[1] but will be defined at each stage by appropriate vertices on the same line. Thus x_2 can be E, F, or G, so that for the sets one has a succession of points corresponding to the variables:

$$
\begin{array}{ll}
x_0: & A \\
x_1: & B, C, D \\
x_2: & E, F, G
\end{array}
\qquad
\begin{array}{ll}
x_3: & H, I, J, K \\
x_4: & L, M \\
x_5: & N
\end{array}
\qquad (11.1)
$$

Let $v_\mathrm{I}(x_0, x_1)$ be the cost of section I. This cost obviously depends on the *values* assigned to x_0 and x_1 (in this case x_0 can only be A), but alternatives would have existed had the starting point of the highway been the subject of choice. In the same way, let $v_\mathrm{II}(x_1, x_2)$ represent the cost of section II, this cost depending on x_1 and x_2. In the same manner one will define $v_\mathrm{III}(x_2, x_3)$, $v_\mathrm{IV}(x_3, x_4)$, and $v_\mathrm{V}(x_4, x_5)$.

---

[1] As a result of the theory of sets we have learned to treat variables in a much more general manner than was customary in "classical" algebra.

The total cost of the highway will be

$$F(x_0, x_1, x_2, x_3, x_4, x_5) = v_I(x_0, x_1) + v_{II}(x_1, x_2) + v_{III}(x_2, x_3)$$
$$+ v_{IV}(x_3, x_4) + v_V(x_4, x_5). \qquad (11.2)$$

Let us first seek the minimum cost of section I for each of its points of termination B, C, or D. In fact, there is no choice to be made. If we call this minimum cost $f_I(x_1)$ we have

$$f_I(B) = v_I(A, B) = 5,$$
$$f_I(C) = v_I(A, C) = 2, \qquad (11.3)$$
$$f_I(D) = v_I(A, D) = 3.$$

Let us now make $f_{I,II}(x_2)$ the minimum cost of sections I and II combined for the different *values* of x_2. We will then have

$$f_{I,II}(E) = \underset{x_1=B,C,D}{\text{MIN}}[f_I(x_1) + v_{II}(x_1, E)],$$
$$f_{I,II}(F) = \underset{x_1=B,C,D}{\text{MIN}}[f_I(x_1) + v_{II}(x_1, F)], \qquad (11.4)$$
$$f_{I,II}(G) = \underset{x_1=B,C,D}{\text{MIN}}[f_I(x_1) + v_{II}(x_1, G)].$$

These equalities are derived from the theorem of optimization, for every optimal path from A to E, to F or to G must be made up of sections for which the cost is minimal.

In considering Fig. 11.2 and giving to x_1, in turn, the values B, C, and D, it follows[1]

$$f_{I,II}(E) = \min[\underbrace{5+11}_{x_1=B}, \underbrace{2+8}_{x_1=C}, \underbrace{3+\infty}_{x_1=D}] = 10, \qquad \text{with} \quad x_1 = C;$$

$$f_{I,II}(F) = \min[\underbrace{5+\infty}_{x_1=B}, \underbrace{2+4}_{x_1=C}, \underbrace{3+6}_{x_1=D}] = 6, \qquad \text{with} \quad x_1 = C; \qquad (11.5)$$

$$f_{I,II}(G) = \min[\underbrace{5+\infty}_{x_1=B}, \underbrace{2+9}_{x_1=C}, \underbrace{3+6}_{x_1=D}] = 9, \qquad \text{with} \quad x_1 = D.$$

Thus, for sections I and II combined, the least costly path is

$$ACE \quad \text{if one stops at } E: \qquad \text{cost} \quad 10,$$
$$ACF \quad \text{if one stops at } F: \qquad \text{cost} \quad 6,$$
$$ADG \quad \text{if one stops at } G: \qquad \text{cost} \quad 9.$$

---

[1] When a vertex is not joined to another, we consider the cost of the link is *infinite*.

Let $f_{I,II,III}(x_3)$ now be the minimal cost for sections I, II, and III combined for the different values of x_3. We shall then have

$$f_{I,II,III}(H) = \underset{x_2=E,F,G}{\text{MIN}} [f_{I,II}(x_2) + v_{III}(x_2, H)],$$

$$f_{I,II,III}(I) = \underset{x_2=E,F,G}{\text{MIN}} [f_{I,II}(x_2) + v_{III}(x_2, I)],$$

$$f_{I,II,III}(J) = \underset{x_2=E,F,G}{\text{MIN}} [f_{I,II}(x_2) + v_{III}(x_2, J)], \tag{11.6}$$

$$f_{I,II,III}(K) = \underset{x_2=E,F,G}{\text{MIN}} [f_{I,II}(x_2) + v_{III}(x_2, K)].$$

These equalities are also the result of the theorem of optimality, since each optimal route going from A to H, to I, to J, or to K must be formed from partial routes from A to E, from A to F, or from A to G for which the costs are minimal.

From Fig. 11.2, it follows:

$$f_{I,II,III}(H) = \min[\underbrace{10+3}_{x_2=E}, \underbrace{6+8}_{x_2=F}, \underbrace{9+\infty}_{x_2=G}] = 13, \qquad \text{with} \quad x_2 = E,$$

$$f_{I,II,III}(I) = \min[\underbrace{10+2}_{x_2=E}, \underbrace{6+11}_{x_2=F}, \underbrace{9+\infty}_{x_2=G}] = 12, \qquad \text{with} \quad x_2 = E,$$

$$f_{I,II,III}(J) = \min[\underbrace{10+\infty}_{x_2=E}, \underbrace{6+5}_{x_2=F}, \underbrace{9+\infty}_{x_2=G}] = 11, \qquad \text{with} \quad x_2 = F, \tag{11.7}$$

$$f_{I,II,III}(K) = \min[\underbrace{10+\infty}_{x_2=E}, \underbrace{6+9}_{x_2=F}, \underbrace{9+4}_{x_2=G}] = 13, \qquad \text{with} \quad x_2 = G.$$

Hence, for the aggregate of sections I, II, and III, the route of least cost is

$$\begin{array}{llll} ACEH & \text{if we stop at} \quad H: & \text{cost} \quad 13, \\ ACEI & \text{if we stop at} \quad I: & \text{cost} \quad 12, \\ ACFJ & \text{if we stop at} \quad J: & \text{cost} \quad 11, \\ ADGK & \text{if we stop at} \quad K: & \text{cost} \quad 13. \end{array}$$

Let us proceed, using a similar notation:

$$f_{I,II,III,IV}(L) = \underset{x_3=H,I,J,K}{\text{MIN}} [f_{I,II,III}(x_3) + v_{IV}(x_3, L)],$$

$$f_{I,II,III,IV}(M) = \underset{x_3=H,I,J,K}{\text{MIN}} [f_{I,II,III}(x_3) + v_{IV}(x_3, M)]. \tag{11.8}$$

Let

$$f_{\text{I,II,III,IV}}(L) = \min[\underbrace{13+9}_{x_3=H},\ \underbrace{12+3}_{x_3=I},\ \underbrace{11+7}_{x_3=J},\ \underbrace{13+\infty}_{x_3=K}] = 15,$$

$$\text{with}\quad x_3 = I;$$

$$f_{\text{I,II,III,IV}}(M) = \min[\underbrace{13+\infty}_{x_3=H},\ \underbrace{12+6}_{x_3=I},\ \underbrace{11+8}_{x_3=J},\ \underbrace{13+5}_{x_3=K}] = 18,$$

$$\text{with}\quad x_3 = I \text{ or } x_3 = K.$$

(11.9)

Thus, for sections I, II, III, and IV combined, the cheapest route is

$$ACEIL \quad \text{if one stops at}\quad L: \quad \text{cost}\quad 15,$$

$$ACEIM \text{ or } ADGKM \quad \text{if one stops at}\quad M: \quad \text{cost}\quad 18.$$

Finally, taking f, as the minimum of $F(x_0, x_1, x_2, x_3, x_4, x_5)$:

$$f = \operatorname*{MIN}_{x_4=L,M}[f_{\text{I,II,III,IV}}(x_4) + v_{\text{V}}(x_4, N)],$$ (11.10)

that is,

$$f = \min[\underbrace{15+4}_{x_4=L},\ \underbrace{18+3}_{x_4=M}] = 19, \quad \text{with}\quad x_4 = L.$$ (11.11)

The cheapest route is therefore $ACEILN$ with a cost of 19 units.

The process of optimization has been carried out from A towards N, and in order that the reader may become fully conversant with the technique of multistage optimization, we propose to repeat the calculation starting from the point N

$$f_{\text{V}}(L) = v_{\text{V}}(L, N) = 4,\quad f_{\text{V}}(M) = v_{\text{V}}(M, N) = 3,$$

$$f_{\text{IV,V}}(H) = \operatorname*{MIN}_{x_4=L,M}[f_{\text{V}}(x_4) + v_{\text{IV}}(H, x_4)],$$ (11.12)

$$f_{\text{IV,V}}(I) = \operatorname*{MIN}_{x_4=L,M}[f_{\text{V}}(x_4) + v_{\text{IV}}(I, x_4)],$$

$$f_{\text{IV,V}}(J) = \operatorname*{MIN}_{x_4=L,M}[f_{\text{V}}(x_4) + v_{\text{IV}}(J, x_4)],$$

$$f_{\text{IV,V}}(K) = \operatorname*{MIN}_{x_4=L,M}[f_{\text{V}}(x_4) + v_{\text{IV}}(K, x_4)],$$ (11.13)

that is,

$$f_{\mathrm{IV,V}}(H) = \min[\underbrace{4 + 9}_{x_4=L}, \underbrace{3 + \infty}_{x_4=M}] = 13, \qquad \text{with} \quad x_4 = L;$$

$$f_{\mathrm{IV,V}}(I) = \min[\underbrace{4 + 3}_{x_4=L}, \underbrace{3 + 6}_{x_4=M}] = 7, \qquad \text{with} \quad x_4 = L; \qquad (11.14)$$

$$f_{\mathrm{IV,V}}(J) = \min[\underbrace{4 + 7}_{x_4=L}, \underbrace{3 + 8}_{x_4=M}] = 11, \qquad \text{with} \quad x_4 = L \text{ or } x_4 = M;$$

$$f_{\mathrm{IV,V}}(K) = \min[\underbrace{4 + \infty}_{x_4=L}, \underbrace{3 + 5}_{x_4=M}] = 8, \qquad \text{with} \quad x_4 = M.$$

Thus, on leaving N, the path of least cost is

NLH	if we stop at	H :	cost	13,
NLI	if we stop at	I :	cost	7,
NLJ or NMJ	if we stop at	J :	cost	11,
NMK	if we stop at	K :	cost	8.

Proceeding:

$$f_{\mathrm{III,IV,V}}(E) = \operatorname*{MIN}_{x_3=H,I,J,K}[f_{\mathrm{IV,V}}(x_3) + v_{\mathrm{III}}(E, x_3)],$$

$$f_{\mathrm{III,IV,V}}(F) = \operatorname*{MIN}_{x_3=H,I,J,K}[f_{\mathrm{IV,V}}(x_3) + v_{\mathrm{III}}(F, x_3)], \qquad (11.15)$$

$$f_{\mathrm{III,IV,V}}(G) = \operatorname*{MIN}_{x_3=H,I,J,K}[f_{\mathrm{IV,V}}(x_3) + v_{\mathrm{III}}(G, x_3)];$$

or,

$$f_{\mathrm{III,IV,V}}(E) = \min[\underbrace{13 + 3}_{x_3=H}, \underbrace{7 + 2}_{x_3=I}, \underbrace{11 + \infty}_{x_3=J}, \underbrace{8 + \infty}_{x_3=K}] = 9, \qquad \text{with } x_3 = I;$$

$$f_{\mathrm{III,IV,V}}(F) = \min[\underbrace{13 + 8}_{x_3=H}, \underbrace{7 + 11}_{x_3=I}, \underbrace{11 + 5}_{x_3=J}, \underbrace{8 + 9}_{x_3=K}] = 16, \qquad \text{with } x_3 = J;$$

$$f_{\mathrm{III,IV,V}}(G) = \min[\underbrace{13 + \infty}_{x_3=H}, \underbrace{7 + \infty}_{x_3=I}, \underbrace{11 + \infty}_{x_3=J}, \underbrace{8 + 4}_{x_3=K}] = 12, \quad \text{with } x_3 = K.$$

$$(11.16)$$

Hence, starting from N, the cheapest road is

$NLIE$	if one stops at	E :	cost	9,
$NLJF$ or $NMJF$	if one stops at	F :	cost	16,
$NMGK$	if one stops at	G :	cost	12.

$$f_{\mathrm{II,III,IV,V}}(B) = \operatorname*{MIN}_{x_2=E,F,G}[f_{\mathrm{III,IV,V}}(x_2) + v_{\mathrm{II}}(B, x_2)],$$

$$f_{\mathrm{II,III,IV,V}}(C) = \operatorname*{MIN}_{x_2=E,F,G}[f_{\mathrm{III,IV,V}}(x_2) + v_{\mathrm{II}}(C, x_2)], \qquad (11.17)$$

$$f_{\mathrm{II,III,IV,V}}(D) = \operatorname*{MIN}_{x=E,F,G}[f_{\mathrm{III,IV,V}}(x_2) + v_{\mathrm{II}}(D, x_2)],$$

that is,

$$f_{\text{II},\text{III},\text{IV},\text{V}}(B) = \min[\underbrace{9 + 11}_{x_2=E}, \underbrace{16 + \infty}_{x_2=F}, \underbrace{12 + \infty}_{x_2=G}] = 20, \quad \text{with } x_2 = E;$$

$$f_{\text{II},\text{III},\text{IV},\text{V}}(C) = \min[\underbrace{9 + 8}_{x_2=E}, \underbrace{16 + 4}_{x_2=F}, \underbrace{12 + 9}_{x_2=G}] = 17, \qquad \text{with } x_2 = E; \quad (11.18)$$

$$f_{\text{II},\text{III},\text{IV},\text{V}}(D) = \min[\underbrace{9 + \infty}_{x_2=E}, \underbrace{16 + 6}_{x_2=F}, \underbrace{12 + 6}_{x_2=G}] = 18, \qquad \text{with } x_2 = G.$$

Thus, with N as one's point of departure, the route of least cost is

$$\begin{array}{llll} NLIEB & \text{if one stops at} & B: & \text{cost} \quad 20, \\ NLIEC & \text{if one stops at} & C: & \text{cost} \quad 17, \\ NMKGD & \text{if one stops at} & D: & \text{cost} \quad 18. \end{array}$$

Lastly, taking f as the minimum of $F(x_0, x_1, x_2, x_3, x_4, x_5)$, if

$$f = \underset{x_1=B,C,D}{\text{MIN}}[f_{\text{II},\text{III},\text{IV},\text{V}}(x_1) + v_{\text{I}}(A, x_1)]; \tag{11.19}$$

We find

$$f = \min[\underbrace{20 + 5}_{x_1=B}, \underbrace{17 + 2}_{x_1=C}, \underbrace{18 + 3}_{x_1=D}] = 19, \quad \text{with } x_1 = C. \tag{11.20}$$

Hence, the path of minimal cost between N and A is $NLIECA$.

B. MANNER IN WHICH OPTIMIZATION IS CARRIED OUT

The method which we have used to find the optimum is that of *dynamic programming*. It consists of finding optimal subpolicies which include an ever increasing number of connected stages until the optimal policy is obtained. The calculations may be started at either end, and the route could even be divided into parts before beginning. For instance, in the example just treated, it would be possible to evaluate the optimal subpolicies for Stages 1 and 2 and those for Stages 3, 4, and 5, and then calculate the optimal policy from the above subpolicies. Given:

Optimal subpolicies for Stages 1 and 2:

$$\begin{array}{llll} \text{From } A \text{ to } E: & ACE, & \text{cost}: & 10, \\ A \text{ to } F: & ACF, & \text{cost}: & 6, \\ A \text{ to } G: & ADG, & \text{cost}: & 9. \end{array}$$

Optimal subpolicies for Stages 3, 4, and 5:

> From N to E : $NLIE$,   cost :   9,
> N to F : $NLJF$ or $NMJF$,   cost :   16,
> N to G : $NMKG$,   cost :   12.

To calculate the optimal policy from A to N we evaluate from the above optimal subpolicies all the paths from A to N which contain these subpolicies.

> $ACEILN$:   cost $10 + 9$,   equals 19;
> $ACFJLN$ or $ACFJMN$:   cost $6 + 16$,   equals 22;
> $ADGKMN$:   cost $9 + 12$,   equals 21.

$ACEILN$ is the optimal path.

In certain problems, calculations can be greatly simplified if the optimization is carried out in a particular fashion, or if one commences at a selected stage, depending on the functional relations which connect the state variables with one another, in the system under consideration. Whatever the manner in which multistage optimization is carried out, the essential thing is that the reasoning employed should depend on the theorem of optimality.

12. Distribution of Investments

A. ENUNCIATION OF A COMBINATORIAL PROBLEM

In this new example, the order in which the different stages are considered is quite immaterial.

Let us imagine four economic regions I, II, III, and IV, in which it has been decided to undertake a sales promotion. A certain sum A is available for allocation among the four regions (Fig. 12.1). It is postulated

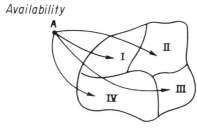

Fig. 12.1.

that the profits which can be realized in each zone are known as a function of the investments (Table 12.1 and Fig. 12.2). Let us suppose that this sum A is equal to 100 millions or preferably 10 units, counting in tens of millions. Thus a policy would consist of allocating, for example, 3 in I, 1 in II, 5 in III, and 1 in IV.

TABLE 12.1

Investments	Profit I	Profit II	Profit III	Profit IV
0	0	0	0	0
1	0.28	0.25	0.15	0.20
2	0.45	0.41	0.25	0.33
3	0.65	0.55	0.40	0.42
4	0.78	0.65	0.50	0.48
5	0.90	0.75	0.62	0.53
6	1.02	0.80	0.73	0.56
7	1.13	0.85	0.82	0.58
8	1.23	0.88	0.90	0.60
9	1.32	0.90	0.96	0.60
10	1.38	0.90	1.00	0.60

From Table 12.1, the profit would be

$$0.65 + 0.25 + 0.62 + 0.20 = 1.72.$$

A problem of this kind is of a combinatorial nature and could be solved by enumeration, but it would be necessary to evaluate a considerable number of policies (286, to be precise). Such a calculation would not be impossibly onerous, but if the indivisible unit selected were a million the number of possible policies would become astronomical.

Let us try to discover which policy for the distribution of investments

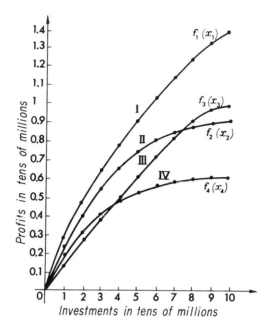

FIG. 12.2.

will provide the maximal profit. To do this we shall make use of the theorem of optimality.

B. MULTISTAGE TREATMENT

Let x_1, x_2, x_3, and x_4 represent the investments in tens of millions (indivisible units) in the regions I, II, III, and IV, and let us call the profits in the corresponding zones $v_1(x_1)$, $v_2(x_2)$, $v_3(x_3)$, and $v_4(x_4)$, and the total profit $F(x_1, x_2, x_3, x_4)$. It follows that

$$F(x_1, x_2, x_3, x_4) = v_1(x_1) + v_2(x_2) + v_3(x_3) + v_4(x_4) \qquad (12.1)$$

with a restriction or constraint:

$$x_1 + x_2 + x_3 + x_4 = 10, \qquad (12.2)$$

it being understood that these four variables can only be assigned the integral values 0 to 10 inclusive.

Let us assume

$$x_1 + x_2 = u_1, \qquad u_1 + x_3 = u_2, \qquad u_2 + x_4 = A, \qquad (12.3)$$

where
$$u_1 \leqslant A, \qquad u_2 \leqslant A;$$

A being given the values 0, 1, 2,..., 10 in turn. Then (12.1) can be written as

$$F(x_1, u_1, u_2, A) = v_1(x_1) + v_2(u_1 - x_1) + v_3(u_2 - u_1) + v_4(A - u_2). \quad (12.4)$$

Thus we can now apply the theorem of optimality and can calculate in stages[1]:

$$f_{1,2}(u_1) = \max_{x_1=0,1,2,\ldots,u_1} [v_1(x_1) + v_2(u_1 - x_1)], \quad (12.5)$$

TABLE 12.2

(1)	(2)	(3)	(4)	(5)
x_1, x_2, u_1	$v_1(x_1)$	$v_2(x_2)$	$f_{1,2}(u_1)$	Optimal subpolicies for I and II
0	0	0	0	(0,0)
1	0.28	0.25	0.28	(1,0)
2	0.45	0.41	0.53	(1,1)
3	0.65	0.55	0.70	(2,1)
4	0.78	0.65	0.90	(3,1)
5	0.90	0.75	1.06	(3,2)
6	1.02	0.80	1.20	(3,3)
7	1.13	0.85	1.33	(4,3)
8	1.23	0.88	1.45	(5,3)
9	1.32	0.90	1.57	(6,3)
10	1.38	0.90	1.68	(7,3)

[1] It should be carefully noted that the order of optimization in this problem is entirely arbitrary. We have chosen I, II, III, and IV, but we might equally, for example, have selected III, I, IV, II, or any other permutation.

$$f_{1,2,3}(u_2) = \underset{u_1=0,1,2,\ldots,u_2}{\text{MAX}} [f_{1,2}(u_1) + v_3(u_2 - u_1)], \tag{12.6}$$

$$f_{1,2,3,4}(A) = \underset{u_2=0,1,2,\ldots,A}{\text{MAX}} [f_{1,2,3}(u_2) + v_4(A - u_2)]. \tag{12.7}$$

The maximum of function F will have as its value $f_{1,2,3,4}(A)$ for $A = 10$. Let us proceed to the numerical calculation.

First, let us calculate (12.5); in order to do this we will construct Table 12.2. Columns (2) and (3) reproduce that portion of Table 12.1 with which we are concerned here. Columns (4) and (5) are completed in the following manner[1]:

$$f_{1,2}(u_1 = 0) = v_1(0) + v_2(0) = 0 + 0 = 0, \tag{12.8}$$

TABLE 12.3

u_1, x_3	$f_{1,2}(u_1)$	$v_3(x_3)$	$f_{1,2,3}(u_2)$	Optimal subpolicy for I and II	Optimal subpolicy for I, II, and III
0	0	0	0	(0, 0)	(0, 0, 0)
1	0.28	0.15	0.28	(1, 0)	(1, 0, 0)
2	0.53	0.25	0.53	(1, 1)	(1, 1, 0)
3	0.70	0.40	0.70	(2, 1)	(2, 1, 0)
4	0.90	0.50	0.90	(3, 1)	(3, 1, 0)
5	1.06	0.62	1.06	(3, 2)	(3, 2, 0)
6	1.20	0.73	1.21	(3, 3)	(3, 2, 1)
7	1.33	0.82	1.35	(4, 3)	(3, 3, 1)
8	1.45	0.90	1.48	(5, 3)	(4, 3, 1)
9	1.57	0.96	1.60	(6, 3)	(5, 3, 1) or (3, 3, 3)
10	1.68	1.00	1.73	(7, 3)	(4, 3, 3)

[1] We shall use the symbols MAX or MIN (in small capital letters) when we are dealing with an analytical expression which is subject to mathematical constraints, and the symbols max or min (in small letters) when it is purely and simply a question of choice between explicit numerical values.

$$f_{1,2}(u_1 = 1) = \max[v_1(0) + v_2(1), v_1(1) + v_2(0)]$$
$$= \max[0 + 0.25; 0.28 + 0]$$
$$= 0.28. \tag{12.9}$$

$$f_{1,2}(u_1 = 2) = \max[v_1(0) + v_2(2), v_1(1) + v_2(1), v_1(2) + v_2(0)]$$
$$= \max[0 + 0.41; 0.28 + 0.25; 0.45 + 0]$$
$$= \max[0.41; 0.53; 0.45] = 0.53, \tag{12.10}$$

and so forth.

Column (5) contains the optimal subpolicies; hence, if a sum of 5 tens of millions is available for investment in I and II, the optimal subpolicy

TABLE 12.4

u_2, x_4	$f_{1,2\,3}(u_2)$	$v_4(x_4)$	$f_{1,2,3,4}(A)$	Optimal subpolicy for I, II, and III	Optimal policy
0	0	0	0	(0, 0, 0)	(0, 0, 0, 0)
1	0.28	0.20	0.28	(1, 0, 0)	(1, 0, 0, 0)
2	0.53	0.33	0.53	(1, 1, 0)	(1, 1, 0, 0)
3	0.70	0.42	0.73	(2, 1, 0)	(1, 1, 0,1 1)
4	0.90	0.48	0.90	(3, 1, 0)	(3, 1, 0, 0) or (2, 1, 0, 1)
5	1.06	0.53	1.10	(3, 2, 0)	(3, 1, 0, 1)
6	1.21	0.56	1.26	(3, 2, 1)	(3, 2, 0, 1)
7	1.35	0.58	1.41	(3, 3, 1)	(3, 2, 1, 1)
8	1.48	0.60	1.55	(4, 3, 1)	(3, 3, 1, 1)
9	1.60	0.60	1.68	(5, 3, 1) or (3, 3, 3)	(4, 3, 1, 1) or (3, 3, 1, 2)
10	1.73	0.60	1.81	(4, 3, 3)	(4, 3, 1, 2)

is $x_1 = 3$ and $x_2 = 2$; if the sum available is 9, the optimal subpolicy
is $x_1 = 6$ and $x_2 = 3$.

With the help of formula (12.6) and the results of Table 12.2, one
can obtain Table 12.3. With the aid of formula (12.7) and the results of
Table 12.3, one can obtain Table 12.4. Hence, one has finally calculated
the optimal policies for $A = 0, 1, 2,..., 10$. For $A = 10$, one finds as
optimal values the following variables:

$$x_1{}^* = 4, \qquad x_2{}^* = 3, \qquad x_3{}^* = 1, \qquad x_4{}^* = 2. \qquad (12.11)$$

The maximal profit is equal to 1.81. One can verify that all the subpolicies for $(4, 3, 1, 2)$ are optimal. For instance, considering I, II,
and III alone, $x_1{}^* = 4$, $x_2{}^* = 3$, $x_3{}^* = 1$ is the correct optimal subpolicy corresponding to $u_2 = 8$; for II, III, and IV, only, $x_2{}^* = 3$,
$x_3{}^* = 1$, $x_4{}^* = 2$ is the correct optimal subpolicy corresponding to
$x_2 + x_3 + x_4 = 6$; for III and IV alone, $x_3{}^* = 1$, $x_4{}^* = 2$ is certainly
the optimal subpolicy corresponding to $x_3 + x_4 = 3$.

13. A Purchasing Problem

A. PROBLEM WITH A CONTINUOUS VARIABLE FORMING A LINEAR DYNAMIC PROGRAM

In the present example, the values considered may vary in a continuous
fashion. Furthermore, this problem is clearly of a multistage character;
the order in which the stages should be dealt with is predetermined,
but optimization can be carried out in various ways.

In order that a certain business may be able to carry out its schedule of
production the purchasing department has to lay in supplies of a
particular primary material at two monthly intervals. The purchasing
price p_i, where $i = 1, 2, 3, 4, 5, 6$, and the demand $d_i(i = 1, 2, 3, 4, 5, 6)$,
are given for the next six two-monthly periods (Table 13.1). The capacity
for storage being limited, the stock can never exceed a certain value S.

TABLE 13.1

Period i	1	2	3	4	5	6
Demand d_i	8	5	3	2	7	4
Purchase price p_i	11	18	13	17	20	10

The initial stock is equal to 2, and the final stock should be null. It is required to decide the quantities to be purchased at the commencement of each period so that the total purchasing cost will be minimal. Let us call

s_i, the stock at time i (end of period $i-1$, beginning of period i), before purchase a_i;

a_i, the stock purchased at time i;

x_i, the amount of stock at time i after the purchase of a_i.

Figure 13.1 provides some kind of solution, but it is clearly not an optimal one.

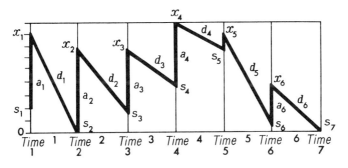

FIG. 13.1.

As the problem is defined, one has

$$x_i = s_i + a_i, \qquad i = 1, 2,..., 6. \tag{13.1}$$

The state of the system at each time i can be described by the value of either s_i or x_i; a_i will be known by the aid of (13.1); it is a matter of indifference whether x_i or s_i is chosen as the variable. We will make an arbitrary choice of x_i as the decision variable.

The nature of the problem imposes the following constraints, with the numerical values which constitute the *bounds* O and S intervening:

$$O \leqslant s_i \leqslant S \qquad i = 1, 2,..., 6, \tag{13.2}$$

$$O \leqslant x_i \leqslant S \qquad i = 1, 2,..., 6, \tag{13.3}$$

$$s_i \leqslant x_i \qquad i = 1, 2,..., 6. \tag{13.4}$$

Again,

$$s_{i+1} = x_i - d_i, \qquad i = 1, 2,..., 6, \tag{13.5}$$

taking $s_1 = 2$.

Let us use (13.5) to eliminate s_i in (13.2) and (13.4):

(1) $O \leqslant s_i$ will give $x_{i-1} - d_{i-1} \geqslant O$, again taking:

$$x_i \geqslant d_i, \qquad i = 1, 2,..., 6. \tag{13.6}$$

(2) $s_i \leqslant x_i$ will give

$$x_i \geqslant x_{i-1} - d_{i-1}, \qquad i = 2, 3,..., 6. \tag{13.7}$$

By combining (13.6) and (13.7) we can state:

$$\max[d_i, x_{i-1} - d_{i-1}] \leqslant x_i. \tag{13.8}$$

(3) We recall that

$$x_i \leqslant S, \qquad i = 1, 2,..., 6. \tag{13.9}$$

(4) Taking (13.7) we can state:

$$x_i \leqslant x_{i+1} + d_i, \qquad i = 1, 2,..., 5. \tag{13.10}$$

By combining (13.9) and (13.10), it follows:

$$x_i \leqslant \min[S, x_{i+1} + d_i], \qquad i = 1, 2,..., 5. \tag{13.11}$$

Lastly, the set of constraints connected with x_i can be written as

$$\max[d_i, x_{i-1} - d_{i-1}] \leqslant x_i \leqslant \min[S, x_{i+1} + d_i], \tag{13.12}$$

where $i = 1, 2,..., 6$, with $x_0 = 2$, $d_0 = 0$, $x_7 = 0$.

B. ENUNCIATION OF A LINEAR PROGRAM

Recalling what was explained in Volume I, Chapter II,[1] one should

[1] Throughout the present work, Volume I refers to A. Kaufmann, *Methods and Models of Operations Research.* Prentice-Hall, Englewood Cliffs, New Jersey, 1963.

recognize that the present problem contains a *linear program*, which can be set out in the following form:

$$
\begin{aligned}
x_1 &\geqslant 8, \\
-x_1 &\geqslant -9, \\
-x_1 + x_2 &\geqslant -8, \\
x_2 &\geqslant 5, \\
-x_2 &\geqslant -9, \\
-x_2 + x_3 &\geqslant -5, \\
x_3 &\geqslant 3, \\
-x_3 &\geqslant -9, \\
-x_3 + x_4 &\geqslant -3, \\
x_4 &\geqslant 2, \\
-x_4 &\geqslant -9, \\
-x_4 + x_5 &\geqslant -2, \\
x_5 &\geqslant 7, \\
-x_5 &\geqslant -9.
\end{aligned}
\tag{13.13}
$$

Linear program with:

$$
\begin{aligned}
S &= 9, & x_0 = s_1 = 2, \\
d_0 &= 0, & x_6 = 4.
\end{aligned}
$$

$$
[\text{MIN}]Z = \sum_{i=1}^{6} p_i a_i = \sum_{i=1}^{6} p_i[x_i - x_{i-1} + d_{i-1}]
$$

$$
= 388 - 7x_1 + 5x_2 - 4x_3 - 3x_4 + 10x_5 . \tag{13.14}
$$

When studying (13.13) observe the *diagonal-type form*[1] of the left-hand member. More often than not, a multistage linear program occurs in a form where the coefficients of the variables become zero above a certain "staircase" running from top left to bottom right.[1]

As we have observed, x_i generally depends on x_{i-1} and x_{i+1}; hence, the state variable of the period following the one under consideration will enter as a *parameter*. One should note that in certain cases this fact can appreciably simplify the process of optimization.

C. SEARCH FOR THE OPTIMAL POLICY

Let us now suppose that $S = 9$. In order to find the optimal policy for purchasing:

$$
x^* = [x_1{}^*, x_2{}^*, x_3{}^*, x_4{}^*, x_5{}^*, x_6{}^*], \tag{13.15}
$$

---

[1] Multistage linear programs occur in a form which Dantzig has termed "block triangular form," of which the diagonal type is a special case (see [I 34]).

which can equally be expressed using a_i :

$$a^* = [a_1{}^*a_2{}^*a_3{}^*a_4{}^*a_5{}^*a_6{}^*], \qquad (13.16)$$

we shall *arbitrarily* begin the optimization process with period 6, although we might equally well have begun in a program of this kind with period 1.[1]

Granted:

$$F(x_1, x_2, x_3, x_4, x_5, x_6) = v_1(x_1) + v_2(x_1, x_2) + v_3(x_2, x_3)$$
$$+ v_4(x_3, x_4) + v_5(x_4, x_5) + v_6(x_5, x_6), \qquad (13.17)$$

where the quantity v_i represents the purchase price for Period i. We will take $x_0 = 2$ as indicated in (13.12).

Period 6. Since $s_7 = 0$, we find, from (13.5):

$$x_6 - d_6 = 0, \qquad \text{let} \quad x_6 = d_6 = 4, \qquad (13.18)$$

from which it necessarily follows that

$$x_6{}^* = 4. \qquad (13.19)$$

The sum expended for period 6 has as its value:

$$\begin{aligned} v_6(x_5, x_6) = p_6 a_6 &= p_6(x_6 - s_6) \\ &= p_6[x_6 - (x_5 - d_5)] \\ &= 10[x_6 - x_5 + 7] \\ &= 110 - 10x_5 \, . \end{aligned} \qquad (13.20)$$

Let us calculate the bounds of x_5 for the optimization of periods 6 and 5. We shall make use of (13.12):

$$d_5 = 7, \qquad d_4 = 2, \qquad x_6 + d_5 = 11, \qquad S = 9, \qquad (13.21)$$

whence:

$$\max[7, x_4 - 2] \leqslant x_5 \leqslant \min[9, 11], \qquad (13.22)$$

that is,

$$\max[7, x_4 - 2] \leqslant x_5 \leqslant 9, \qquad (13.23)$$

---

[1] Or again, for example, we might have optimized for stages 1, 2, 3, 4, on the one hand, and then for stages 5 and 6, on the other, and have next sought the optimal policy for the set of stages; indeed, any "dividing up" into connected sets of stages would be equally suitable.

but in the particular case with which we are dealing, where x_i is always less than or equal to 9:

$$\max[7, x_4 - 2] = 7. \tag{13.24}$$

Lastly

$$7 \leqslant x_5 \leqslant 9. \tag{13.25}$$

Periods 6 and 5 Combined. Let us assume:

$$f_{6,5}(x_4) = \underset{7 \leqslant x_5 \leqslant 9}{\text{MIN}}[v_5(x_4, x_5) + v_6(x_5, x_6)]. \tag{13.26}$$

We have

$$
\begin{aligned}
v_5(x_4, x_5) &= p_5 a_5 \\
&= 20[x_5 - (x_4 - d_4)] \\
&= 20[x_5 - x_4 + 2] \\
&= 40 + 20x_5 - 20x_4 .
\end{aligned}
\tag{13.27}
$$

$$
\begin{aligned}
v_5(x_4, x_5) + v_6(x_5, x_6) &= 110 - 10x_5 + 40 + 20x_5 - 20x_4 \\
&= 150 + 10x_5 - 20x_4 ,
\end{aligned}
\tag{13.28}
$$

and

$$f_{6,5}(x_4) = \underset{7 \leqslant x_5 \leqslant 9}{\text{MIN}}[150 + 10x_5 - 20x_4]. \tag{13.29}$$

But $150 + 10x_5 - 20x_4$ is a *monotone[1] increasing function* of x_5, and it is therefore necessary to take the lower bound of x_5. Let

$$x_5{}^* = 7, \quad \text{whence } s_6{}^* = 0 \quad \text{and} \quad a_6{}^* = 4 \quad \text{(Fig. 13.2)} \tag{13.30}$$

and

$$f_{6,5}(x_4) = 220 - 20x_4 . \tag{13.31}$$

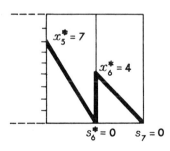

FIG. 13.2.

---

[1] We must remember that a function is *monotone in an interval* if its variation retains the same sign during this interval.

Let us calculate the bounds of x_4 :

$$\max[d_4, x_3 - d_3] \leqslant x_4 \leqslant \min[S, x_5 + d_4] \qquad (13.32)$$

gives

$$\max[2, x_3 - 3] \leqslant x_4 \leqslant \min[9, 9], \qquad (13.33)$$

so that we have

$$\max[2, x_3 - 3] \leqslant x_4 \leqslant 9. \qquad (13.34)$$

Periods 6, 5, and 4 Combined.   Let us assume

$$f_{6,5,4}(x_3) = \min_{\max[2,x_3-3] \leqslant x_4 \leqslant 9} [v_4(x_3, x_4) + f_{6,5}(x_4)]. \qquad (13.35)$$

We have

$$\begin{aligned}
v_4(x_3, x_4) &= p_4 a_4 \\
&= 17[x_4 - (x_3 - d_3)] \\
&= 17[x_4 - x_3 + 3] \\
&= 51 + 17x_4 - 17x_3,
\end{aligned} \qquad (13.36)$$

$$\begin{aligned}
v_4(x_3, x_4) + f_{6,5}(x_4) &= 51 + 17x_4 - 17x_3 + 220 - 20x_4 \\
&= 271 - 3x_4 - 17x_3,
\end{aligned} \qquad (13.37)$$

and

$$f_{6,5,4}(x_3) = \min_{\max[2,x_3-3] \leqslant x_4 \leqslant 9} [271 - 3x_4 - 17x_3]. \qquad (13.38)$$

Since $271 - 3x_4 - 17x_3$ is a monotone decreasing function of x_4, we must take the upper bound, whence

$$x_4{}^* = 9, \qquad s_5{}^* = 7, \qquad a_5{}^* = 0 \quad \text{(Fig. 13.3)} \qquad (13.39)$$

and

$$f_{6,5,4}(x_3) = 244 - 17x_3. \qquad (13.40)$$

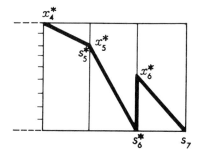

FIG. 13.3.

Let us calculate the bounds of x_3

$$\max[d_3, x_2 - d_2] \leqslant x_3 \leqslant \min[S, x_4 + d_3], \tag{13.41}$$

$$\max[3, x_2 - 5] \leqslant x_3 \leqslant \min[9, 12], \tag{13.42}$$

$$\max[3, x_2 - 5] \leqslant x_3 \leqslant 9. \tag{13.43}$$

Periods 6, 5, 4, and 3 Combined. Let us assume

$$f_{6,5,4,3}(x_2) = \underset{\max[3,x_2-5]\leqslant x_3\leqslant 9}{\text{MIN}} [v_3(x_2, x_3) + f_{6,5,4}(x_3)]. \tag{13.44}$$

We have

$$
\begin{aligned}
v_3(x_2, x_3) &= p_3 a_3 \\
&= 13[x_3 - (x_2 - d_2)] \\
&= 13[x_3 - x_2 + 5] \\
&= 65 + 13x_3 - 13x_2,
\end{aligned} \tag{13.45}
$$

$$
\begin{aligned}
v_3(x_2, x_3) + f_{6,5,4}(x_3) &= 65 + 13x_3 - 13x_2 + 244 - 17x_3 \\
&= 309 - 4x_3 - 13x_2,
\end{aligned} \tag{13.46}
$$

and

$$f_{6,5,4,3}(x_2) = \underset{\max[3,x_2-5]\leqslant x_3\leqslant 9}{\text{MIN}} [309 - 4x_3 - 13x_2]. \tag{13.47}$$

Since $309 - 4x_3 - 13x_2$ is a monotone decreasing function of x_3, one must take the upper bound, whence

$$x_3{}^* = 9, \qquad s_4{}^* = 6, \qquad a_4{}^* = 3 \quad \text{(Fig. 13.4)}, \tag{13.48}$$

and

$$f_{6,5,4,3}(x_2) = 273 - 13x_2. \tag{13.49}$$

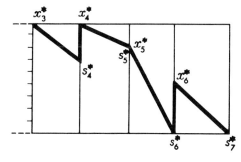

FIG. 13.4.

II. DYNAMIC PROGRAMMING

Let us calculate the bounds of x_2

$$\max[d_2, x_1 - d_1] \leqslant x_2 \leqslant \min[9, x_3 + d_2], \qquad (13.50)$$

$$\max[5, x_1 - 8] \leqslant x_2 \leqslant \min[9, 14], \qquad (13.51)$$

$$5 \leqslant x_2 \leqslant 9. \qquad (13.52)$$

Periods 6, 5, 4, 3 and 2 Combined. Assuming

$$f_{6,5,4,3,2}(x_1) = \min_{5 \leqslant x_2 \leqslant 9} [v_2(x_1, x_2) + f_{6,5,4,3}(x_2)], \qquad (13.53)$$

we have

$$\begin{aligned}
v_2(x_1, x_2) &= p_2 a_2 \\
&= 18[x_2 - (x_1 - d_1)] \\
&= 18[x_2 - x_1 + 8] \\
&= 144 + 18x_2 - 18x_1, \qquad (13.54)
\end{aligned}$$

$$\begin{aligned}
v_2(x_1, x_2) + f_{6,5,4,3}(x_2) &= 144 + 18x_2 - 18x_1 + 273 - 13x_2 \\
&= 417 + 5x_2 - 18x_1, \qquad (13.55)
\end{aligned}$$

and

$$f_{6,5,4,3,2}(x_1) = \min_{5 \leqslant x_2 \leqslant 9} [417 + 5x_2 - 18x_1]. \qquad (13.56)$$

Since $417 + 5x_2 - 18x_1$ is a monotone increasing function of x_2, one must take the lower bound, whence

$$x_2{}^* = 5, \qquad s_3{}^* = 0, \qquad a_3{}^* = 9 \quad \text{(Fig. 13.5)}, \qquad (13.57)$$

and

$$f_{6,5,4,3,2}(x_1) = 442 - 18x_1. \qquad (13.58)$$

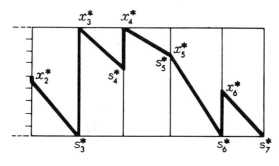

FIG. 13.5.

Let us calculate the bounds of x_1 :

$$\max[d_1, s_1] \leqslant x_1 \leqslant \min[S, x_2 + d_1], \tag{13.59}$$

$$\max[8,2] \leqslant x_1 \leqslant \min[9,13] \tag{13.60}$$

$$8 \leqslant x_1 \leqslant 9. \tag{13.61}$$

Periods 6, 5, 4, 3, 2, 1 Combined.

$$F^* = \underset{8 \leqslant x_1 \leqslant 9}{\text{MIN}} [v_1(x_0, x_1) + f_{6,5,4,3,2}(x_1)]. \tag{13.62}$$

We have, with $x_0 = 2$, $d_0 = 0$,

$$
\begin{aligned}
v_1(x_0, x_1) &- p_1 a_1 \\
&= 11[x_1 - (x_0 - d_0)] \\
&= 11[x_1 - 2] \\
&= -22 + 11x_1,
\end{aligned}
\tag{13.63}
$$

$$
\begin{aligned}
v_1(x_0, x_1) + f_{6,5,4,3,2}(x_1) &= -22 + 11x_1 + 442 - 18x_1 \\
&= 420 - 7x_1
\end{aligned}
\tag{13.64}
$$

$$F^* = \underset{8 \leqslant x_1 \leqslant 9}{\text{MIN}} (420 - 7x_1). \tag{13.65}$$

Since $420 - 7x_1$ is a monotone decreasing function of x_1, one must take the upper bound, whence

$$x_1{}^* = 9, \qquad s_2{}^* = 1, \qquad a_2{}^* = 4 \quad \text{(Fig. 13.6)}, \tag{13.66}$$

and

$$F^* = 357. \tag{13.67}$$

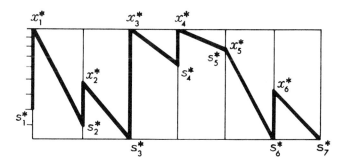

Fig. 13.6.

Lastly

$$a_1{}^* = x_1{}^* - s_1{}^* = 7. \tag{13.68}$$

Hence, the optimal policy (Fig. 13.6) is given by

$$x^* = [9, 5, 9, 9, 7, 4] \tag{13.69}$$

or by

$$a^* = [7, 4, 9, 3, 0, 4], \tag{13.70}$$

and the minimal expenditure for the 6 periods is $F^* = 357$.

It would be incorrect to assume that dynamic programming can always be carried out automatically with the aid of a general algorithm, as can be done in the case of linear programs by means of the simplex method (see Vol. I, Chap. VII).[1] In point of fact, the difficulty in applying the method of dynamic programming in the case of variables with continuous values,[2] frequently arises from the problem of carrying forward the relations introduced by the bounds of the decision variables from one stage to the following ones. In order that the reader may convince himself of this, we suggest that he rework the present problem, taking $S = 20$. He will find

$$x^* = [20, 12, 12, 9, 7, 4], \tag{13.71}$$

$$a^* = [18, \ \ 0, \ \ 5, 0, 0, 4], \tag{13.72}$$

with

$$F^* = 303. \tag{13.73}$$

In certain problems it is possible to obtain a speedy optimal solution by means of intuitive reasoning; this is true, for instance, in the case of the present one if we make $S = 20$.[3] The method of dynamic programming may not always be the easiest approach, but it can be used in a great variety of problems.

14. Decisions in the Face of Uncertainty

A. STUDY OF A GAME OF CHANCE

To demonstrate the manner in which dynamic programming can be applied in the face of uncertainty let us consider the following game of chance.

---

[1] See footnote on p. 94.

[2] The appropriate calculations are given in Section 52 of Part II.

[3] All these difficulties disappear if the problem can be treated by means of *discrete variables*, with the reservation that each variable is to be assigned such discrete values as will be suitable for the method of calculation (manual or electronic) which is available. If this is possible, the problem becomes one of seeking an optimal path as in the example introduced in Section 11.

The game will take place on four squares, numbered A, B, C, and D, and will be divided into a preliminary stage followed by three further stages. The player will receive a counter which will be placed in one of the squares (see Fig. 14.1).

FIG. 14.1.

Stage 0. The player is told in which of the squares his counter is to be placed initially.

Stage 1. Starting from the square in question, the player can move his counter in accordance with the broken arrows shown on Fig. 14.2. Thus, if he is initially in B, he may move to either A, B, or D.

After this move has been made, chance intervenes, and a suitable lottery wheel then moves the counter to a different square in accordance with the various laws of probability shown in the column marked "chance" for Stage 1. Thus, if the decision results in the counter being in square D, any of the three following events may take place: it is moved to A with a gain of 3 and a probability of 0.5; to C with a gain of 2 and a probability of 0.3; or to D with a gain of 1 and a probability of 0.2.

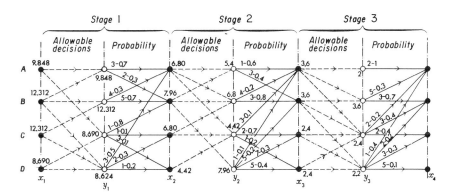

FIG. 14.2.   Graph of optimal strategies. The numbers beside the arcs represent the value of a move followed by its probability; the numbers in column y_1 represent the expected value at time i after the decision taken at that time; the numbers in x_1 represent the optimal expected value at time i.

Stage 2. The same process takes place, but the set of allowable decisions is different, as well as the gains and laws of probability.

Stage 3. Again, the same process takes place, but with a new set of allowable decisions, gains, and probabilities.

The game ends in Stage 3, and the total profit for the three stages is then calculated.

It is our intention to determine the *maximum expected gain*, that is to say, the *strategy*[1] (or strategies) giving the expected value of the maximal profit.

To work out this problem by means of dynamic programming, we shall consider the 3 *variables* y_1, y_2, y_3 which can be equal to A, B, C, or D, and represent a decision of the player in the successive Stages 1, 2, and 3. We shall take x_1, x_2, x_3, and x_4 as "the variables" of positions which may be equal to A, B, C, D at the commencement of Stages 1, 2, and 3, and at the conclusion of Stage 3. Figure 14.2 shows the moments at which these different variables are to be considered.

Let $\bar{z}_3(x_3, y_3)$ represent the expected value of gain when one decides to move from x_3 to y_3. In point of fact, this quantity depends solely on y_3. One now has

$$\bar{z}_3(A, A) = (2)(1) = 2,$$

$$\bar{z}_3(A, B) = \bar{z}_3(B, B) = (5)(0.3) + (3)(0.7) = 3.6,$$

$$\bar{z}_3(A, C) = \bar{z}_3(B, C) = \bar{z}_3(C, C) = \bar{z}_3(D, C)$$
$$= (2)(0.2) + (3)(0.4) + (2)(0.4) = 2.4, \tag{14.1}$$

$$\bar{z}_3(B, D) = \bar{z}_3(C, D) = \bar{z}_3(D, D)$$
$$= (1)(0.4) + (2)(0.2) + (3)(0.3) + (5)(0.1) = 2.2.$$

By calling $f_3(x_3)$ the maximum expected value when we are in x_3, it follows that

$$f_3(A) = 3.6, \qquad \text{with} \quad y_3 = B;$$

$$f_3(B) = 3.6, \qquad \text{with} \quad y_3 = B;$$

$$f_3(C) = 2.4, \qquad \text{with} \quad y_3 = C; \tag{14.2}$$

$$f_3(D) = 2.4, \qquad \text{with} \quad y_3 = C.$$

Stages 3 and 2 Combined. Let $\bar{z}_2(x_2, y_2)$ be the expected value of gain for Stages 2 and 3 combined when one is in x_2 at the commencement

---

[1] When dealing with statistical processes it is preferable to replace the term *policy* by the term *strategy*; it is now a question of a set of determining decisions to meet every possible eventuality.

of Stage 2. We shall apply the theorem of optimality by finding the sum of the expected value of gain in Stage 2 and of the maximal gain in Stage 3 for the values of x_2, y_2, and x_3 under consideration. The optimal expected values are given by (14.4).

It follows that

$$\bar{z}_2(A, A) = \bar{z}_2(B, A) = (1 + 3.6)(0.6) + (3 + 3.6)(0.4) = 5.40,$$

$$\bar{z}_2(A, B) = \bar{z}_2(B, B) = \bar{z}_2(C, B)$$
$$= (4 + 3.6)(0.2) + (3 + 3.6)(0.8) = 6.80,$$

$$\bar{z}_2(A, C) = \bar{z}_2(B, C) = \bar{z}_2(C, C) = \bar{z}_2(D, C) \qquad (14.3)$$
$$= (3 + 3.6)(0.1) + (2 + 2.4)(0.7) + (1 + 2.4)(0.2) = 4.42,$$

$$\bar{z}_2(B, D) = (1 + 3.6)(0.1) + (5 + 3.6)(0.2)$$
$$+ (7 + 2.4)(0.3) + (5 + 2.4)(0.4) = 7.96.$$

Taking $f_{3,2}(x_2)$ to represent the maximal expected value when we are in x_2, it follows that

$$
\begin{aligned}
f_{3,2}(A) &= 6.80, &\text{with} \quad y_2 &= B; \\
f_{3,2}(B) &= 7.96, &\text{with} \quad y_2 &= D; \\
f_{3,2}(C) &= 6.80, &\text{with} \quad y_2 &= B; \\
f_{3,2}(D) &= 4.42, &\text{with} \quad y_2 &= C.
\end{aligned}
\qquad (14.4)
$$

Stages 3, 2, and 1 Combined. Let $\bar{z}_1(x_1, y_1)$ be the expected value of gain for these stages when we are in x_1 at the commencement of Stage 1. We shall again apply the theorem of optimality in the same manner as before.

$$\bar{z}_1(A, A) = \bar{z}_1(B, A)$$
$$= (3 + 6.80)(0.7) + (2 + 7.96)(0.3) = 9.848,$$

$$\bar{z}_1(B, B) = \bar{z}_1(C, B)$$
$$= (4 + 6.80)(0.3) + (5 + 7.96)(0.7) = 12.312,$$

$$\bar{z}_1(A, C) = \bar{z}_1(D, C) \qquad (14.5)$$
$$= (1 + 7.96)(0.8) + (1 + 6.80)(0.1) + (3 + 4.42)(0.1) = 8.690,$$

$$\bar{z}_1(B, D) = \bar{z}_1(C, D) = \bar{z}_1(D, D)$$
$$= (3 + 6.80)(0.5) + (2 + 6.80)(0.3) + (1 + 4.42)(0.2) = 8.624.$$

Lastly, if $F_{\max}(x_1)$ represents the maximal value, we shall have as the different values of y_1 :

$$
\begin{aligned}
x_1 &= A, F_{\max}(A) = \ 9.848, &\text{with}\quad y_1{}^* = A; \\
x_1 &= B, F_{\max}(B) = 12.312, &\text{with}\quad y_1{}^* = B; \\
x_1 &= C, F_{\max}(C) = 12.312, &\text{with}\quad y_1{}^* = B; \\
x_1 &= D, F_{\max}(D) = \ 8.690, &\text{with}\quad y_1{}^* = C.
\end{aligned}
\qquad (14.6)
$$

B. OPTIMAL STRATEGIES

We should obtain the above maximal values as average values by repeating the game a sufficient number of times, and after 100 games the average values would differ but very slightly from those obtained in our calculations.

Hence the optimal strategy will be

$$
\begin{array}{llll}
\text{In Stage 1:} & \text{if } x_1 = A, & \text{take } y_1 = A, \\
 & = B, & = B, \\
 & = C, & = B, \\
 & = D, & = C; \\
\text{In Stage 2:} & \text{if } x_2 = A, & \text{take } y_2 = B, \\
 & = B, & = D, \\
 & = C, & = B, \\
 & = D, & = C; \\
\text{In Stage 3:} & \text{if } x_3 = A, & \text{take } y_3 = B, \\
 & = B, & = B, \\
 & = C, & = C, \\
 & = D, & = C.
\end{array}
\qquad (14.7)
$$

These optimal strategies are shown on Fig. 14.3.

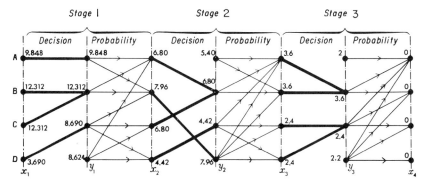

FIG. 14.3.   Graph of optimal strategies.

Important observation. In the face of an uncertain future, if we take for criterion the optimization of the expected value of the total value, such optimization can only be carried out in one direction: by proceeding from the future towards the past. This rule is the consequence of the multistage method by which this expected value is calculated.

15. Two Examples of Dynamic Programming in the Face of an Uncertain Future

A. A PROBLEM OF RESTOCKING

Restocking of a certain article takes place every three months, and the annual policy has to be planned. The initial stock is nil.

The quarterly demand u_i for the three-month period i, $i = 1, 2, 3, 4$, is uncertain and its law of probability $\phi(u_i)$ is the same for each of the four quarters. The cost of stocking up with the article for a three-month period is C_1; when the amount of stock is insufficient to meet demand (condition of scarcity) there are *lost* sales resulting in a loss evaluated at C_2 per article.

On the other hand, for various reasons, it is impossible to restock beyond an amount S. Let us make use of the following notation:

s_i,   stock at the end of the three-month period $i - 1$ (before the restocking a_i);

σ_i,   stock at the beginning of the three-month period i (after the restocking a_i);

$a_i = \sigma_i - s_i$,   the amount of restocking at the start of the quarter i;

u_i,   the amount of the uncertain demand[1] in the three-month period i.

We intend to show how it is possible to obtain the minimal expected value of the total annual cost of restocking, by making use of dynamic programming. We shall introduce numerical assumptions which will make it easy to follow the calculations, will assume that the demand u_i can consist of 0, 1, 2, or 3 articles, and that the stock must never exceed 3 articles. The law of probability $\phi(u_i)$ for the demand u_i will be assumed to be the one shown in Table 15.1, in which the cumulative probabilities

---

[1] It is not known in advance on what dates of the quarter the demand for the articles will occur, and one has to suppose that the probability of demand on a certain day j of this quarter will be the same as on any other day.

$\Phi(u_i)$ are also included. Lastly, we shall take the numerical values: $C_1 = 4$, $C_2 = 3$, and $C_1 = 12$.

TABLE 15.1

u	$\phi(u)$	$\Phi(u)$
0	0.2	0.2
1	0.3	0.5
2	0.4	0.9
3	0.1	1
>3	0	1

Figure 15.2 represents a possible *history* of this problem of stock when the strategy for purchasing has been $a_1 = 2$, $a_2 = 1$, $a_3 = 3$, $a_4 = 0$, while the demand has been $u_1 = 1$, $u_2 = 3$, $u_3 = 1$, $u_4 = 2$. In this figure, broken lines represent the mean variation of stock in each quarter, the assumption being that each day possesses an *equiprobability* for the arrival of orders. The cost of restocking in situations (a) and (b), which are described later, will therefore be taken as being proportional to the shaded areas in Fig. 15.1; it is this assumption which will enable us to write the two formulas for cost (15.2) and (15.4).

It is evident that two distinct situations may arise in the course of a quarter:

Situation (a):

$$u_i \leqslant \sigma_i ; \tag{15.1}$$

then

$$s_{i+1} = \sigma_i - u_i = s_i + a_i - u_i , \qquad \text{there is no shortage.}$$

The average cost for quarter i is

$$C_1(\sigma_i - u_i/2) \qquad \text{if the demand is } u_i . \tag{15.2}$$

Situation (b):

$$u_i \geqslant \sigma_i ; \tag{15.3}$$

then

$$s_{i+1} = 0, \qquad \text{there is a shortage.}$$

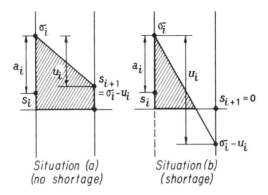

FIG. 15.1.

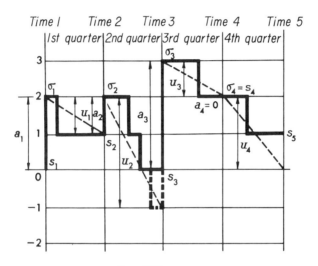

FIG. 15.2.

The average cost for quarter i is

$$\tfrac{1}{2}C_1 \frac{\sigma_i^2}{u_i} + C_2(u_i - \sigma_i) \qquad \text{if the demand is } u_i . \tag{15.4}$$

The following constraints intervene:

$$s_i \leqslant \sigma_i \leqslant S; \tag{15.5}$$

let

$$s_i \leqslant a_i + s_i \leqslant S, \tag{15.6}$$

or

$$0 \leqslant a_i \leqslant S - s_i . \tag{15.7}$$

The expected value of the cost of the stock for a quarter i, taking into account the probabilities of situations (a) and (b), is

$$\bar{z}(\sigma_i) = \sum_{u_i=0}^{\sigma_i-1} C_1\left(\sigma_i - \frac{u_i}{2}\right) \cdot \phi(u_i) + \sum_{u_i=\sigma_i}^{\infty} \left[\tfrac{1}{2}C_1\frac{\sigma_i^2}{u_i} + C_2(u_i - \sigma_i)\right] \cdot \phi(u_i)$$

(15.8)

We shall successively evaluate the optimum for the three-month period 4, then for quarters 4 and 3 combined, then for 4, 3, 2, and finally for 4, 3, 2, and 1. Let us remember that in a case of an uncertain future optimization has to be carried out in a reverse direction in relation to time.

Taking the numerical values which we assigned earlier, and taking into account the fact that each of the quarters has the same law of probability in regard to demand, we can state

$$\bar{z}(\sigma_i) = \sum_{u_i=0}^{\sigma_i-1} 4\left(\sigma_i - \frac{u_i}{2}\right) \cdot \phi(u_i) + \sum_{u_i=\sigma_i}^{3} \left[2\frac{\sigma_i^2}{u_i} + 12(u_i - \sigma_i)\right] \cdot \phi(u_i). \quad (15.9)$$

Let us begin by evaluating the minimal cost for the fourth quarter.

Quarter 4. Let $f_4(s_4)$ be the minimal cost[1] for the fourth quarter, allowing of course for the fact that this minimal cost depends on s_4.

$$f_4(s_4) = \min_{s_4 \leqslant \sigma_4 \leqslant 3} \bar{z}(\sigma_4). \quad (15.10)$$

We have in succession

$$\bar{z}(0) = \sum_{u_4=0}^{3} 12u_4\phi(u_4)$$

$$= 12[(1)(0.3) + (2)(0.4) + (3)(0.1)] = 16.80, \quad (15.11)$$

$$\bar{z}(1) = 4 \cdot \phi(0) + \sum_{u_4=1}^{3} [2/u_4 + 12(u_4 - 1)] \cdot \phi(u_4)$$

$$= (4)(0.2) + (2)(0.3) + (13)(0.4) + (74/3)(0.1) = 9.07, \quad (15.12)$$

$$\bar{z}(2) = \sum_{u_4=0}^{1} 4\left(2 - \frac{u_4}{2}\right) \cdot \phi(u_4) + \sum_{u_4=2}^{3} [8/u_4 + 12(u_4 - 2)] \cdot \phi(u_4)$$

$$= (8)(0.2) + (6)(0.3) + (4)(0.4) + (44/3)(0.1) = 6.47, \quad (15.13)$$

---

[1] When there can be no risk of confusion we shall frequently use "cost" instead of "expected value of the cost" to shorten our explanations.

$$\bar{z}(3) = \sum_{u_4=0}^{2} 4(3 - u_4/2) \cdot \phi(u_4) + \sum_{u_4=3}^{3} [18/u_4 + 12(u_4 - 3)] \cdot \phi(u_4)$$

$$= (12)(0.2) + (10)(0.3) + (8)(0.4) + (6)(0.1) = 9.20. \qquad (15.14)$$

Beginning with the values $\bar{z}(0)$, $\bar{z}(1)$, $\bar{z}(2)$, and $\bar{z}(3)$ above, we now have to evaluate the quantities $f_4(0)$, $f_4(1)$, $f_4(2)$, and $f_4(3)$; we have

$$f_4(0) = \underset{0 \leqslant \sigma_4 \leqslant 3}{\text{MIN}} \bar{z}(\sigma_4) = \bar{z}(2) = 6.47, \qquad (15.15)$$

$$f_4(1) = \underset{1 \leqslant \sigma_4 \leqslant 3}{\text{MIN}} \bar{z}(\sigma_4) = \bar{z}(2) = 6.47, \qquad (15.16)$$

$$f_4(2) = \underset{2 \leqslant \sigma_4 \leqslant 3}{\text{MIN}} \bar{z}(\sigma_4) = \bar{z}(2) = 6.47, \qquad (15.17)$$

$$f_4(3) = \underset{\sigma_4=3}{\text{MIN}} \bar{z}(\sigma_4) = \bar{z}(3) = 9.20. \qquad (15.18)$$

Let $\bar{v}_4(\sigma_3)$ be the expected value of the cost for the fourth quarter, taking into account all possible values of s_4 which depend on σ_3 (see 15.1 and 15.3):

$$s_4 = \sigma_3 - u_3 \quad \text{if} \quad u_3 \leqslant \sigma_3$$
$$= 0 \quad \text{if} \quad u_3 \geqslant \sigma_3, \qquad (15.19)$$

that is

$$s_4 = \max[\sigma_3 - u_3, 0]. \qquad (15.20)$$

We have

$$\bar{v}_4(\sigma_3 = 0) = f_4(0) \cdot (1) = 6.47, \qquad (15.21)$$

$$\bar{v}_4(\sigma_3 = 1) = f_4(1) \cdot \phi(0) + f_4(0)[\phi(1) + \phi(2) + \phi(3)]$$
$$= (6.47) \cdot (0.2) + (6.47) \cdot (0.8) = 6.47, \qquad (15.22)$$

$$\bar{v}_4(\sigma_3 = 2) = f_4(2) \cdot \phi(0) + f_4(1) \cdot \phi(1) + f_4(0)[\phi(2) + \phi(3)]$$
$$= (6.47)(0.2) + (6.47)(0.3) + (6.47)(0.5) = 6.47, \qquad (15.23)$$

$$\bar{v}_4(\sigma_3 = 3) = f_4(3) \cdot \phi(0) + f_4(2) \cdot \phi(1) + f_4(1) \cdot \phi(2) + f_4(0) \cdot \phi(3)$$
$$= (9.2) \cdot (0.2) + (6.47) \cdot (0.8) = 7.02. \qquad (15.24)$$

Quarters 4 and 3 Combined. Letting $f_{4,3}(s_3)$ be the minimal cost, we have

$$f_{4,3}(s_3) = \underset{s_3 \leqslant \sigma_3 \leqslant 3}{\text{MIN}} [\bar{z}(\sigma_3) + \bar{v}_4(\sigma_3)]. \qquad (15.25)$$

Taking account of (15.11 to 14) on one hand, and (15.21 to 24), on the other, it follows that

$$\bar{z}(0) + \bar{v}_4(0) = 16.80 + 6.47 = 23.27, \tag{15.26}$$
$$\bar{z}(1) + \bar{v}_4(1) = 9.07 + 6.47 = 15.54, \tag{15.27}$$
$$\bar{z}(2) + \bar{v}_4(2) = 6.47 + 6.47 = 12.94, \tag{15.28}$$
$$\bar{z}(3) = \bar{v}_4(3) = 9.20 + 7.02 = 16.22, \tag{15.29}$$

whence,

$$f_{4,3}(0) = \min_{0 \leqslant \sigma_3 \leqslant 3} [\bar{z}(\sigma_3) + \bar{v}_4(\sigma_3)] = 12.94, \qquad \text{with} \quad \sigma_3{}^* = 2, \tag{15.30}$$

$$f_{4,3}(1) = \min_{1 \leqslant \sigma_3 \leqslant 3} [\bar{z}(\sigma_3) + \bar{v}_4(\sigma_3)] = 12.94, \qquad \text{with} \quad \sigma_3{}^* = 2, \tag{15.31}$$

$$f_{4,3}(2) = \min_{2 \leqslant \sigma_3 \leqslant 3} [\bar{z}(\sigma_3) + \bar{v}_4(\sigma_3)] = 12.94, \qquad \text{with} \quad \sigma_3{}^* = 2, \tag{15.32}$$

$$f_{4,3}(3) = \min_{\sigma_3 = 3} [\bar{z}(\sigma_3) + \bar{v}_4(\sigma_3)] = 16.22, \qquad \text{with} \quad \sigma_3{}^* = 3. \tag{15.33}$$

Let $\bar{v}_{4,3}(\sigma_2)$ be the expected value of the cost for the fourth and third quarters together, taking account of all the possible values of s_3, which depend on σ_2:

$$s_3 = \max[\sigma_2 - u_2, 0]. \tag{15.34}$$

We have

$$\bar{v}_{4,3}(\sigma_2 = 0) = f_{4,3}(0) \cdot (1) = 12.94, \tag{15.35}$$

$$\bar{v}_{4,3}(\sigma_2 = 1) = f_{4,3}(1) \cdot \phi(0) + f_{4,3}(0) \cdot [\phi(1) + \phi(2) + \phi(3)] = 12.94, \tag{15.36}$$

$$\bar{v}_{4,3}(\sigma_2 = 2) = f_{4,3}(2) \cdot \phi(0) + f_{4,3}(1) \cdot \phi(1) \\ + f_{4,3}(0)[\phi(2) + \phi(3)] = 12.94, \tag{15.37}$$

$$\bar{v}_{4,3}(\sigma_2 = 3) = f_{4,3}(3) \cdot \phi(0) + f_{4,3}(2) \cdot \phi(1) \\ + f_{4,3}(1) \cdot \phi(2) + f_{4,3}(0) \cdot \phi(3) = 13.60. \tag{15.38}$$

Quarters 4, 3, and 2 Combined. Taking $f_{4,3,2}(s_2)$ for the minimal cost, we have

$$f_{4,3,2}(s_2) = \min_{s_2 \leqslant \sigma_2 \leqslant 3} [\bar{z}(\sigma_2) + \bar{v}_{4,3}(\sigma_2)]. \tag{15.39}$$

Taking into account (15.11 to 14) and (15.35 to 38), it follows that

$$\bar{z}(0) + \bar{v}_{4,3}(0) = 16.80 + 12.94 = 29.74, \tag{15.40}$$
$$\bar{z}(1) + \bar{v}_{4,3}(1) = 9.07 + 12.94 = 22.01, \tag{15.41}$$
$$\bar{z}(2) + \bar{v}_{4,3}(2) = 6.47 + 12.94 = 19.41, \tag{15.42}$$
$$\bar{z}(3) + \bar{v}_{4,3}(3) = 9.20 + 13.60 = 22.80; \tag{15.43}$$

thence,

$$f_{4,3,2}(0) = \underset{0 \leqslant \sigma_2 \leqslant 3}{\text{MIN}} [\bar{z}(\sigma_2) + \bar{v}_{4,3}(\sigma_2)] = 19.41, \qquad \text{with} \quad \sigma_2{}^* = 2, \quad (15.44)$$

$$f_{4,3,2}(1) = \underset{1 \leqslant \sigma_2 \leqslant 3}{\text{MIN}} [\bar{z}(\sigma_2) + \bar{v}_{4,3}(\sigma_2)] = 19.41, \qquad \text{with} \quad \sigma_2{}^* = 2, \quad (15.45)$$

$$f_{4,3,2}(2) = \underset{2 \leqslant \sigma_2 \leqslant 3}{\text{MIN}} [\bar{z}(\sigma_2) + \bar{v}_{4,3}(\sigma_2)] = 19.41, \qquad \text{with} \quad \sigma_2{}^* = 2, \quad (15.46)$$

$$f_{4,3,2}(3) = \underset{\sigma_2 = 3}{\text{MIN}} [\bar{z}(\sigma_2) + \bar{v}_{4,3}(\sigma_2)] = 22.80, \qquad \text{with} \quad \sigma_2{}^* = 3. \quad (15.47)$$

Taking $\bar{v}_{4,3,2}(\sigma_1)$ for the expected value of the cost for quarters 4, 3, and 2 together and taking account of all the values of s_2, which depend on σ_1; we have

$$s_2 = \max[\sigma_1 - u_1, 0], \qquad (15.48)$$

$$\bar{v}_{4,3,2}(\sigma_1 = 0) = f_{4,3,2}(0) \cdot (1) = 19.41, \qquad (15.49)$$

$$\bar{v}_{4,3,2}(\sigma_1 = 1) = f_{4,3,2}(1) \cdot \phi(0)$$
$$+ f_{4,3,2}(0)[\phi(1) + \phi(2) + \phi(3)] = 19.41, \qquad (15.50)$$

$$\bar{v}_{4,3,2}(\sigma_1 = 2) = f_{4,3,2}(2) \cdot \phi(0) + f_{4,3,2}(1) \cdot \phi(1)$$
$$+ f_{4,3,2}(2)[\phi(2) + \phi(3)] = 19.41, \qquad (15.51)$$

$$\bar{v}_{4,3,2}(\sigma_1 = 3) = f_{4,3,2}(3) \cdot \phi(0) + f_{4,3,2}(2) \cdot \phi(1)$$
$$+ f_{4,3,2}(1) \cdot \phi(2) + f_{4,3,2}(0) \cdot \phi(3) = 20.09. \quad (15.52)$$

Quarters 4, 3, 2, and 1 Combined. Taking $f_{4,3,2,1}(s_1)$ for the minimal cost, we have

$$f_{4,3,2,1}(s_1) = \underset{s_1 \leqslant \sigma_1 \leqslant 3}{\text{MIN}} [\bar{z}(\sigma_1) + \bar{v}_{4,3,2}(\sigma_1)]. \qquad (15.53)$$

Taking account of (15.11 to 14) and (15.49 to 52), it follows that

$$\bar{z}(0) + \bar{v}_{4,3,2}(0) = 16.80 + 19.41 = 36.21, \qquad (15.54)$$

$$\bar{z}(1) + \bar{v}_{4,3,2}(1) = 9.07 + 19.41 = 28.48, \qquad (15.55)$$

$$\bar{z}(2) + \bar{v}_{4,3,2}(2) = 6.47 + 19.41 = 25.88, \qquad (15.56)$$

$$\bar{z}(3) + \bar{v}_{4,3,2}(3) = 9.20 + 20.09 = 29.29. \qquad (15.57)$$

Thence

$$f_{4,3,2,1}(0) = \underset{0 \leqslant \upsilon_1 \leqslant 3}{\text{MIN}} [\bar{z}(\sigma_1) + \bar{v}_{4,3,2}(\sigma_1)] = 25.88, \qquad \text{with} \quad \sigma_1{}^* = 2,$$
$$(15.58)$$

$$f_{4,3,2,1}(1) = \underset{1 \leqslant \sigma_1 \leqslant 3}{\text{MIN}} [\bar{z}(\sigma_1) + \bar{v}_{4,3,2}(\sigma_1)] = 25.88, \qquad \text{with} \quad \sigma_1{}^* = 2,$$
$$(15.59)$$

$$f_{4,3,2,1}(2) = \underset{2 \leqslant \sigma_1 \leqslant 3}{\text{MIN}} [\bar{z}(\sigma_1) + \bar{v}_{4,3,2}(\sigma_1)] = 25.88, \qquad \text{with} \quad \sigma_1{}^* = 2,$$
$$(15.60)$$

$$f_{4,3,2,1}(3) = \underset{\sigma_1 = 3}{\text{MIN}} [\bar{z}(\sigma_1) + \bar{v}_{4,3,2}(\sigma_1)] = 29.28, \qquad \text{with} \quad \sigma_1{}^* = 3.$$
$$(15.61)$$

Lastly, as $s_1 = 0$ is required,

$$F^* = f_{4,3,2,1}(0) = 25.88,$$ (15.62)

with the following strategy[1]:

Quarter 1.

$$s_1 = 0, \qquad \sigma_1{}^* = 2 \qquad \text{hence} \quad a_1{}^* = 2.$$ (15.63)

Quarter 2.

$$
\begin{array}{rlllll}
\text{if} \quad s_2 = 0 & \text{take} & \sigma_2{}^* = 2, & \text{whence} & a_2{}^* = 2, \\
= 1 & & = 2, & & = 1, \\
= 2 & & = 2, & & = 0, \\
= 3 & & = 3, & & = 0.
\end{array}
$$ (15.64)

Quarter 3.

$$
\begin{array}{rlllll}
\text{if} \quad s_3 = 0 & \text{take} & \sigma_3{}^* = 2, & \text{whence} & a_3{}^* = 2, \\
= 1 & & = 2, & & = 1, \\
= 2 & & = 2, & & = 0, \\
= 3 & & = 3, & & = 0.
\end{array}
$$ (15.65)

Quarter 4.

$$
\begin{array}{rlllll}
\text{if} \quad s_4 = 0 & \text{take} & \sigma_4{}^* = 2, & \text{whence} & a_4{}^* = 2, \\
= 1 & & = 2, & & = 1, \\
= 2 & & = 2, & & = 0. \\
= 3 & & = 3, & & = 0.
\end{array}
$$ (15.66)

Figure 15.3 illustrates a sequence of optimal decisions corresponding to the demand:

$$u_1 = 1, \qquad u_2 = 3, \qquad u_3 = 0, \qquad u_4 = 1.$$

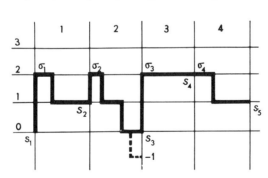

FIG. 15.3.

[1] If $s_1 = 0$; $s_2 = 3$, $s_3 = 3$, and $s_4 = 3$ cannot be worked out by taking the strategy we have just chosen. On the other hand, if s_1 can have other values such as 1, 2, or 3, the values $s_2 = 3$, $s_3 = 3$, $s_4 = 3$ can be found (see Fig. 15.6).

If, at the beginning of the second quarter, we had known that $u_1 = 3$, we would obviously have taken $\sigma_2 = 3$. But, with the available information, the decision $u_2 = 2$ is the best average one.

It should be pointed out that the stock must always remain less than or equal to 2. The optimal strategy lies in bringing the stock up to this value; hence the expected value of the minimal cost is equal to 25.88.

It should also be noted that in this problem the optimal decisions for quarters 1, 2, 3, and 4 are the same, but this must not lead us to generalize and to think that this will always be the case, for the decisions depend on the nature of the problem under consideration and its implicit limits.

In problems where the data is certain, a policy may consist of a set of N numbers. In those where the future is uncertain, if N is the number of stages or periods, a strategy takes the form of a series of N decisions, each comprising M values (with the possibility that M will not remain constant during the process). Hence, in the present problem, an optimal strategy assumes the following aspect, the values shown being those of $a_1{}^*$ (Fig. 15.4):

$$\varpi^* = \left[\begin{pmatrix} 2 \\ 1 \\ 0 \\ 0 \end{pmatrix} \begin{pmatrix} 2 \\ 1 \\ 0 \\ 0 \end{pmatrix} \begin{pmatrix} 2 \\ 1 \\ 0 \\ 0 \end{pmatrix} \begin{pmatrix} 2 \\ 1 \\ 0 \\ 0 \end{pmatrix} \right] \quad \begin{matrix} s_i = 0 \\ s_i = 1 \\ s_i = 2 \\ s_i = 3. \end{matrix} \qquad (15.67)$$

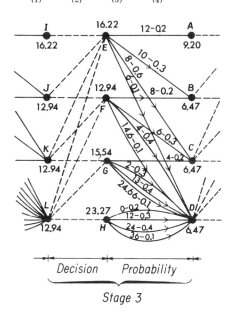

FIG. 15.4.

B. GRAPHICAL REPRESENTATION OF A DISCRETE MULTISTAGE PROCESS IN THE FACE OF UNCERTAINTY

In general, a multistage process with discrete values in a statistic situation may be represented by a graph such as the one in Fig. 15.5,

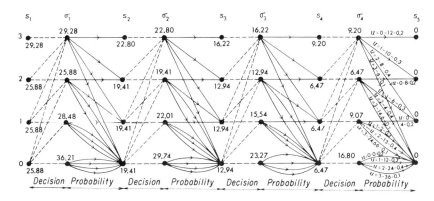

FIG. 15.5. The numbers placed in the σ columns give the expected values brought back to the period for the value of σ decided on; the numbers in the s columns give the optimal values brought back for each value of s. The numbers on the arcs of the probability show, in this order: the demand, the variation of the cost, and the probability of transition.

which corresponds to the example with which we have just dealt. It is possible to envisage more complicated problems in which the formulating of a decision and/or chance may intervene several times in the same stage. It is recommended that the graph of the changes of state should be drawn as we have just drawn it,[1] in order to establish the multistage mathematical pattern for which the graph itself will provide the structure. A representation of the optimal strategy or strategies, shown by heavy lines on the same graph (Fig. 15.6), is generally interesting to study.

In point of fact, the representation of a problem in the form of a graph, whenever it is possible to do so, can prove an extremely convenient method of carrying out the calculations: all that is required is that the expected values of the total subsequent values (if it is a matter of probability), or the optimal value (if it is a matter of decision), should be brought back to the same point. We begin by writing beside the arcs

---

[1] In accordance with convention, the arcs representing the decisions will not be arrowed and will be shown by dotted lines. In addition, as for example GD in Fig. 15.4, there may be several arcs between two points: this is an instance of a "p-colored graph" (see Section 34). It should be noted that it is simple in the graph in question to replace several arcs joining two vertices by a single arc between these vertices, after calculating the value and probability of this arc.

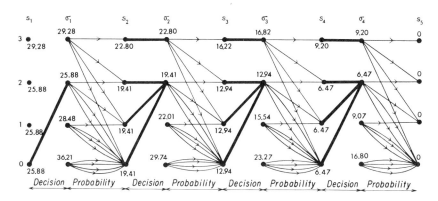

FIG. 15.6. Showing the optimal strategy.

which represent chance the values of the transition and the probabilities associated with them, doing so for all the hypotheses implicit in the problem. In the same way, we write beside the arcs of decision the transitional values associated with them (the present problem does not include any such values, but it is possible to imagine transportation costs proportionate to the number of units ordered in restocking). Next, we write the final values, beside the vertices at the end of the last stage; these values being nil in the present case, since the remaining stock is supposed to be liquidated without any return; though in a different problem, it might have been sold.

To clarify how we should proceed, let us select Stage 3 (the third quarter) of the problem represented in Fig. 15.1; this stage is shown in Fig. 15.4. To simplify the explanations, we have used a letter to denote each vertex of the multigraph.

Let us find out, in the first place, how the values of transition are assigned to the arcs of chance. Why, for example, should the value 10 have been placed beside arc EB? For this arc, we have $\sigma_i = 3$, $u_i = 1$, so that $\sigma_i > u_i$. From (15.2) we obtain

$$4(3 - \tfrac{1}{2}) = 4 \times 2.5 = 10.$$

To take another example: What value should we give to the arc GD, for which $\sigma_i = 1$, $u_i = 3$? In this case $\sigma_i < u_i$ and we must use (15.4). From this we obtain

$$(\tfrac{1}{2})(4)(\tfrac{1}{3}) + (12)(3 - 1) = 0.66 + 24 = 24.66.$$

We shall proceed in the same way for all the arcs of chance in this problem. Having now obtained the values associated with the vertices

A, B, C, and D; What values should we now place beside the vertices E, F, G, and H? Taking E, the cost at E has as its expected value:

$$(12)(0.2) + (10)(0.3) + (8)(0.4) + (6)(0.1),$$

on the one hand, and on the other,

$$(9.20)(0.2) + (6.47)(0.3) + (6.47)(0.4) + (6.47)(0.1),$$

which, through the use of the distributive property of multiplication, becomes

$$(12 + 9.20)(0.2) + (10 + 6.47)(0.3)$$
$$+ (8 + 6.47)(0.4) + (6 + 6.47)(0.1) = 16.216;$$

hence, the value 16.22 has been assigned to E, the value being given correct to two decimal places.

Taking G, the expected value of the cost is

$$(4)(0.2) + (2)(0.3) + (13)(0.4) + (24.66)(0.1),$$

on the one hand, and on the other,

$$(6.47)(0.2) + (6.47)(0.3 + 0.4 + 0.1),$$

or, by regrouping the products:

$$(4 + 6.47)(0.2) + (2 + 6.47)(0.3)$$
$$+ (13 + 6.47)(0.4) + (24.66 + 6.47)(0.1) = 15.536,$$

which is the reason we have given the value 15.54 to vertex G. Vertices F and H are then calculated in the same way.

We will now consider vertices I, J, K, and L by comparing the extremities of the arcs of decision drawn from these points. From I there is only one arc, IE, and the value 16.22 is therefore assigned to I. From J there are two arcs, JE and JF, and as the value of JE is 16.22 and that of JF is 12.94, the smaller of the two numbers is written below J, the same procedure being followed[1] for K and L. We now know the optimal values brought back to the points I, J, K, and L, and all that we now have to do is to calculate for Stage 2 by the same method, and continue from that point.

---

[1] If the nature of the problem implies values for the arcs of decision, the comparison which enables us to determine the optimal decision will be found from the sums of these values added to those of the corresponding vertices.

As will now be realized, the use of a graph is a very convenient method of calculating the results needed for a dynamic program.

It need scarcely be added that the extreme simplicity of the process greatly facilitates programming with an electronic computer. Hence, dynamic programming with discrete values provides a very convenient instrument for scientific planning, even if it should be necessary to consider tens or even hundreds[1] of vertices or changes of state in each stage, in the shape of the descriptive graph.

C. REPLACEMENT OF A SET OF MACHINES

The problem which concerns us is the administration of a set of machines over a five-year period. A machine may either be kept in service or replaced annually by a new one of the same kind. The decision which has to be made annually can be expressed as follows:

(a)   To replace a machine by a new one: R,

(b)   Not to replace it: NR.

The graph of survival[2] is given in Fig. 15.7; Table 15.2 gives the same information in greater detail.

Table 15.3 gives the maintenance cost and the value at each time n. The operation of these machines can be expressed by a dynamic program *decision probability* which is shown on Fig. 15.8. The state variable which

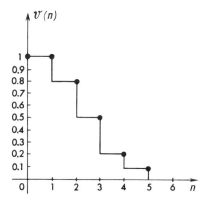

FIG. 15.7.

---

[1] In this case it would not be necessary to draw the graph, since all the calculations would be carried out by means of perforated cards and the computer.

[2] See Vol. I, Section 43. *Note*: see footnote on p. 94, this volume.

TABLE 15.2

(1)	(2)[a]	(3)[b]
n	$v(n)$	$p_c(n) = \dfrac{v(n-1)-v(n)}{v(n-1)}$
0	1	0
1	1	0
2	0.8	$\frac{1}{5}$
3	0.5	$\frac{3}{8}$
4	0.2	$\frac{3}{5}$
5	0.1	$\frac{1}{2}$
$\geqslant 6$	0	1

TABLE 15.3

n	Cost of maintenance	Value at time n
0	0	100
1	4	60
2	10	40
3	17	30
4	26	20
5	38	10
$\geqslant 6$	—	0

[a] The probability that a machine will still be in sound condition after time n.

[b] The conditional probability of wear in the interval $n-1$ to n, such wear necessitating replacement at the end of the period.

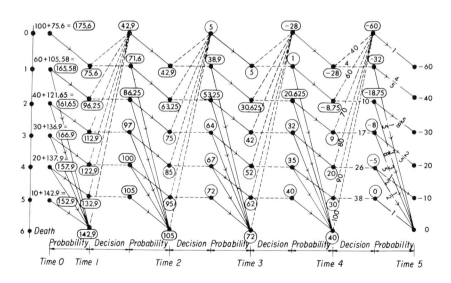

FIG. 15.8.

expresses the condition of the equipment at each time n is its age a. At the beginning of the following period there is a choice of either retaining the equipment (in which case its age a remains unchanged), or of replacing it (in which event the age drops to 0). Chance now leads either to age 6, representing the deterioration, or to age $(a + 1)$. From this point the reader will find little difficulty in continuing the calculations.

The optimal strategy is shown in Table 15.4.

TABLE 15.4

OPTIMAL DECISION

If we are in	Time 1	Time 2	Time 3	Time 4
1	NR	NR	NR	NR
2	NR	NR	NR	NR
3	R	R	R	NR
4	R	R	R	R
5	R	R	R	R
6	R	R	R	R

The results given above have been shown in Fig. 15.9, and it will be seen that beginning with time 3, and returning towards time 0, a permanent strategy is disclosed.

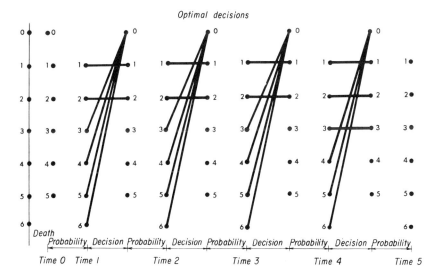

FIG. 15.9.

D. A More Complex Problem of Replacement

We will now study a similar problem of a more complex and realistic kind. The equipment is assumed to have the same curve of survival (Fig. 15.7 and Table 15.2), but a machine may either be replaced by a newer machine or it may be retained: for example, a machine of age 3 may be replaced by a new machine or by one of age 1 or 2, or it may be retained. In addition, the selling price is always lower than the purchase price of a machine of the same age; replacement involves a loss and retaining a machine in service entails the cost of its maintenance. In Fig. 15.10 the different expenses or costs are shown for each possible decision. The purchase price at time 0 is given in the left-hand column of this figure, and the resale price at time 6 appears in the column on the right. In addition, whenever a machine "dies," a loss of production equal to 35 is assumed. The economic period involved in this case is one of six years.

The dynamic program for this new problem is shown in Fig. 15.11. The results are recorded on this figure and on Fig. 15.12, and the optimal multistage decisions are given on the latter. It will be observed that from time 5 a cyclic change of strategies enters, with the result that the strategy followed will depend on whether $2k$ or $(2k + 1)$ is chosen as the economic horizon.

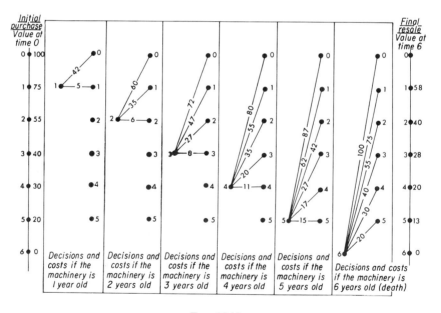

FIG. 15.10.

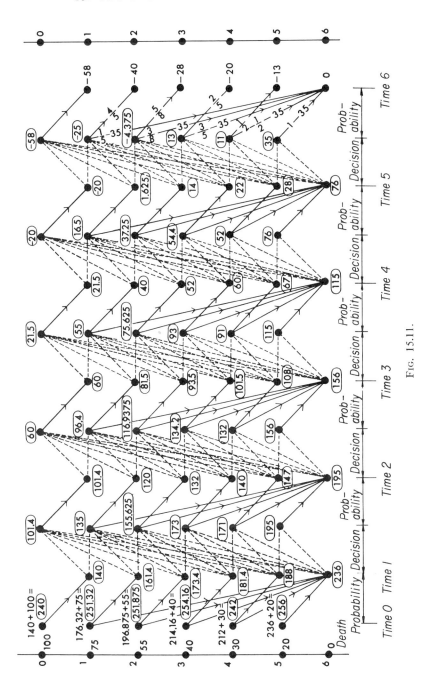

Fig. 15.11.

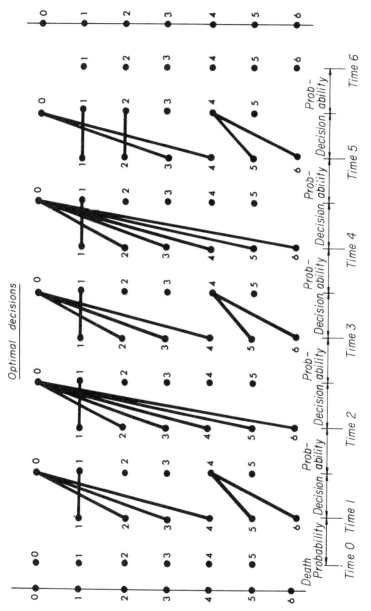

FIG. 15.12.

In mathematical terminology (see Part II, Sections 53 and 58) the two problems form *Markovian decision chains*, but the first is an *ergodic* process, which means that there is a permanent optimal policy which is not dependent on the economic duration[1]; this is not true of the problem in which a *cycle* appears.

E. Evaluation of Risk in an Optimal Strategy

The discovery of the optimal strategy may be usefully supplemented in a case involving uncertainty by a knowledge of the worst and best results provided by this strategy, and of the distribution of values (the law of probability) which may be obtained with this optimal strategy.

To show how these supplementary calculations should be carried out, we shall return to the operational problem shown in Fig. 15.8, the optimal strategy of which is given in Fig. 15.9. To begin with, we will draw the graph of the dynamic programming, retaining only the arcs of optimal decisions and all the arcs of probability. Instead of giving the latter probabilities, we shall now ignore them, so that the program becomes a deterministic one, in which we shall first evaluate the minimal value (the most favorable possible result of the optimal strategy in Fig. 15.13) and then the maximal value (the least favorable possible result of the optimal strategy in Fig. 15.14).

An analytical calculation can be made of the distributive law for the results of an optimal strategy. Considering the same example, and

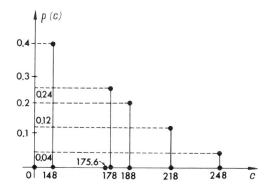

FIG. 15.13. Distribution of the costs in the case of new machinery at time 0 using the optimal strategy.

[1] Beginning with a certain number of stages and going backwards in relation to time. In the example shown in Fig. 15.9, a steady-state strategy appears, beginning at the last stage, starting at the end.

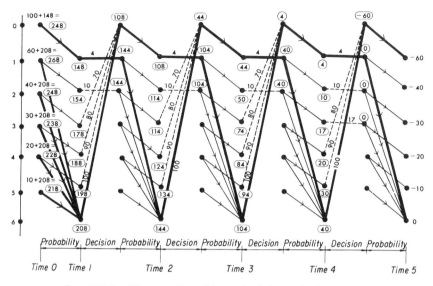

FIG. 15.14.   The most favorable result of the optimal strategy.

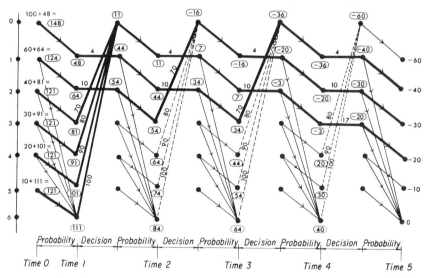

FIG. 15.15.   The most unfavorable result of the optimal strategy.

starting with a new machine at time 0, what is the probability of the least favorable result, namely a cost of 248 (Fig. 15.14)? To discover this, all that need be done is multiply the values carried by each arc of probability forming a part of the path by one another. If there were several paths of equal cost starting from the same origin, the sum of

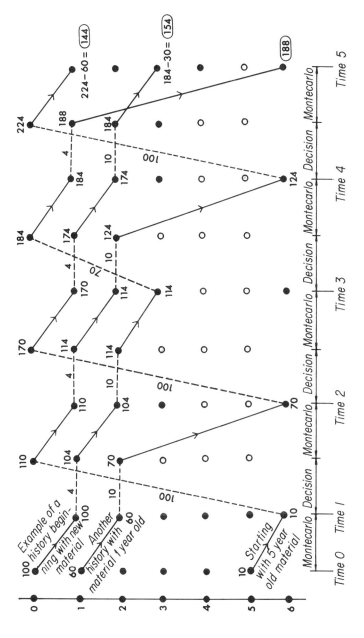

Fig. 15.16.   Representation of three histories.

the products would be calculated. Thus for the path with a cost of 248, starting with a machine which was new at time 0, the probability is

$$1 \times \tfrac{1}{5} \times 1 \times \tfrac{1}{5} \times 1 = \tfrac{1}{25}.$$

It is in this way, by studying the different states at time 0, that we can obtain the distributive laws, an example of which is given in Fig. 15.13 for the case of a machine which was new at time 0.

This enables us to evaluate the risk. Since the expected value of the cost is 175.6, it follows that the probability of obtaining a higher value than 200 is 0.16 in the case of a single machine.

By using the laws of distribution for the different initial states under consideration, the conclusion may be reached that it is not necessarily essential to start with equipment of an age corresponding to the *optimum optimorum*, but that the risk of passing a critical cost should always be taken into account.

F. ALMOST-OPTIMAL STRATEGIES (*n*th OPTIMAL STRATEGIES)

The interesting question may be asked: the optimal strategy (or strategies) having been obtained, what strategy or strategies correspond to the value immediately below the optimal value? In the same way, in a deterministic case, it would be interesting to discover the policies corresponding to values immediately below that of the optimal policy or policies. Such strategies are called *almost optimal*,[1] and the same term is applied to policies.

The method of calculating almost-optimal strategies or policies is known,[2] and their value will be realized if one considers questions of operation in which the criterion of expected value cannot be accepted for a number of reasons. This applies in particular when the decisions are only made once, each in its respective stage, or when they are not made often enough to allow the accepted value to be regarded as *a priori* reliable. In such cases it becomes necessary to calculate the 1-optimal,

---

[1] Some authors prefer to define these strategies as: (1) an element or set which renders the function optimal, and (2) an element or set which renders the function optimal if the element or set 1-optimal is ignored; hence, *n*-optimal.

[2] See, for example:

Bellman, R. and Kalaba, R., The *k*th best policy, *J. Soc. Industrial and Appl. Math.* No. 4, 582–588 (1960).

Pollack, M., Solutions of the *k*th best route through a network, a review, *J. Math. Anal. Appl.* No. 3, 547–559 (1961).

Hoffman, N. and Pavley, R., A method for the solution of the *N*th best path problem, *J. Assoc. Computing Machinery* No. 4, 506–514 (1959).

2-optimal,..., k-optimal strategies (these eventually becoming multiple),[1] and to determine their respective laws of probability, whence we learn the probability of realizing the expected value and, where it is useful, the variance of this law. We can imagine that the following problem of decision might arise, though the calculations might, as the articles quoted will show, prove complicated due to the highly combinatorial nature of the transitions to be considered:

(a) 1-minimal strategy:
       average cost (expected value): 1000,
       probability of exceeding 1500: 0.2;

(b) 2-minimal strategy:
       average cost (expected value): 1100,
       probability of exceeding 1500: 0.1.

A poor person might prefer the 2-minimal strategy to the other if the risk of exceeding 1500 seems too great for him. It is this entrance of the amount of the risk which justifies the search for the almost-optimal strategies in certain cases.

16. Theorem of Optimality in the Case of Uncertainty. Discrete Systems

A. ENUNCIATION OF THE THEOREM

Given a system capable of changing its state at each stage $n(n = 0, 1, 2, 3,..., N)$, the number of states being either finite or infinite, but denumerable.[2] We will take $p_{ij}^{(r)}$ as the probability of passing from a state E_i to a state E_j when the decision is r (the number of possible decisions being either finite or infinite, but denumerable); hence, the probability will depend on i, j, r, and finally n. To every change of state $E_i \rightarrow E_j$, and to each decision r, is attached a value $R_{ij}^{(r)}$ which may also depend on n.

The term "decision vector at stage n" is applied to the set of decisions taken at stage n; the term "strategy" to a consecutive sequence of decision vectors from time 0 to time N. A "substrategy" is a

---

[1] There is a possibility of the r-optimal not existing; for instance, if there are two $(r-1)$-optimal strategies, the next is the $(r+1)$-optimal.

[2] The popular meaning of the word is "which can be counted": that is, the states can be numbered by consecutive integers. It should be noted, however, that the present theorem can be extended to cases of continuous magnitudes (cf. the example in Part II, Section 64, and also our reference [I.1]).

consecutive sequence of decision vectors from time $N - n$ to time N. With these definitions we can now enunciate the theorem of optimality:

An Optimal Substrategy from $N - n$ to N Must Consist of
Optimal Substrategies from $N - n + 1$ to N.
This Must Hold for Every n from 1 to N.

This theorem can be proved very simply by the indirect method which we used to prove the deterministic case.

It will be relevant at this point to recall the more synthetic enunciation given by Bellman, and quoted in Section 10, in which we have replaced the word "policy" by "strategy":

"A strategy[1] is optimal if at a given period (stage), whatever the previous decisions have been, the decisions still to be taken form an optimal strategy when the previous ones are taken into account."

Luckily, the reader has learned from the examples in the previous section how to apply the theorem of optimality in a case of uncertainty, and will in consequence find it easier to understand the enunciation given above. Indeed, sometimes it pays to put the cart before the horse!

Let us also remind ourselves that the use of the criterion of the optimal expected value forces us to move from the future towards the present, hence the definition of the substrategies as $N - n$ to N, and not as 0 to n.

A further example will enable us to convert the enunciation into formulas, and to examine what happens when the phenomenon is stationary while the number of stages becomes increasingly large. It will also serve as an introduction to the concept of Markovian decision chains, which are treated more fully in Section 58 of Part II.

B. ANOTHER EXAMPLE[2]

A factory produces an article and every month introduces a variant which is tested for customer approval. If the product has been successful in month n (a month between time $n - 1$ and time n), the system is considered as being in state E_1 ; if the product has been a failure it is in state E_2 .

---

[1] Anglo-Saxon writers tend to use the word *policy* for both *politique* or *stratégie*. We prefer to make a distinction: a struggle with chance implies decisions which must take into account different states and which thus form "a decision vector"; the set of these vectors comprises a strategy.

[2] We have chosen the example given in Reference [16].

After evaluating what has happened during the last month, a decision is taken as follows:

$D_1^{(1)}$: not to advertise if the past month was successful one.

$D_1^{(2)}$: to do a certain amount of advertising if the past month was a successful one.

$D_2^{(1)}$: not to alter the price if the past month was a failure.

$D_2^{(2)}$: to reduce the price by 10% if the past month was a failure.

All the hypotheses and data of the present problem have been greatly simplified to avoid complicated numerical calculations, and in an actual case it would be impossible to make such systematic hypotheses and deductions, so that there would be a very large number of states and possible decisions; the calculations would then be much longer, but the method would remain the same.

Hence, in the simplified problem which we have just outlined, there are two possible states E_1 (success) and E_2 (failure), and four possible decisions $D_1^{(1)}$ or $D_1^{(2)}$ for E_1, and $D_2^{(1)}$ and $D_2^{(2)}$ for E_2. The possible preservations or changes of state are represented in Fig. 16.1.

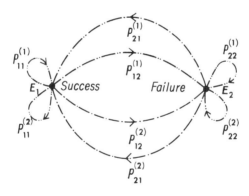

FIG. 16.1.

It is assumed that a statistic study has made it possible to evaluate the probabilities corresponding to change or nonchanges of state from one month to the next (from time n to time $n - 1$), and that these probabilities do not vary according to time. In addition, it is assumed that it is possible to calculate the profit which will be made from each of these changes of state, and that these profits do not vary according to time.

Table 16.1 gives the probabilities and profits corresponding to the states and decisions.

TABLE 16.1

State at time n	Decision at time n	State at time $n+1$	Probability	Monthly profit
E_1	$D_1^{(1)}$	E_1	$p_{11}^{(1)} = 0.5$	$R_{11}^{(1)} = 500$
E_1	$D_1^{(1)}$	E_2	$p_{12}^{(1)} = 0.5$	$R_{12}^{(1)} = 150$
E_1	$D_1^{(2)}$	E_1	$p_{11}^{(2)} = 0.6$	$R_{11}^{(2)} = 400$
E_1	$D_1^{(2)}$	E_2	$p_{12}^{(2)} = 0.4$	$R_{12}^{(2)} = 200$
E_2	$D_2^{(1)}$	E_1	$p_{21}^{(1)} = 0.7$	$R_{21}^{(1)} = 200$
E_2	$D_2^{(1)}$	E_2	$p_{22}^{(1)} = 0.3$	$R_{22}^{(1)} = -400$
E_2	$D_2^{(2)}$	E_1	$p_{21}^{(2)} = 0.8$	$R_{22}^{(2)} = 100$
E_2	$D_2^{(2)}$	E_2	$p_{22}^{(2)} = 0.2$	$R_{21}^{(2)} = -800$

Figure 16.2 shows how the situation may develop over a period of N months.

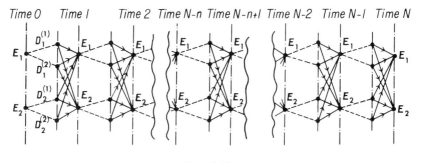

FIG. 16.2.

The first step is to calculate the expected value of the returns for a month corresponding to each of the decisions. $\bar{q}_i^{(r)}$ will represent the expected value of the montly profit when the state is E_i at the beginning of the month and the decision is not to make any change ($r = 1$), or to make a change $(r = 2)$ of commercial activity. Using the above notation, we have

$$
\begin{aligned}
D_1^{(1)}: \quad \bar{q}_1^{(1)} &= p_{11}^{(1)} R_{11}^{(1)} + p_{12}^{(1)} R_{12}^{(1)} \\
&= (0.5)(500) + (0.5)(150) = 325, \\
D_1^{(2)}: \quad \bar{q}_1^{(2)} &= p_{11}^{(2)} R_{11}^{(2)} + p_{12}^{(2)} R_{12}^{(2)} \\
&= (0.6)(400) + (0.4)(200) = 320, \\
D_2^{(1)}: \quad \bar{q}_2^{(1)} &= p_{21}^{(1)} R_{21}^{(1)} + p_{22}^{(1)} R_{22}^{(1)} \\
&= (0.7)(200) + (0.3)(-400) = 20, \\
D_2^{(2)}: \quad \bar{q}_2^{(2)} &= p_{21}^{(2)} R_{21}^{(2)} + p_{22}^{(2)} R_{22}^{(2)} \\
&= (0.8)(100) + (0.2)(-800) = -80.
\end{aligned}
\tag{16.1}
$$

Let $\bar{v}_i(N - n, N)$ be the optimal expected value of the total profit over n months from time $N - n$ to time N, when at time $N - n$, the system is in state E_i ; then, by the theorem of optimality, we can state

$$
\bar{v}_i(N - n, N) = \max_{r=1,2} \left[\bar{q}_i^{(r)} + \sum_{j=1}^{2} p_{ij}^{(r)} \bar{v}_j(N - n + 1, N) \right] \tag{16.2}
$$

with

$$
\bar{v}_i(N, N) = 0, \qquad i = 1, 2. \tag{16.3}
$$

In this case it is assumed that the factory ceases production at time N, and is handed over *gratis*, whatever its condition may then be. Without this assumption, we should be confronted with another problem which might, under certain conditions, lead to different optimal strategies.

Formula (16.2) is rather complicated, and in order to explain its content more fully, we shall employ numerical calculation to show how the optimal strategy or strategies can be worked out.

C. Finding the Permanent Optimal Strategy

(1) Optimal administration for 1 month (from time $N - 1$ to time N).

$$
\bar{v}_1(N - 1, N) = \max_{r=1,2}[\bar{q}_1^{(r)}] = \max[325, 320] = 325; \tag{16.4}
$$

hence, if at time $N - 1$ we are in state E_1, the best decision at this time is $D_1^{(1)}$.

$$
\bar{v}_2(N - 1, N) = \max_{r=1,2}[\bar{q}_2^{(r)}] = \max[20, -80] = 20; \tag{16.5}
$$

hence, if at time $N - 1$ we are in state E_2, the best decision is $D_2^{(1)}$.

(2) Optimal administration for 2 months (from time $N - 2$ to time N).

We have

$$\bar{v}_1(N - 2, N) = \underset{r=1,2}{\text{MAX}}[\bar{q}_1^{(r)} + p_{11}^{(r)}\bar{v}_1(N - 1, N) + p_{12}^{(r)}\bar{v}_2(N - 1, N)]$$

$$= \max[325 + (0.5)(325) + (0.5)(20),$$
$$\qquad\qquad 320 + (0.6)(325) + (0.4)(20)]$$

$$= \max[497.5; 523]$$

$$= 523; \tag{16.6}$$

hence, if at time $N - 2$, we are in state E_1, the best decision is $D_1^{(2)}$.

$$\bar{v}_2(N - 2, N) = \underset{r=1,2}{\text{MAX}}[q_2^{(r)} + p_{21}^{(r)}\bar{v}_1(N - 1, N) + p_{22}^{(r)}\bar{v}_2(N - 1, N)]$$

$$= \max[20 + (0.7)(325) + (0.3)(20),$$
$$\qquad\qquad -80 + (0.8)(325) + (0.2)(20)]$$

$$= \max[253.5, 184]$$

$$= 253.5; \tag{16.7}$$

hence, if at time $N - 2$, we are in state E_2, the best decision is $D_2^{(1)}$.

(3) Optimal administration for 3 months (from time $N - 3$ to time N).

We have

$$\bar{v}_1(N - 3, N) = \underset{r=1,2}{\text{MAX}}[q_1^{(r)} + p_{11}^{(r)}\bar{v}_1(N - 2, N) + p_{12}^{(r)}\bar{v}_2(N - 2, N)]$$

$$= \max[325 + (0.5)(523) + (0.5)(253.5),$$
$$\qquad\qquad 320 + (0.6)(523) + (0.4)(253.5)]$$

$$= \max[713.25, 735.2]$$

$$= 735.2; \tag{16.8}$$

hence, if at time $N - 3$, we are in state E_1, the best decision is $D_1^{(2)}$.

$$\bar{v}_2(N - 3, N) = \underset{r=1,2}{\text{MAX}}[\bar{q}_2^{(r)} + p_{21}^{(r)}\bar{v}_1(N - 2, N) + p_{22}^{(r)}\bar{v}_2(N - 2, N)]$$

$$= \max[20 + (0.7)(523) + (0.3)(253.5),$$
$$\qquad\qquad -80 + (0.8)(523) + (0.2)(253.5)]$$

$$= \max[462.15, 389.1]$$

$$= 462.15; \tag{16.9}$$

hence, if at time $N - 3$, we are in state E_2, the best decision is $D_2^{(1)}$.

(4) By following the same process, and by assuming, for instance, that $N = 20$, we find

Time	Average revenue		Optimal decision	
$n = 4$	$\bar{v}_1(N-4, N) =$	945.98	$D_1^{(2)}$	
	$\bar{v}_2(N-4, N) =$	673.28	$D_2^{(1)}$	(16.10)
$n = 5$	$\bar{v}_1(N-5, N) =$	1156.90	$D_1^{(2)}$	
	$\bar{v}_2(N-5, N) =$	884.17	$D_2^{(1)}$	(16.11)
$\vdots$	$\vdots$		$\vdots$	
$n = 20$	$\bar{v}_1(0.20) =$	4320.33	$D_1^{(2)}$	
	$\bar{v}_2(0.20) =$	4047.60	$D_2^{(1)}$	(16.12)

To conclude, the optimal administration for 20 months from time 0 to time N (the business being handed over free at time N, whatever its condition), will be

1. For all times 0, 1, 2,..., $N - 2$:
   (a) If the previous month has been successful: choose $D_1^{(2)}$.
   (b) If the previous month has been unsuccessful: choose $D_2^{(1)}$.

2. For time $N - 1$:
   (a) If the previous month has been successful: choose $D_1^{(1)}$.
   (b) If the previous month has been unsuccessful: choose $D_2^{(1)}$.
   This strategy is shown in Fig. 16.3.

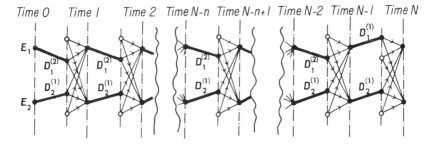

FIG. 16.3. Optimal strategy.

By following this optimal strategy, the expected value of revenue will be maximal and as given in Table 16.2.

TABLE 16.2

n	1	2	3	4	5	6	7		20
Administrative period Income	19 to 20	18 to 20	17 to 20	16 to 20	15 to 20	14 to 20	13 to 20		0 to 20
$\bar{v}_1(N-n, N)$	325	523	735.2	945.98	1156.90	1367.80	1578.61		4320.33
$\bar{v}_2(N-n, N)$	20	253.5	462.15	673.15	884.17	1095.07	1305.88		4047.60

By calling:

$$\bar{r}_i(N) = \frac{\bar{v}_i(0, N)}{N}, \tag{16.13}$$

the average monthly revenue when, at time 0, we are in state E_1, it can be shown[1]:

$$\lim_{N \to \infty} \bar{r}_1(N) = \lim_{N \to \infty} \bar{r}_2(N) = \bar{r}^* = 210.92. \tag{16.14}$$

From Table 16.2 we are forced to deduce that $r_1(20)$ and $r_2(20)$ are approaching r^*:

$$\bar{r}_1(20) = \frac{4320.33}{20} = 216.02; \quad \bar{r}_2(20) = \frac{4047.60}{20} = 202.38. \tag{16.15}$$

Indeed, as N increases, the difference $\bar{r}_1(N) - \bar{r}_2(N)$ approaches zero.

17. Interval of Anticipation. Dynamic Program with Adaptation

A. EFFECT OF THE INTERVAL OF ANTICIPATION

As a general rule, we do not have a precise knowledge of the future, nor even of its probability, over a sufficiently long period. The further ahead we plan, the less we know of the laws of probability. To understand

---

[1] Part II, Section 60.

how dynamic programming is carried out under such uncertain conditions, we shall take an example with a deterministic future, but we must first of all define as "the interval of anticipation" the series of consecutive stages beginning with the present time, during which we possess a formal knowledge of the future.

In considering the discrete decision process shown in Fig. 17.1, it will

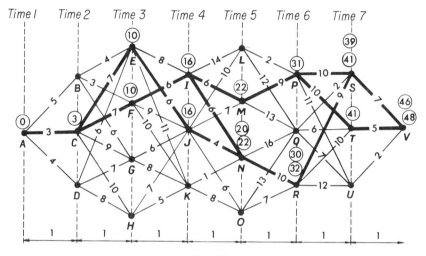

FIG. 17.1.

be assumed that the interval of anticipation θ is equal to one stage only; hence, the vision which "the decision maker"[1] has of the future is limited to one stage, and at each time he will select his path by the following criterion: assuming he wishes to find the path of minimal cost, he will choose the link with minimal cost leading from the point at which he is then situated. Using this criterion, vertex V will be reached with a cost of 46 or 48, depending on the path which is followed, since at certain points different choices are open to him.

Let us now assume that $\theta = 2$, in other words that "the decision maker" knows the future for a distance of two stages. From each point at which he finds himself at time n, he will seek the best path for stages n and $n + 1$, which are known to him, and will decide to go to the $n + 1$ point which is on this route (if there are several he will select one from an urn where each has equal probability).

He will also be required, at each time, to decide on a dynamic program for two stages. By these means, point V will be reached with a cost of 44. With $\theta = 3$, the decision to choose the point with time n on the best

---

[1] This is another neologism for the person who makes or is responsible for the decision.

of the three known stages n, $n + 1$, and $n + 2$, would have a minimal cost of 42 or 43. Continuing this process with increasing values for θ, we find that for $\theta = 4, 5, 6$, and 7, the minimal costs are, respectively, 41 or 42, 42, 40, and 40.

It is now possible to imagine other criteria: for example, if θ is the interval of anticipation, we could place ourselves in stage $n + \alpha$ $(0 < \alpha < \theta)$ on the optimal path or paths of stage n at stage $n + \theta$. Stated differently, instead of deciding at each time a dynamic program with stages, we should limit our work to reviewing, at every stage, the decisions which have already been made. While this method will clearly not provide such good results, it is also a simpler one. But whichever criteria are chosen, and however the data are obtained and utilized, the optimal value is a monotone decreasing one (that is it diminishes or remains constant) as the interval of anticipation increases, which is the direct consequence of the theorem of optimality (see Figs 17.1–17.7).

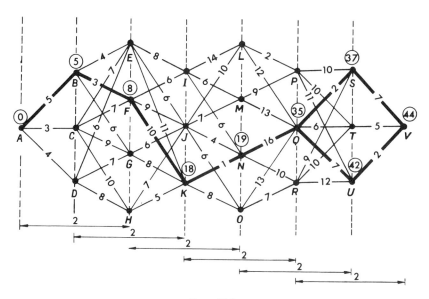

FIG. 17.2.

In practice, the information is not obtained gratuitously, and the difficulty of obtaining it increases the more distant the future time considered. In such cases, and under certain conditions, adapted optimization can lead to a convergent process for finite values of α, even when $\theta \to \infty$.

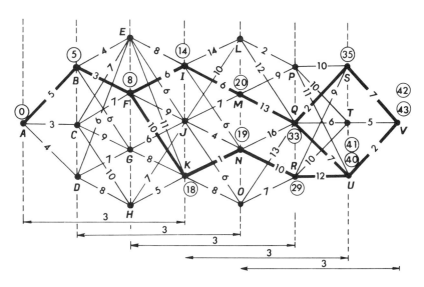

FIG. 17.3.

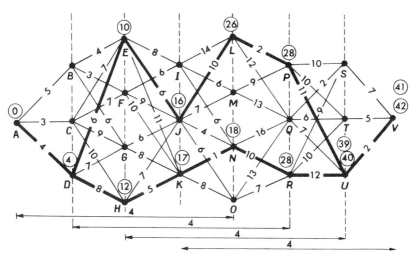

FIG. 17.4.

B. ADAPTATION UNDER UNCERTAINTY. PERIOD OF LEARNING

The concepts which have just been briefly examined are only of practical use if applied to actual cases which are subject to chance and uncertainty. Rarely is the future known with any degree of certitude, though the economic horizon is sometimes described with the help of

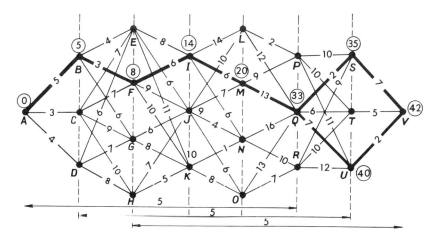

FIG. 17.5.

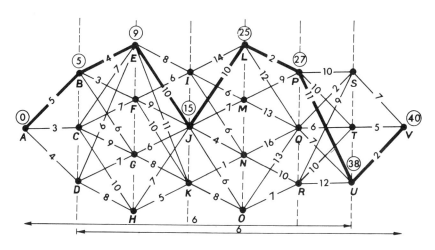

FIG. 17.6.

laws of probability based on statistics or oversimplified assumptions. The further ahead we attempt to anticipate developments, the weaker and more diffused the laws of probability become, and it would be unrealistic to accept these laws without admitting the need for readjustment and adaptation.

To understand the processes of adaptation, let us consider a *cinematic* analogy. The subjects in the example consist of a man and his dog, the latter of whom continuously proceeds at a fixed pace in the direction of

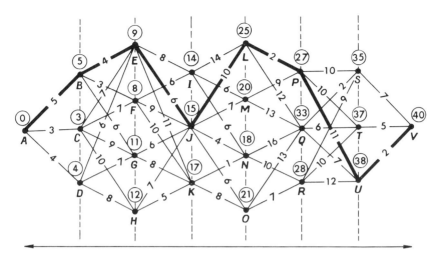

FIG. 17.7.

his master, who is following a trajectory T_M, also at a constant speed. The dog's path T_C can easily be expressed as an equation, the classic "dog's curve," so well known to students of higher mathematics (Fig. 17.8).

Let us further assume that the dog is an unusually wise and intelligent one who is capable of anticipating the future position of his master; that is, of being able at any given moment to foresee his position a moment later. With this power of anticipation he would then be able to improve his trajectory and overtake his master more quickly (if that is possible) (Fig. 17.9). It is by such methods that the trajectory of pursuit missiles can be improved if the estimate of the future position of the moving object can be obtained under satisfactory conditions.

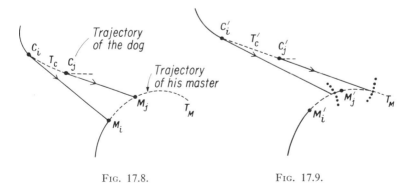

FIG. 17.8.   FIG. 17.9.

Indeed, in the phenomena of organization, the knowledge of the laws of probability controlling the particular system increases as information accumulates, provided the phenomenon is static; the outcome is a period of learning about which it will be useful to provide some details.

C. AN EXAMPLE OF ADAPTIVE BEHAVIOR

In the *discrete* process to be studied, a system may take four different states E_1, E_2, E_3, and E_4, the changes in which occur at times 1, 2, 3,..., $N - 1$ and N. At time $N + 1$ the process is stopped, and the system can be in any of the four states. The probabilities and revenues are given in Table 17.1, which also contains two possible decisions D_1 or D_2.

TABLE 17.1

	D_1		D_2	
	Probability	Revenue	Probability	Revenue
$E_1 \rightarrow E_1$	$p^{(1)}$	10	$p^{(2)}$	13
$E_1 \rightarrow E_2$	$q^{(1)} = 1 - p^{(1)}$	5	$q^{(2)} = 1 - p^{(2)}$	-4
$E_2 \rightarrow E_1$	$p^{(1)}$	7	$p^{(2)}$	6
$E_2 \rightarrow E_3$	$q^{(1)} = 1 - p^{(1)}$	-2	$q^{(2)} = 1 - p^{(2)}$	5
$E_3 \rightarrow E_2$	$p^{(1)}$	4	$p^{(2)}$	-3
$E_3 \rightarrow E_4$	$q^{(1)} = 1 - p^{(1)}$	1	$q^{(2)} = 1 - p^{(2)}$	10
$E_4 \rightarrow E_3$	$p^{(1)}$	12	$p^{(2)}$	6
$E_4 \rightarrow E_4$	$q^{(1)} = 1 - p^{(1)}$	-5	$q^{(2)} = 1 - p^{(2)}$	4

Figures 17.10 and 11 provide an illustration of this process, and it will be assumed that the number of stages N is sufficiently large, and also that the probabilities $p^{(1)}$ and $p^{(2)}$ are not known *a priori*.

For the commencement in Stage 1, at time 1, we shall asume (since we do not have any information) that the probabilities are those shown:

$$\text{decision 1:} \quad p^{(1)} = \tfrac{1}{2}, \quad q^{(1)} = \tfrac{1}{2}$$
$$\text{decision 2:} \quad p^{(2)} = \tfrac{1}{2}, \quad q^{(2)} = \tfrac{1}{2}. \tag{17.1}$$

A quick calculation for several stages shows us that in Stage 1 the following optimal decisions should be applied:

$$
\begin{array}{lll}
\text{time 1:} & \text{in } E_1 & \text{choose } D_1, \\
& E_2 & D_2, \\
& E_3 & D_3, \\
& E_4 & D_4.
\end{array} \tag{17.2}
$$

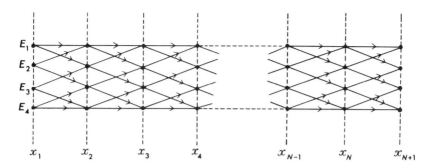

FIG. 17.10.

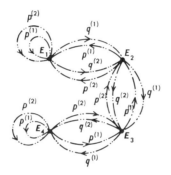

FIG. 17.11.

Let us now suppose that we are initially in E_1, and that we have chosen D_1. Chance, of which the real laws are unknown to us (since the changes of state have introduced unknown probabilities $p^{(1)}, q^{(1)}$ and $p^{(2)}, q^{(2)}$ which do not depend on the state being considered, but only on the decision) has then led us to to E_2. Choosing probabilities *a posteriori* according to the hypothesis of Laplace-Bayes,[1] it follows that

$$q^{(1)} = (1 + 1)/(1 + 2) = \tfrac{2}{3}, \qquad p^{(1)} = 1 - \tfrac{2}{3} = \tfrac{1}{3} \qquad (17.3)$$

---

[1] This assumes an urn containing numerous balls of r colors c_1, c_2,..., c_r. A succession of draws of 1 ball is made of a part of the contents of the urn. By representing the time of a draw as i, and the time of the first one as 0, the sequence defined below constitutes the set of the "Laplace-Bayes probabilities" of the emergence of the colors c_1, c_2,..., c_r:

$$p_1(0) = 1/r, \quad p_2(0) = 1/r, ..., \quad p_r(0) = 1/r,$$

$$p_1(i) = \frac{m_1 + 1}{i + r}, \quad p_2(i) = \frac{m_2 + 1}{i + r}, ..., \quad p_r(i) = \frac{m_r + 1}{i + r}$$

the other probabilities being unchanged:

$$p^{(2)} = \tfrac{1}{2}, \qquad q^{(2)} = \tfrac{1}{2}. \tag{17.4}$$

A fresh calculation shows that the optimal decision in Stage 2 is then

$$
\begin{array}{lll}
\text{time 2:} & \text{in } E_1 & \text{choose } D_2, \\
& E_2 & D_2, \\
& E_3 & D_1, \\
& E_4 & D_2.
\end{array} \tag{17.5}
$$

In Stage 2, probability will bring about another transition; hence a readjustment and a search for the optimal strategy at time 3, and so on. By assuming that the real, but unknown probabilities are

$$\pi^{(1)} = 0.2, \qquad 1 - \pi^{(1)} = 0.8, \tag{17.6}$$

$$\pi^{(2)} = 0.7, \qquad 1 - \pi^{(2)} = 0.3, \tag{17.7}$$

there will be the following corresponding permanent optimal strategy, provided N is very large:

$$
\begin{array}{lll}
\text{time } n \ll N: & \text{in } E_1 & \text{choose } D_2, \\
& E_2 & D_2, \\
& E_3 & D_2, \\
& E_4 & D_2.
\end{array} \tag{17.8}
$$

By carrying out a series of inexhaustive draws from an urn containing two red balls and eight blue balls for D_1, seven red and three blue balls for D_2, and using the following formulas:

$$D_1: \quad p^{(1)}(i) = \frac{m_1^{(1)} + 1}{i^{(1)} + 2}, \qquad q^{(1)}(i) = \frac{m_2^{(1)} + 1}{i^{(1)} + 2}, \tag{17.9}$$

$$D_2: \quad p^{(2)}(i) = \frac{m_1^{(2)} + 1}{i^{(2)} + 2}, \qquad q^{(2)}(i) = \frac{m_2^{(2)} + 1}{i^{(2)} + 2}, \tag{17.10}$$

---

if, at time i, starting from time 0, the following draws have been made: m_1 draws of c_1, m_2 draws of c_2 ,..., m_r draws of c_r.

If the urn contains n_1 balls of color c_1, n_2 of color c_2 ,..., n_r of color c_r

$$n_1 + n_2 + \cdots + n_r = n,$$

then, in the direction of converging probability, we shall have

$$\lim_{i \to \infty} \frac{m_k + 1}{i + r} = \frac{n_k}{n}, \qquad k = 1, 2, ..., r.$$

In other words, when i is large enough:

$$\frac{m_k}{i} \approx \frac{n_k}{n}, \qquad k = 1, 2, ..., r.$$

where
$$m_1^{(1)} + m_2^{(1)} = i^{(1)}, \qquad m_1^{(2)} + m_2^{(2)} = i^{(2)}, \qquad i^{(1)} + i^{(2)} = i,$$

$m_1^{(1)}$: number of red balls for $i^{(1)}$ decisions D_1,

$m_2^{(1)}$: number of blue balls for decisions D_1,

$m_1^{(2)}$: number of red balls for $i^{(2)}$ decisions D_2,

$m_2^{(2)}$: number of blue balls for decisions D_2,

we should discover how to vary the optimal decisions to reach (17.8) at the end of a sufficient number of stages.

If, at time 0, we possessed a knowledge of a certain pattern of past events, as, for example,

$$m_{1,0}^{(1)}, \qquad m_{2,0}^{(1)}, \qquad i_0^{(1)}, \qquad m_{1,0}^{(2)}, \qquad m_{2,0}^{(2)}, \qquad i_0^{(2)}, \qquad i_0,$$

the convergence would be more rapid and formulas (17.9) and (17.10) would then become

$$D_1: \quad p_i^{(1)} = \frac{m_1^{(1)} + m_{1,0}^{(1)} + 1}{i^{(1)} + i_0^{(1)} + 2}, \qquad q_i^{(1)} = \frac{m_2^{(1)} + m_{2,0}^{(1)} + 1}{i^{(1)} + i_0^{(1)} + 2}, \quad (17.11)$$

$$D_2: \quad p_i^{(2)} = \frac{m_1^{(2)} + m_{1,0}^{(2)} + 1}{i^{(2)} + i_0^{(2)} + 2}, \qquad q_i^{(2)} = \frac{m_2^{(2)} + m_{2,0}^{(2)} + 1}{i^{(2)} + i_0^{(2)} + 2}. \quad (17.12)$$

Obviously, a process of this kind could not be accepted unless the process were stationary or if the number of stages N of the formal anticipation of revenue were too small.

It is possible to think of other cases where the form of the laws of probability (for instance, one of Laplace-Gauss's in a continuous case) remains constant while the average varies and the standard deviation increases, as we anticipate further ahead; indeed, an analytical study can be carried out in certain problems. What we frequently encounter in practical problems is an approximative knowledge of a law of probability and of a reasonable period of anticipation. Optimization is then carried out with a progressive increase of our knowledge of the law or laws of probability involved. Various methods can be used: decreasing squares, line of regression, method of the x^2, etc. Each of these processes is open to criticism, and we have to admit that the methods employed in an adaptive process still lack a general theory which might well be entitled "the mathematical theory of learning." Nevertheless, a number of recent works have shown the obvious importance which is attached to these concepts.

18. Effect of Introducing a Rate of Interest

A. OPTIMAL POLICY FOR THE DISCOUNTED VALUE

Let us return to the problem of the motor route which we studied in Section 12, using a graph and different values (Fig. 18.1), and assume that one section is completed annually, so that the knowledge of the duration of the work leads to a discounting of the cost. Taking α for the annual rate of interest, formula (11.2) will become

$$F(x_0, x_1, x_2, x_3, x_4, x_5) = v_I(x_0, x_1) + \frac{v_{II}(x_1, x_2)}{1 + \alpha} + \frac{v_{III}(x_2, x_3)}{(1 + \alpha)^2}$$

$$+ \frac{v_{IV}(x_3, x_4)}{(1 + \alpha)^3} + \frac{v_V(x_4, x_5)}{(1 + \alpha)^4}. \quad (18.1)$$

The optimal policy or policies will be found in the same way as in Section 11, but the following observation is important:

The optimal policy for the discounted values may differ from the optimal policy which does not take the rate of interest into account.

This conclusion will be confirmed if we observe Figs. 18.1 and 18.2. On the former, the optimal policy without discounting is shown, and on the latter the policy corresponding to a rate of interest $\alpha = 0.10$, so that these policies are not the same.

The formulas for multistage optimization given in Section 11, and used in the direction of the future towards the past, now become

$$f_{n, n-1, \ldots, n-r}(x_n, x_{n-r-1}) = \underset{x_{n-r}}{\text{MIN}} \left[v_{n-r}(x_{n-r-1}, x_{n-r}) \right.$$

$$\left. + \frac{1}{1 + \alpha} f_{n, n-1, \ldots, n-r+1}(x_{n-r}, x_n) \right], \quad (18.2)$$

or similar formulas if we are finding the maximum.

It is easy to prove that an introduced rate can modify the optimal policy.

B. MODIFICATIONS INTRODUCED BY DISCOUNTING

Let us take three numbers A, B, B' with $B > B'$ and a coefficient $0 < k < 1$, and show that the values of x may be such that

$$A + B > x + B' \quad (18.3)$$

and

$$A + kB < x + kB'. \quad (18.4)$$

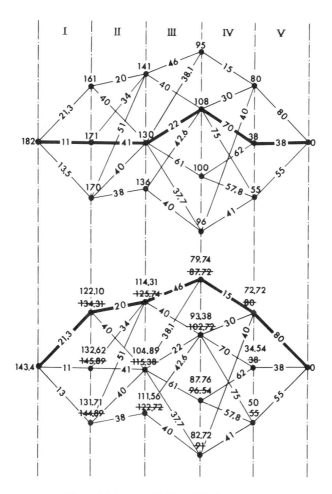

FIG. 18.1 (top) and FIG. 18.2 (bottom).

By combining (18.3) and (18.4), it follows that

$$A + k(B - B') < x < A + B - B'. \tag{18.5}$$

If $B > B'$, there must always be a number such as x. The inequalities in (18.3) and (18.4) represent situations which can occur when we compare (18.2) with $\alpha = 0$ and (18.2) with $\alpha = 0$, assuming $k = 1/(1 + \alpha)$. The proof given in (18.3)–(18.5) can be used with changed signs for the case when the maximum is to be found.

Formula (18.2) can easily be transformed into a generalization for cases where the x_i terms are continuous variables and also for those containing uncertain variables of a discrete or continuous nature. It is thus that we have returned to the data of the problem in Fig. 15.9 and have introduced a rate of $\alpha = 0.1$ [given $r = 1(1 + \alpha) = 0.9090$]. From this example, it can be seen that introducing a rate modifies the strategy (different decision at time 1). By increasing the number of stages, it would be possible to observe the development of the strategy.

In some problems the second member of (18.1) has to be multiplied by $1/(1 + \alpha)$ if the first expenditure or revenue, as well as those that follow, occur at the end of the period.

It should be noted that in a deterministic case, optimization can be carried out in either direction, even when discounting enters; but, of course, in cases of uncertainty we must work from the future in the direction of the past.

C. DISCOUNTING AND INFLATION

To conclude, we will consider the case where the economy is in a state of inflation, so that it becomes necessary to impose an inflationary rate of interest β which may be regarded as a negative rate $(-\beta')$. β and β' can be connected by the equation

$$1 + \beta = 1/(1 - \beta').$$

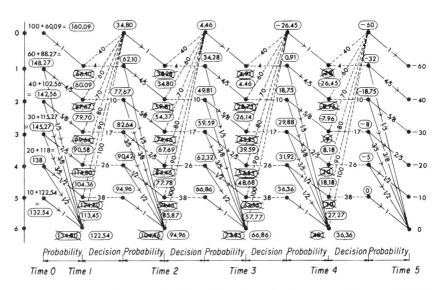

FIG. 18.3. The problem of Fig. 15.8 with an introduced rate $\alpha = 0.1$ ($r = 0.9090$).

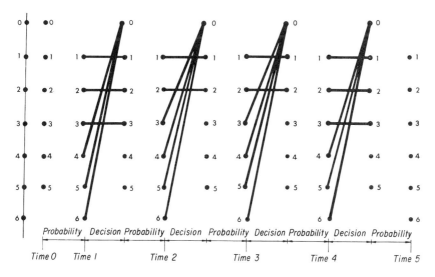

FIG. 18.4.   Optimal strategy for the problem of Fig. 15.8 with an introduced rate $\alpha = 0.1$.

Lastly, taking account of the rate of interest α and the inflationary rate β, we can conclude that:

$$k = (1 + \beta)/(1 + \alpha).$$

If $k < 1$, which means that $\beta < \alpha$, the process may be convergent; if $k > 1$, which means that $\beta > \alpha$, it is divergent.

Obviously, if a multistage process is divergent, optimization cannot have any meaning for a nonfinite economic period. For a study of convergence in a nonfinite economic period, the reader should refer to Section 63 of Part II.

CHAPTER III

THE THEORY OF GAMES OF STRATEGY[1]

19. Introduction

Certain problems can be presented in the form of a game in which we are confronted by an opponent (opponents) whose reactions to our decisions are unknown, or at best dimly known to us. A battle, a duel, or an economic struggle are examples of such problems. How are we to make rational decisions when we are unaware how our adversaries will act, even if the rules of warfare or of the competition are known and will be accepted by all the participants?

The theory of games of strategy, first studied by Borel[2] and von Neumann,[3] provides an answer to this question. Different aspects of this theory, which is of much earlier origin than would be imagined, had already received the attention of a number of writers such as Cardan, Kepler, Galileo, Huygens, Pascal, Fermat, and Bernoulli. The popularity of such games as chess, checkers, and cards stems from the fact that they confront the participants with situations where a wide choice of action is open to them, and thereby provide what might be termed artificial adventures.

Unfortunately, real problems, such as economic or military ones, are too strongly combinatorial to provide satisfactory analytical solutions if the theory of games of strategy is applied to them. It is in providing an approach to these problems, rather than in their systematic treatment, that the theory is valuable. By analyzing games played on rectangular "boards," it is possible to define the best strategy in reponse to an opponent's decisions, and to produce a state of equilibrium which is of basic importance.

---

[1] The reader who wishes to gain a deeper knowledge of the theory of games might profitably study *Games and Decisions* by Duncan, Luce, and Raiffa.

[2] Address to the *Académie des Sciences* in 1921 on the theory of games of chance and integral equations with left symmetrical nucleus.

[3] J. von Neumann and O. Morgenstern, *Theory of Games and Economic Behavior*, 1948.

20. Game on a Rectangular Matrix

A. CHOICE

Let us consider the following game to be played by two persons A and B. The board to be used has three columns and four lines (Table 20.1), and the spaces formed by their intersections contain either positive or negative numbers or zeros, which are known to the contestants. B is asked to choose a column, and A has to select a line, each being ignorant of the other's choice. Whether or not the selections are made simultaneously, B will give A the amount shown at the intersection of the column and line which they have selected. Thus, if B chooses column (2) and A line (3), B will give A an amount equal to 2; if B elects (1) and A (3), B will allow A the sum of (-4), which means that he will gain 4. (We are not concerned as to whether the game is a fair one or not, nor whether A or B was anxious to take part in it; we merely assume they have agreed to do so.)

TABLE 20.1

	B			
	(1)	(2)	(2)	.
(1)	3	-2	1	-2
(2)	3	3	2	2
(3)	-4	2	5	-4
(4)	4	6	0	0
	4	6	5	

A labels the lines (1), (2), (3), (4).

We can now visualize any number of moves which are open to the players. For instance, B might choose column (2) in the hope of gaining 2, though this would be unwise, for there is the risk of his losing 6. Alternatively, B might calculate the average of his possible gains and losses in each column and choose the one with the least average loss. For column (1) this would be:

$$\tfrac{1}{4}(3 + 3 - 4 + 4) = 1.5;$$

for column (2) 2.25; for column (3) 2. Hence, B would be led to choose column (1), but we can imagine many other more or less reasonable methods of play for B, and also for A as regards his gains.

B. PLAYING WITH INTELLIGENCE AND PRUDENCE

We will assume that A and B are "intelligent and prudent"[1] players, and will translate this description into mathematical language in the following arbitrary manner:

B finds the largest number in each column (Table 20.1), namely 4,6,5, and decides to choose column (1), for he is then *certain* of not losing more than 4, whichever line his opponent chooses.

A finds the smallest number in each line, $-2,2,-4,0$, and selects line (2), for he is then *certain* to gain at least 2, whatever his opponent's choice may be.

It is assumed that each player is unaware of his opponent's method.

Hence, by taking a_{ij} ($i = 1, 2, 3, 4$; $j = 1, 2, 3$), for the board on which the game is being played,

$$B \quad \text{chooses} \quad \min_{j}(\max_{i} a_{ij}), \tag{20.1}$$

$$A \quad \text{chooses} \quad \max_{i}(\min_{j} a_{ij}). \tag{20.2}$$

C. SYMMETRICAL PLAY

A preliminary question arises as to whether the systems of A and B may be asymmetrical, since one of the contestants chooses the column corresponding to the minimum of the maxima, and the other the line representing the maximum of the minima. Such is not the case, for this asymmetry is simply the consequence of the type of board chosen; indeed, A should be given a board having the same numbers with different signs, in which the columns would replace the lines (Table 20.2).

TABLE 20.2

		A				
		(1)	(2)	(3)	(4)	
	(1)	-3	-3	4	-4	**—4**
B	(2)	2	-3	-2	-6	-6
	(3)	-1	-2	-5	0	-5
		2	**—2**	4	0	

---

[1] *Intelligence*: adopting a rational system of play; *prudence*: its meaning in the present context will become clearer when we have explained the concept of "minimax." For the moment, we can say that it signifies "logical" under certain conditions.

A and B now make their choice by taking the minimum of the maxima as their criterion. It would also be possible to use the two boards related in such a way that A and B would each choose the maximum of the minima on their respective "tables."

D. DEFINING A RECTANGULAR GAME WITH TWO PLAYERS AND A NULL SUM

To generalize what we have written about this example, we shall define a game played on a rectangular board showing the losses of one opponent to another "a game of strategy on a rectangular table a_{ij}," or a "game with two players and a null sum on a rectangular table." One player called B will choose a column and the other called A will choose a line, each being ignorant of the decisions of his opponent. If we show the table of B's losses to A to each of the opponents, then B's system: choose $\min_j (\max_i a_{ij})$ will correspond to A's: choose $\max_i (\min_j a_{ij})$.

Player B will be called "the minimizing player" and A will be referred to as "the maximizing player."

We shall prove (Part II, Section 66) that for any table

$$a_{ij} \qquad (i = 1, 2,..., m; \quad j = 1, 2,..., n):$$

$$\min_j [\max_i a_{ij}] \geqslant \max_i [\min_j a_{ij}]. \tag{20.3}$$

Thus, in the rectangular game of Table 20.1:

$$\min_j [\max_i a_{ij}] = 4; \qquad \max_i [\min_j a_{ij}] = 2: \quad 4 > 2.$$

The important special case where

$$\min_j [\max_i a_{ij}] = \max_i [\min_j a_{ij}]$$

will now be considered.

21. Point of Equilibrium of a Rectangular Game

A. SADDLE POINT AND VALUE OF THE PLAY

Let us study the play in Table 21.1, where we have

$$\min_j [\max_i a_{ij}] = 3, \tag{21.1}$$

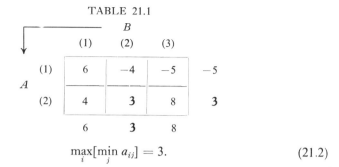

TABLE 21.1

$$\max_i[\min_j a_{ij}] = 3. \tag{21.2}$$

This case of equality plays a very important role in the present theory. When Table a_{ij} of a rectangular game is such that

$$\max_i[\min_j a_{ij}] = \min_j[\max_i a_{ij}] = v, \tag{21.3}$$

we say that the game has a *point of equilibrium* or *saddle point*.[1] The quantity v is called the *value of the game*.

A rectangular game may possess one or more points of equilibrium, but the game has never more than one value. Table 21.2 (a)–(i) gives some examples.

B. FINDING A SADDLE POINT, IF ONE EXISTS

There is a very simple method of finding whether a game has a point of equilibrium:

"An element which is *the smallest in its line* and *the largest in its column* is a point of equilibrium."

With this definition in mind, the reader can now try to find the points of equilibrium in the matrices shown in Table 21.2.

Why are such points given this name? If a game has a point of equilibrium, and if the opponents play "intelligently and prudently," the result of the game will correspond to the hypotheses made by each player, and their efforts will balance each other.

Further on, we shall see how important this concept can be in any problem involving competition.

---

[1] Let us consider a surface such as, for instance, a hyperbolic paraboloid with the shape of a horse's saddle; there is a certain point in it which is the minimum of the curve obtained by cutting the surface by a plane, and which is the maximum of another curve obtained by cutting the surface by a second plane. It is what is also referred to as a *collar*.

TABLE 21.2

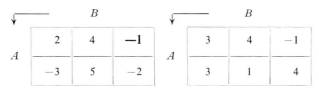

	B		
	2	4	**−1**
A			
	−3	5	−2

a_{13} is a point of equilibrium.

(a)

	B		
	3	4	−1
A			
	3	1	4

No point of equilibrium.

(b)

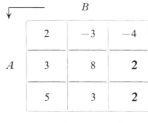

	B		
	2	−3	−4
A	3	8	**2**
	5	3	**2**

a_{23} and a_{33} are points of equilibrium.

(c)

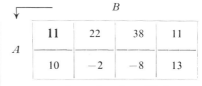

	B			
	11	22	38	11
A	10	−2	−8	13

a_{11} is a point of equilibrium.

(d)

	B	
	2	8
A		
	4	1

No point of equilibrium.

(e)

	B	
	11	13
A		
	7	5

a_{11} is a point of equilibrium.

(f)

	B	
	0	1
A		
	−1	0

a_{11} is a point of equilibrium.

(g)

	B	
	0	1
A		
	1	0

No point of equilibrium.

(h)

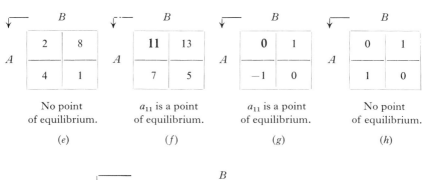

	B					
	2	1	2	1	−1	−3
A	−3	−1	2	0	3	−2
	6	5	3	0	−1	**−2**

a_{36} is a point of equilibrium.

(i)

22. Pure Strategy and Mixed Strategy

A. BEHAVIOR WHEN THERE IS NO SADDLE POINT

In a rectangular game it is customary to call the choice of a column or a line "pure strategy"; this word "strategy" could lead to some confusion due to the meaning usually given it, but in the present section it is to be understood in the restricted sense given to it above.

When a game has a point of equilibrium, the line (or the column) that passes through this point constitutes a *pure optimal strategy* for player A (or player B).

We must now ask whether it is possible to introduce a concept of equilibrium, and hence of optimality, into a game where such a point is lacking. The answer is affirmative, and leads us to examine the theory introduced by Borel and von Neumann.

We will assume that the opponents in a rectangular game are allowed a fixed number of moves, for instance 100. In this game, where one chooses a line and the other a column, 100 times, respectively, how are we to introduce strategies when there is no point of equilibrium?

B. AVERAGE GAIN

We shall take as an example, the game in Table 22.1 where there is no point of equilibrium. If A chooses line (1) 25 times and line (2) 75 times, that is to say if he gives himself relative frequencies of 0.25 and 0.75 for

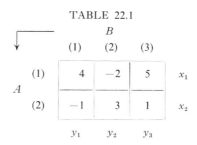

TABLE 22.1

		B			
		(1)	(2)	(3)	
A	(1)	4	-2	5	x_1
	(2)	-1	3	1	x_2
		y_1	y_2	y_3	

the decisions he is going to make, then if B chooses column (1), A's average gain will be

$$(4)(0.25) + (-1)(0.75) = 0.25; \qquad (22.1)$$

if B chooses column (2), A's average gain will be

$$(-2)(0.25) + (3)(0.75) = 1.75; \qquad (22.2)$$

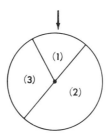

FIG. 22.1.

if B chooses column (3), A's average gain will be

$$(5)(0.25) + (1)(0.75) = 2. \tag{22.3}$$

If B were certain of the weighting chosen by A, he would then select column (1) 100 times with a total loss of 25; but if B is ignorant of A's intentions, he will, as an intelligent and prudent player, try to avoid the worst. We shall show how he can do this, but we first wish to introduce an important definition.

C. MIXED STRATEGY

The term "weighted strategy" or "mixed strategy" or even "strategy" alone, is applied to the choice of a weighting of lines (for A) or columns (for B), that is to say, in a game formed of m lines and n columns:

for A: m quantities x_1, x_2 ,..., x_m nonnegative and such that

$$x_1 + x_2 + \cdots + x_m = 1$$

for B: n quantities y_1, y_2 ,..., y_n nonnegative and such that

$$y_1 + y_2 + \cdots + y_n = 1$$

To employ a mixed strategy in a game of N plays, it is useful to select the lines or columns by a chance draw, using an urn or a roulette wheel or a similar device which will provide the numbers of lines or columns corresponding to the law of probability of the weighting selected. Without this precaution, the opponent would be in possession of information which might assist him unfairly. For instance, if B has chosen the weighting $y_1 = 0.2$, $y_2 = 0.5$, and $y_3 = 0.3$ in the example in Table 22.1, he could employ a roulette wheel like the one in Fig. 22.1, and by using it to make his draws, he might, for example, obtain

$$(3), (2), (2), (3), (1), (2), (2), (1),\ldots .$$

The frequencics of a draw of this kind will correspond to the given frequencies y_1, y_2, y_3, provided the game includes a sufficient number of moves, and in this way, the secret of B's behavior will be preserved, while A will take similar precautions to preserve his intentions. Alternatively, the results of the plays might not be announced until the end of the game, in which case secrecy would no longer be necessary, and each player would then be free to carry out his mixed strategy in an arbitrary order.

Let us return to the game in Table 22.1 and assume that B has chosen a weighting y_1, y_2, y_3, of which A is ignorant. If A systematically chooses line (1), his average gain will be

$$4y_1 - 2y_2 + 5y_3,$$

and if he systematically chooses line (2) it will be

$$-y_1 + 3y_2 + y_3.$$

If A chooses x_1, x_2, his average gain will then be

$$(4y_1 - 2y_2 + 5y_3)x_1 + (-y_1 + 3y_2 + y_3)x_2,$$

and a rational system for A would be to choose x_1, x_2, for he is then assured, on average, of gaining at least a certain value v, which is the value of the game, as we shall prove later by von Neumann's theorem (cf. Part II, Section 67).

In the same way, let us assume that A has chosen a weighting x_1, x_2, of which B is unaware. If B always chooses column (1), his average loss will be $4x_1 - x_2$; if he chooses column (2) it will be $-2x_1 + 3x_2$; if he chooses column (3) it will be $5x_1 + x_2$. Finally, if B chooses a weighting of y_1, y_2, y_3, the average loss will be

$$(4x_1 - x_2)y_1 + (-2x_1 + 3x_2)y_2 + (5x_1 + x_2)y_3$$

so that a rational behavior for B is to choose y_1, y_2, y_3, since his greatest average loss cannot then be larger than the value v of the game. We have

$$
\begin{aligned}
(4y_1 - 2y_2 + 5y_3)x_1 &+ (-y_1 + 3y_2 + y_3)x_2 \\
&= (4x_1 - x_2)y_1 + (-2x_1 + 3x_2)y_2 + (5x_1 + x_2)y_3.
\end{aligned}
\tag{22.4}
$$

Let F be this function of x_1, x_2, y_1, y_2, y_3 :

$$F = 4x_1y_1 - 2x_1y_2 + 5x_1y_3 - x_2y_1 + 3x_2y_2 + x_2y_3. \tag{22.5}$$

D. VON NEUMANN'S THEOREM;
   THE CORRESPONDENCES GOVERNING BEHAVIOR

We will now enunciate the main proposition of von Neumann's theorem as it applies to the given example:

"There exists for A a mixed optimal strategy $x_1{}^*$, $x_2{}^*$ for which his average gain $F(x_1{}^*, x_2{}^*)$ will be greater than, or equal to, a quantity v which is unique and is called "the value of the game"; and for B a mixed optimal strategy $y_1{}^*$, $y_2{}^*$, $y_3{}^*$ for which the average loss $F(y_1{}^*, y_2{}^*, y_3{}^*)$ will be less than or equal to v."

Expressing this in a different form, we can say that there is a quantity v such that

$$4x_1 - x_2 \geqslant v,$$
$$-2x_1 + 3x_2 \geqslant v,$$
$$5x_1 + x_2 \geqslant v, \qquad x_1, x_2 \geqslant 0. \qquad (22.6)$$
$$x_1 + x_2 = 1,$$

on the one hand, and on the other,

$$4y_1 - 2y_2 + 5y_3 \leqslant v,$$
$$-y_1 + 3y_2 + y_3 \leqslant v, \qquad y_1, y_2 \geqslant 0. \qquad (22.7)$$
$$y_1 + y_2 + y_3 = 1,$$

The optimal strategies $x_1{}^*$, $x_2{}^*$, on the one hand, and $y_1{}^*$, $y_2{}^*$, $y_3{}^*$, on the other, are the respective solutions for (22.6) and (22.7). The set of magnitudes $x_1{}^*$, $x_2{}^*$, $y_1{}^*$, $y_2{}^*$, $y_3{}^*$ and v constitute what is called "the solution of the game." There can be more than one optimal strategy for $x_1{}^*$, $x_2{}^*$ and $y_1{}^*$, $y_2{}^*$, $y_3{}^*$, in which case every linear combination of the multiple optimal strategies (either for A or for B) which has nonnegative coefficients and a sum equal to 1, is also an optimal strategy. We shall prove all these properties in Part II, and in particular, that every rectangular game has a solution.

Hence, to find the solution of the game taken as our example (Table 22.1) we shall resolve the inequalities (22.6) and (22.7).

E. SOLUTION OF A GAME

As a rule this is not easy; as we shall explain in the Part II, Section 70, it is necessary to resolve a linear program in order to find the mixed optimal strategies and the value of the game. But in the present example, the strategies of A contain only three unknowns x_1, x_2, and v which must be found, and as we shall see, this can be done without much difficulty.

Substituting $x_2 = 1 - x_1$ in the three inequalities of (22.6), we find:

$$
\begin{align}
(1) \quad & v \leqslant 5x_1 - 1, \\
(2) \quad & v \leqslant 3 - 5x_1, \\
(3) \quad & v \leqslant 4x_1 + 1, \\
\text{with} \quad & 0 \leqslant x_1 \leqslant 1.
\end{align}
\tag{22.8}
$$

We now construct the perpendiculars which limit the domains defined by the inequalities (22.8) in the system of reference $x_1 O v$ (Fig. 22.2).

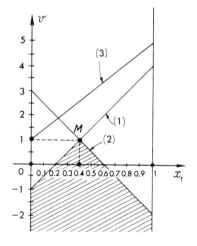

FIG. 22.2.

Each point in the hatched area provides a possible strategy and a value for v. But since A wishes this value to be as large as possible, he chooses point M, for which $x_1^* = 0.4$ and $v = 1$; and since $x_1^* + x_2^* = 1$, $x_2^* = 0.6$. Hence, by choosing the optimal strategy $x_1^* = 0.4$, $x_2^* = 0.6$, A will have an average gain equal to or greater than 1, whatever strategy his opponent chooses.

Is it possible for us to find B's optimal strategy without too complicated calculations? To do this let us make use of another proposition of von Neumann's theorem (cf. Part II, Section 67):

"If an inequality of A becomes strict (in other words if it is satisfied by replacing $\leqslant$ by $<$), after the optimal strategy has been found, the corresponding frequency of B's optimal strategy should be equal to zero. The converse is not always true, and a zero frequency in B's strategy does not necessarily produce a corresponding strict inequality for A.

"If an inequality of B becomes strict (is satisfied by $>$ instead of $\geqslant$) when the optimal strategy has been found, the corresponding frequency of A's optimal strategy should be zero, though the converse is not always true."

Hence, in (22.8) the inequality (3) becomes strict when we take the mixed strategy $x_1{}^* = 0.4$, $x_2{}^* = 0.6$, and the value of the game $v = 1$:

$$1 < (4)(0.4) + 1, \qquad \text{or} \quad 1 < 2.6. \tag{22.9}$$

From this we conclude that the frequency y_3 is zero, and system (22.7) becomes

$$\begin{aligned} 4y_1 - 2y_2 &\leqslant 1, \\ -y_1 + 3y_2 &\leqslant 1. \end{aligned} \tag{22.10}$$

Substituting $y_2 = 1 - y$, in these equations, if follows that

$$y_1 \leqslant \tfrac{1}{2}, \qquad y_1 \geqslant \tfrac{1}{2}; \tag{22.11}$$

hence;

$$y_1{}^* = \tfrac{1}{2} \qquad \text{and} \qquad y_2{}^* = \tfrac{1}{2}.$$

Assuming that the value of v is not known, but only that $y_3{}^* = 0$, we have to determine its value from the right-hand members of the inequalities (22.7). Proceeding as before, we construct Fig. 22.3 in which we find a point M' for which $v = 1$.

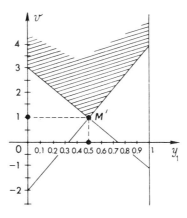

FIG. 22.3.

Thus, by choosing the strategy $y_1{}^* = 0.5$, $y_2{}^* = 0.5$, $y_3{}^* = 0$, B will have an average loss less than or equal to 1, whatever mixed strategy is chosen by A.

To conclude:

If A chooses $x_1{}^* = 0.4$, $x_2{}^* = 0.6$, his average gain will be greater than or equal to 1.

If B chooses $y_1 = 0.5$, $y_2 = 0.5$, $y_3 = 0$, his average loss will be less than or equal to 1.

We have returned to the concept of equilibrium introduced in Section 21, but in the form of a generalization that is applicable to every form of rectangular game. As we shall prove in Part II, Sections 67 and 68, von Neumann's theorem, enunciated in the above propositions, holds true whatever the matrix of the rectangular game which is considered.

It is possible for the optimal strategy of one of the players to be a pure one, but this is a special case.

23. Various Properties

A. A FAIR GAME

A rectangular game is called "fair" if its value is zero. Table 23.1 shows two examples of fair games.

TABLE 23.1

Fair Game without a Point of Equilibrium

3	−3	4
−2	2	0

(a)

Fair Game with a Point of Equilibrium

7	−7	1
2	−1	−3
4	**0**	1

(b)

B. INVARIANCE OF THE OPTIMAL SOLUTIONS

The reader should have little difficulty in finding proofs for the following properties:

(1) If the same quantity α is added to each element in the matrix of a rectangular game, the optimal strategies remain unaltered but the value of the game becomes $v + \alpha$. In particular, by adding $\alpha = -v$ to all the elements of a game, one renders it fair.

(2)   If all the elements in the matrix of a rectangular game are multiplied by a quantity $k > 0$, the optimal strategies remain the same, but the value of the game becomes kv.

This property can be used, in certain cases, to get rid of awkward fractions.

(3)   Properties (1) and (2) can be combined. If a_{ij} is the element of line i and column j, and if v is the value of the game, the rectangular game with elements $\alpha + Ka_{ij}$ $(K > 0)$ has the same optimal strategies as the former, and the value of the game is $\alpha + Kv$.

(4)   By multiplying all the elements in a matrix by (-1) we produce a new game in which the players have changed places. As we saw in Section 20, we are really dealing with the same game, but the value has a different sign because the cashier has changed.

C. SYMMETRICAL GAME

A game is said to be *symmetrical* if the matrix is an antisymmetrical square in which all the elements of the main diagonal are zero and the other elements are antisymmetrical in relation to this diagonal.

TABLE 23.2

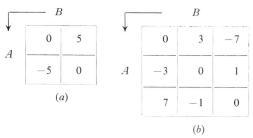

(a)

(b)

$$a_{ii} = 0 \quad (i = 1, 2, ..., n); \qquad a_{ij} = -a_{ji} \quad (i \neq j, \quad i, j = 1, 2, ..., n). \quad (23.1)$$

Table 23.2 shows two examples of symmetrical games.
If a game is symmetrical:

(a)   It is fair; in other words, the value of the game is zero.

(b)   Any strategy which is optimal for A is optimal for B, and conversely.

D. REDUCTION BY DOMINATION

The calculation of mixed strategies can often be simplified by the process of *"reduction by domination"* which we shall explain with the

III. THE THEORY OF GAMES OF STRATEGY

aid of examples; it can easily be shown how the properties and the process derived from them can be made into generalizations.

Turning to the game in Table 23.3, we notice that all the elements of the third column are greater than the corresponding ones in the first column, since $4 > 2$, $9 > 7$, and $11 > 0$. It is obvious that B, who is supposed to be an intelligent player, will not bring column (3) into his strategy; hence, $y = 0$, and the game can be reduced to that of Table 23.4.

TABLE 23.3   TABLE 23.4   TABLE 23.5

TABLE 23.3

		B	
	(1)	(2)	(3)
(1)	2	5	4
A (2)	7	3	9
(3)	0	4	11

TABLE 23.4

	(1)	(2)
(1)	2	5
(2)	7	3
(3)	0	4

TABLE 23.5

	(1)	(2)
(1)	2	5
(2)	7	3

A, in turn, observes that line (3) in the new table is composed of elements of less value than the corresponding ones in line (1), so that he will avoid bringing line (3) into his strategy.

In the last stage we are left with Table 23.5, the solution of which can easily be found:

$$x_1 = \tfrac{4}{7}, \quad x_2 = \tfrac{3}{7}; \qquad y_1 = \tfrac{2}{7}, \quad y_2 = \tfrac{5}{7}; \qquad v = \tfrac{29}{7}; \qquad (23.2)$$

whence, for the complete game of Table 23.2:

$$x_1 = \tfrac{4}{7}, \qquad x_2 = \tfrac{3}{7}, \qquad x_3 = 0;$$
$$y_1 = \tfrac{2}{7}, \qquad y_2 = \tfrac{5}{7}, \qquad y_3 = 0; \qquad v = \tfrac{29}{7}. \qquad (23.2')$$

The generalization for the reduction by domination is carried out as follows, in the case where a column (or line) is greater (or smaller if a line is concerned) than a convex linear combination[1] of other columns or lines. For example, in Table 23.6 we can verify that column (3) is such that:

$$\frac{1}{5}\begin{Bmatrix} 5 \\ -3 \end{Bmatrix} + \frac{2}{5}\begin{Bmatrix} 2 \\ 8 \end{Bmatrix} + \frac{2}{5}\begin{Bmatrix} -4 \\ 1 \end{Bmatrix} < \begin{Bmatrix} 1 \\ 11 \end{Bmatrix}. \qquad (23.3)$$
$$\quad (1) \qquad\quad (2) \qquad\quad (4) \qquad (3)$$

---

[1] Linear form $\lambda_1 x_1 + \lambda_2 x_2 + \cdots + \lambda_n x_n$ of the variables x_1, x_2, ..., x_n, of which the λ_i terms are nonnegative and with a sum equal to 1.

TABLE 23.6   TABLE 23.7

		B							
		(1)	(2)	(3)	(4)		(1)	(2)	(4)
A	(1)	5	2	1	−4	(1)	5	2	−4
	(2)	−3	8	11	1	(2)	−3	8	1

Hence, the rectangular game in Table 23.7 has the same optimal strategies as the one in Table 23.6.

In the same way, in the game shown in Table 23.8 it can be seen that

$$\frac{1}{8}[12 \quad -5] + \frac{7}{8}[0 \quad 4] > [1 \quad 2].$$ (23.4)

TABLE 23.8   TABLE 23.9

		A					
		(1)	(2)			(1)	(2)
	(1)	1	2		(1)	12	−5
B	(2)	12	−5		(2)	0	4
	(3)	0	4				

From this we conclude that the first line can be eliminated and this game can be replaced by the one in Table 23.9.

Let us now give some mathematical definitions:

A vector {**b**} *dominates* a vector {**c**} when all the elements of {**b**} are greater than or equal to the corresponding elements of {**c**}. This is denoted as:

$$\{\mathbf{b}\} \geqslant \{\mathbf{c}\}.$$

A vector {**b**} *strictly dominates* a vector {**c**} when all the elements of {**b**} are strictly greater than the corresponding elements of {**c**}.

$$\left\{ \begin{matrix} 5 \\ 2 \\ -1 \end{matrix} \right\} \text{ dominates } \left\{ \begin{matrix} 3 \\ 2 \\ -7 \end{matrix} \right\} \text{ and strictly dominates } \left\{ \begin{matrix} 3 \\ 1 \\ -2 \end{matrix} \right\}$$

The domination is *transitive*[1] if $\{\mathbf{a}\} \geqslant \{\mathbf{b}\}$ and $\{\mathbf{b}\} \geqslant \{\mathbf{c}\}$, whence $\{\mathbf{a}\} \geqslant \{\mathbf{c}\}$, and the same applies to strict domination.

In cases where reduction by domination does not appear in the strict sense, certain precautions are necessary. With this purpose in mind, we shall now consider the following theorem, the proof of which is given in reference [J16].

(1)   Let J represent a rectangular game in which the elements are a_{ij} ($i = 1, 2,..., m; j = 1, 2,..., n$), and in which the line k is *dominated* by a convex linear combination of the other lines; let J' be the new game obtained when line k is eliminated.

The value v' of the new game J' is the same as the value v of J, and any optimal strategy of B in J' is also an optimal strategy in J. If $\{x'\}$ is an optimal strategy of A in J', the strategy obtained by adding a term[2] such that $k = 0$ in $\{x'\}$ is also optimal for A in J. However, if line k of game J is *strictly dominated* by a convex linear combination of the other lines, then any solution of J can be obtained in this way beginning with a solution of J'.

(2)   Let us assume that the first column *dominates* a convex linear combination of the other columns, and let J'' be the new game obtained by eliminating column l. The value v'' of J'' is the same as v in game J, and any optimal strategy for A in J'' is also optimal in J. If $\{y''\}$ is an optimal strategy for B in J'', the strategy $\{y\}$, obtained by adding a term[2] such that $1 = 0$ in $\{y''\}$, is also optimal for A in J. However, if column l in J strictly dominates a convex linear combination of the other columns, any solution can be obtained in the same way beginning with J''.

To sum up:

If the domination is *strict*, the value of the game is the same, and any solution for J is found in J' (or J''); if it is *not strict*, the value of the game is the same, and an optimal solution for J' (or J'') is optimal for J', though there may be others in J.

Since the above may seem rather complicated for the reader who is not a mathematician, we shall attempt to make it clearer with the help of some examples.

Example 1.   In the game shown in Table 23.10, line (1) is strictly dominated by line (3) and can therefore be eliminated, with the result that $x_1 = 0$. We now obtain Table 23.11 in which column (3) strictly

---

[1] For the property of transitivity, see also Part II, Section 31.

[2] After the eventual "unwedging" of the others.

dominates column (2), and also dominates a linear combination of columns (1) and (2). In fact:

$$(\tfrac{1}{3})(2) + (\tfrac{2}{3})(5) < 5,$$
$$(\tfrac{1}{3})(7) + (\tfrac{2}{3})(3) < 6.$$

(23.5)

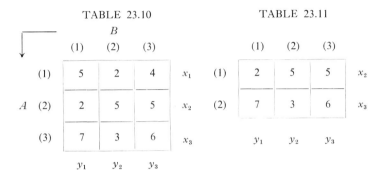

TABLE 23.10

		B		
	(1)	(2)	(3)	
(1)	5	2	4	x_1
A (2)	2	5	5	x_2
(3)	7	3	6	x_3
	y_1	y_2	y_3	

TABLE 23.11

	(1)	(2)	(3)	
(1)	2	5	5	x_2
(2)	7	3	6	x_3
	y_1	y_2	y_3	

We therefore eliminate column (3), concluding that $y_3 = 0$. The solution of the game in Table 23.12 is

$$x_2 = \tfrac{4}{7}, \qquad x_3 = \tfrac{3}{7};$$
$$y_1 = \tfrac{2}{7}, \qquad y_2 = \tfrac{5}{7}; \qquad v = \tfrac{29}{7};$$

(23.6)

TABLE 23.12

	(1)	(2)	
(1)	2	5	x_2
(2)	7	3	x_3
	y_1	y_2	

hence (as this is a case of reduction by strict domination), the only optimal solution of the game in Table 23.10 is

$$x_1 = 0, \qquad x_2 = \tfrac{4}{7}, \qquad x_3 = \tfrac{3}{7};$$
$$y_1 = \tfrac{2}{7}, \qquad y_2 = \tfrac{5}{7}, \qquad y_3 = 0; \qquad v = \tfrac{29}{7}.$$

(23.7)

Example 2. The game shown in Table 23.13 can have any optimal solution such that

$$x_1 = \alpha, \qquad x_2 = \beta, \qquad x_3 = \beta,$$
$$y_1 = \alpha, \qquad y_2 = \beta, \qquad y_3 = \beta,$$

(23.8)

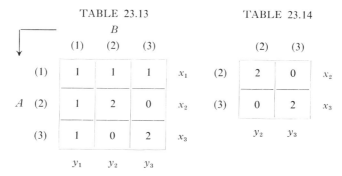

TABLE 23.13   TABLE 23.14

with $v = \frac{1}{2}$.

There are an infinite number of optimal solutions; for instance:

$$x_1 = \tfrac{2}{3}, \qquad x_2 = \tfrac{1}{6}, \qquad x_3 = \tfrac{1}{6};$$
$$y_1 = \tfrac{2}{3}, \qquad y_2 = \tfrac{1}{6}, \qquad y_3 = \tfrac{1}{6}, \tag{23.9}$$

is an optimal solution, and

$$x_1 = 1/100, \qquad x_2 = 99/200, \qquad x_3 = 99/200;$$
$$y_1 = 1/100, \qquad y_2 = 99/200, \qquad y_3 = 99/200, \tag{23.10}$$

is another.

It can easily be verified that line (1) is dominated, but not strictly, by a convex linear combination of lines (2) or (3), so that it will be eliminated. In the table thus obtained, column (1) dominates, though not strictly, a convex linear combination of columns (2) and (3), and it will also be eliminated. Finally, Table 23.13 is obtained, in which there is one and only one solution:

$$x_2 = \tfrac{1}{2}, \qquad x_3 = \tfrac{1}{2};$$
$$y_2 = \tfrac{1}{2}, \qquad y_3 = \tfrac{1}{2}; \qquad v = \tfrac{1}{2}. \tag{23.11}$$

Using this solution, we go back to that of the initial game and find

$$x_1 = 0, \qquad x_2 = \tfrac{1}{2}, \qquad x_3 = \tfrac{1}{2};$$
$$y_1 = 0, \qquad y_2 = \tfrac{1}{2}, \qquad y_3 = \tfrac{1}{2}; \qquad v = \tfrac{1}{2}. \tag{23.12}$$

This solution clearly belongs to the group given in (23.8) and is not the only one.

The discovery of convex linear combinations which reveal domination is not always easy. It will not be discussed here, but the reader will find various theorems relating to this problem in Part II.

24. Use in Concrete Cases of Competition

Within the scope of the present volume, which is intended to be elementary, it is difficult to do justice to the role which the theory of games of strategy may play in the field of economics. Moreover, the space which is devoted to such games is too limited to allow us to consider real examples, which are generally played by more than two people and with a nonzero sum (see Sections 75 and 76). In the present section the reader must excuse the fact that only two examples, which are obviously of an artificial nature, are discussed, though they will be useful as an introduction to further study of the subject.

The first example deals with "a business game" of a very simplified kind in which the behavior of the opponents, who are commercial rivals, is based on the theory of games of strategy. The assumptions will be such that only pure strategies can have a concrete meaning, and we shall compare the intuitive methods of two actual teams of players to a system based on the theoretical concepts which have been discussed in the preceding pages. The second example, which is of a completely different type, will give substance to the concept of a mixed strategy in the case of a duel: one between the police and thieves.

A. EXAMPLE 1: A SIMPLIFIED BUSINESS GAME[1]

Let us remember that "a business game" is the artificial reproduction of the activities of several firms who are competitors in the same market. In it, the relation of cause and effect, both between the firms, and also between them and the market, is reproduced by means of a mathematical model. In a business game the teams are allowed to receive certain information and have to take decisions for a series of stages, each of which may be of 3–6 months' duration. In our example the duration is of one year.

The reader should not be surprised at the choice of a very simplified example, since we are more concerned with explaining how a certain form of reasoning can be used than with applying it to cases which are more general and closer to reality. Where business games are detailed and complex enough to give a simulation of real situations,[2] the use of

---

[1] This example is taken from the thesis by Simon-Pierre Jacot, *Stratégie et concurrence*, Faculté de Droit et des Sciences économiques de Lyon, April, 1961. In this work, one of the best on the subject, the reader will find a very detailed study of the employment of the theory of games in the analysis of competition.

[2] See, for example, the model OMNILOG in A. Kaufmann, R. Faure, and A. Le Garff, *Les jeux d'entreprises*, Series *Que sais-je?*, Presses Universitaires de France, 1950.

the theory of games of strategy for the purpose of obtaining optimal strategies has often proved difficult, and even impossible. The attempt to attain a close resemblance to actual situations requires such a highly combinatorial form that it is impossible to describe such games in terms of matrices; besides, they do not have a zero sum and usually include more than two firms (often four). Hence, from a theoretical standpoint they are games with n people $(n > 2)$ with a nonzero sum.

In the present example, two firms A and B are producing and selling the same product. We shall deal exclusively with the sales problems, taking the cost price as a given constant, and shall only consider the possible variations of the selling price and of the promotion expenses.

At the beginning of each stage the players or teams representing firm A and firm B will receive certain information about their competitor and about the market; they will then make their decisions on the basis of this information, and these decisions will be transferred to the mathematical model in the form of corresponding variables, the results being used in making new decisions, and so on. Since the mathematical model only provides results in the last stage, it will not be given until the end of this subsection.

Describing the market as a firm of several years' experience might do, we can say that the sales price will affect both the size of the market and the proportional share of it obtainable by the two firms. In this game the competitor's price will not be made known to his rival until the conclusion of a stage or "exercise."

The proportion in which the market is shared will also depend on the promotion expenses of each firm, and these will further affect the size of the market, though it cannot be either increased or diminished indefinitely.

The basic difference between the effect of the sales price and of promotion expenses, both on the size of the market and on the proportions in which it is shared, is that the former is experienced immediately, whereas the result of an advertising campaign is only partial during the following year, and may not be fully effective until nearly a year later. After two years its effect becomes nil, unless further advertising has been launched during this period.

Each firm has a qualitative idea of the other's annual expenditure on advertising.

When fixing A's selling price, and the amount the firm will spend on promotion during the next financial year, the following figures are taken into account:

(1) The number of articles A sold during the last year, 18,600;

(2) B's latest selling price, $280;

(3) A's selling price during the last financial year, $300;

(4) A's cost of production per article, $200;

(5) A's promotion expenses during the last financial year $200,000:

Lastly, information is obtained about B's promotion expenses, which shows that they are roughly the same as A's.

In order to decide A's selling price for the next financial year, we draw up Table 24.1, which shows the gross profit (number of articles sold multiplied by the gross margin of profit) from various situations which take into account different selling prices for B.

TABLE 24.1

Sales price of competitor B

		$300	$280	$260
	$300	20,460 × 100	18,600 × 100	16,740 × 100
Sales price of competitor A	$280	22,320 × 80	20,460 × 80	18,600 × 80
	$260	24,180 × 60	22,320 × 60	20,460 × 60

The figures in the "boxes" have been obtained on the assumption that the results of the second financial year can be used as a basis for confirmation or change of policy: the loss in sales due to the $20 increase in the selling price amounts to 10% of A's sales, with the result that his competitor's now amount to 22,320 out of a total of 40,920.

This matrix of gross profits is used as the basis for the first game of strategy.

Game 1.

TABLE 24.2

Customers

		$300	B's price $280	$260
	$300	$2,046,000	$1,860,000	**$1,674,000**
A's price	$280	$1,786,000	$1,637,000	$1,488,000
	$260	$1,451,000	$1,339,000	$1,228,000

While it is obvious that the pure strategy of maintaining the higher price ($300) is the best for A, since it dominates the two others and ensures a minimal gain of $1,674,000, B's course is less clear. If we follow the rules for games of strategy introduced by von Neumann, the pure strategy of a low price dominates the two others, but this result seems unrealistic, since A's reasoning should be equally valid for B, who sells in the same market.

The reason for this apparent contradiction is that we are not dealing with two players (as in the games previously discussed), but with three: A, B, and the market. In other words, A and B are not the real opponents in the game shown in Table 24.2, for it is an extension of A's struggle with the market, and A's gains are obtained from the customers. This is what the arrow connecting the customers with A is intended to express.

The real contest between A and B is over the number of customers each can take away from the other by their respective policies for the selling price. This is shown in Table 24.3.

Game 2. The game has a point of equilibrium, since the minimum for B is equal to the maximum for A and the value of the game is 0. In the contest in which they are engaged, the aim of both competitors is the lowering of prices.

A characteristic feature in the present problem of concurrence appears in Table 24.3, for the same value of the game (0) can be obtained with high prices. In other words, at the expense of the market.

TABLE 24.3

A ↓	B $300	$280	$260
$300	0	−1860	−3720
$280	1860	0	−1860
$260	3720	1860	**0**

If we exclude the possibility of an understanding between the two firms, which would result in a monopoly and the end of competitiveness, the choice between a high or a low price largely depends on the loyalty of the customer. Earlier we made the assumption that a reduction in price would have an immediate and proportional influence, and we must now make the further assumption that there is an almost entire absence of customer loyalty as far as the article in question is concerned. Further, it would be of no advantage to A or B to lower their prices in order to attract new customers, for these new customers would be lost as soon as the price was increased. In other words, a combined study of Game 1 (A against the consumers—Table 24.2) and Game 2 (A against B—Table 24.3) shows that only the former corresponds to the assumptions which have so far been made.

The question of *expenditure on promotion* is of a different kind, for this only has an influence at a later date. Despite this, it is possible to include it within the same framework by artificially compressing the year's expenditure on advertising and the profits of the two subsequent years into a one-year period. As an initial hypothesis, A calculates that, if his promotion expenses are found to exceed his rival's, he can expect his business index to rise by 10% the next year and by 20% the following one, in accordance with Table 24.4.

TABLE 24.4

	Evaluation of A's expenditure	A's sales	
		Next year	Year after
Expenses of A = 4 × expenses of B	Very much higher	+20%	+40%
Expenses of A = 3 × expenses of B	Much higher	+15%	+30%
Expenses of A = 2 × expenses of B	Higher	+10%	+20%
Expenses of A = 1.5 × expenses of B	Slightly higher	+5%	+10%
Expenses of A = expenses of B	Equal	0	0
Expenses of B = 1.5 × expenses of A	Slightly lower	−5%	−10%
Expenses of B = 2 × expenses of A	Lower	−10%	−20%
Expenses of B = 3 × expenses of A	Much lower	−15%	−30%
Expenses of B = 4 × expenses of A	Very much lower	−20%	−40%

On this basis it is possible to formulate a game of strategy for the net profits obtained from the different expenditures on promotion, with the proviso that it must be revised annually and that the policy of a stable price, which was agreed upon earlier, will be maintained.

Game 3. Expenditure for Promotion. Supplementary profits obtained during the two years following the advertising campaign (gross profits less the corresponding expenditure) are shown in Table 24.4a.

TABLE 24.4a

	$100,000	$200,000	$300,000	$400,000
$100,000	−10	−65.8	−93.7	−121.6
$200,000	35.8	−20	−47.9	−75.8
$300,000	53.7	−2.1	−30	−48.6
$400,000	71.6	15.8	−21.4	**− 40**

The numbers in the boxes (Table 24.4a) have been obtained as follows: for example, for the box in the second line of the first column, we find

$$\text{supplementary gross profit} = (10 + 20\%) \text{ of } \$1,860,000$$
$$= \$558,000$$
$$\text{supplementary net profit} = \$558,000 - \$200,000$$
$$= \$358,000$$

The above numbers are given in units of $10,000.

A's strategy is clearly defined: whatever action *B* takes, it is to *A*'s advantage to increase his promotion expenditure to $400,000.

Once more, in this game *A*'s opponent is not exclusively *B*. To be precise, it is not a game with a zero sum, a prerequisite if the theory of rectangular games is to apply. In fact, the numbers in each box certainly represent *A*'s profits (or losses, if they are negative), but it is only *B* who experiences the corresponding losses (or gains) since each of these numbers is derived from the combination of the profit (or loss) resulting from the increase (or decrease) of the sales index, and of the advertising expenditure of *A* alone.

To find the best pure strategy for B, it is necessary to reconstruct his game, which will be similar in all respects to A's; hence, the conclusion is reached that the prudent course for each competitor is to raise his promotion expenses, since the loss in either case is then limited to $400,000 (the value of the game).

The above reasoning is based on the assumption that neither competitor has any idea of his rival's intentions. It is obvious that any disclosure of them alters the premises of the game and *will always increase the value of the game*.[1] For instance, if A's enquiries lead to the conclusion that he may assign the following probabilities to B's plans:

Advertising expenditure	Probabilities
$100,000	0.10
$200,000	0.15
$300,000	0.50
$400,000	0.25

we shall, as a result of multiplying the columns in Table 24.4a by the corresponding probabilities, produce Table 24.5.

Game 4.

TABLE 24.5

A	B 100,000	200,000	300,000	400,000	Total
100,000	−1	−9.9	−46.8	−30.4	−88.1
200,000	3.6	−3	−23.9	−18.9	−42.2
300,000	5.4	−0.3	−15	−12.1	−22
400,000	7.2	2.4	−10.7	−10	**−11.1**

A's best strategy is still to spend $400,000 on advertising, but the information obtained (and consequent changes) has increased the value of the game from −40 to −11.1.

In fact, as we have no present information about the other firm's intentions, our decisions will remain unchanged: new selling price $300; new expenditure on promotion, $400,000.

---

[1] See Section 28.

In the case of the competitor, the decisions will be: $250 and "slight increase" of promotion expenditure to $500,000. Used in conjunction with the mathematical model giving the relations between cause and effect, these decisions produce a net profit of $1,390,000 for *A* for 17,900 articles sold.

We can now check and improve our first assumption about the effect of the selling price (though not of that of the outlay on advertising, since its influence is only felt a year later). An increase of $50 instead of $20 in the selling price has caused a decrease of 700 in the sales. Hence, the estimates can be made for the new financial year:

TABLE 24.6   TABLE 24.7

A	┌─── *B* $280	$250
$300	18,600	**17,900**
$280	19,066	18,366
$250	19,766	19,066

A	┌─── Customers $280	$250
$300	$1,860,000	**$1,790,000**
$280	$1,503,000	$1,470,000
$250	$ 990,000	$ 950,000

Game 5.   Articles Sold.   After obtaining Table 24.6 we can express it as gross profits (Table 24.7). The sale price of $300 will be maintained.

As regards the advertising expenditure, all we know is that *B*'s is slightly higher than *A*'s, and as we have already shown that it is not to the advantage of either firm to be outpaced in this field, we shall increase *A*'s outlay to $500,000.

The results of the new financial year, as given by the mathematical model, and taking into account *B*'s decisions to fix its sale price at $240 and to spend the same amount on advertising as *A*, are:

A's profit $1,180,000 for 16,848 articles sold.

By comparing these figures with the estimates, we can now verify our assumptions as to the effect of advertising.

According to the estimates, *A* should have sold 17,667 articles (if we take only the effect of a competitiveness in price into account) and 16,784 (if we include the effect of promotion expenditure). Our earlier assumptions about the reaction of the market can therefore be regarded as sufficiently accurate.

Teams were picked from people with a knowledge of competitive business to represent *A* and *B*. On the basis of the information which was available for our games, they came to various empirical decisions which are compared below with those obtained by the theory of games of strategy. The two sets of decisions differ so materially that the value of the theory is obvious in cases where the phenomena are simple enough to permit its use.

	Empirical decisions	Decisions obtained by the theory of games
1st Year		
Selling price	$ 270	$ 300
Cost of advertising	$ 200,000	$ 400,000
Gross profit (after deducting cost of advertising)	$1,240,000	$1,390,000
2nd Year		
Selling price	$ 260	$ 300
Cost of advertising	$ 250,000	$ 500,000
Gross profit (after deducting cost of advertising)	$ 870,000	$1,180,000

Model which Simulates the Market. The very simple model which has been used to simulate the reactions of the market can be summarized as follows:

(a) *Cost of advertising and its effect on the market:*

At the outset the available market is for 40,000 articles, and each firm invests $200,000 on different kinds of promotion. Each time that the two firms invest a further $200,000 in advertising, the market is increased by 2% the following year and by 4% after two years, then decreases to its original size if no further sums have been spent on promotion in the meanwhile.

Floor: 92% (no expenditure on promotion);

Ceiling: 120% of the original market.

(b) *Expenditure on promotion and its effect on the proportional division of the market:*

The following formula determines the percentage by which the initial equal division is modified in favor of one or other of the competitors:

$$\pm p\% = 13.5 \frac{A\text{'s outlay} - B\text{'s outlay}}{A\text{'s outlay} + B\text{'s outlay}}, \tag{24.1}$$

for the following year, and double this percentage two years later.

(c) *Effect of the selling price on the volume of the market:*

The minimal price, fixed by one or other of the competitors, alone has an effect on the volume of the market during the current year. The linear effect is expressed as:

Minimal price: $320:   84% of the market

$300:   92% of the market

$280:   100% of the market

$260:   108% of the market

$250:   112% of the market (ceiling)

(d) *Effect of the sales price on the share of the market*:

The original equal division is modified for the current financial year by the formula:

$$\pm a\% = 50\,\frac{A\text{'s price} - B\text{'s price}}{\text{minimal price}} \qquad (24.2)$$

It must be understood, of course, that both the selling price and the expenditure on promotion have a cumulative effect.

(e) *Evaluation of the commercial expenditure of the compititor*:

The relation

$$k = \frac{\text{competitor's expenditure}}{\text{total expenditure}}$$

enables us to make the following evaluations:

k	Competitor's expenditure
>0.6	Much higher,
0.57 − 0.6	Higher,
0.53 − 0.56	Slightly higher,
0.48 − 0.52	Equal,
0.44 − 0.47	Slightly lower,
0.40 − 0.43	Lower,
<0.40	Much lower.

B. EXAMPLE 2: DETECTIVE STRATEGY

This example was suggested by a process used by a naval operations research group to determine the optimal strategies of attack and defense for submarines. The contest to be considered is between thieves and detectives in a department store (see Table 24.8).

TABLE 24.8

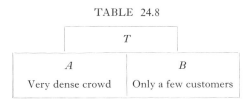

T	
A	B
Very dense crowd	Only a few customers

Two areas, *A* and *B*, are on the same floor of a department store; the first, on account of the type of merchandise sold there, is densely crowded with customers, the latter has comparatively few people in it. The store employs two plain-clothes detectives, who are free to take up positions in *A* or *B*; and television sets, situated in *T*, can cover both areas. Hence, the two detectives can be in either *A*, *B*, or *T*, but the thieves in *A* or *B* only (see Table 24.9).

TABLE 24.9

		Thief	
		A	B
	T	0.3	0.5
Detective	A	0.4	0.2
	B	0.1	0.7

The detectives have given an estimate of the probability of finding and arresting a thief, and have assumed that it is rare for several thieves to operate in the store at the same time. As a result of experience, they estimate that if a thief is in *A* and a detective at *T*, the probability of an arrest is 0.3; if the detective is in *A*, it is 0.4, and so on, these results being shown in Table 24.9. But if there are two detectives, they are free to operate separately or together in *A*, *B*, or *T*.

It is a simple matter to calculate the total probabilities of arrest for cases *TT* (with the detectives in *T*), for *AA* (when they are in *A*) and for *TA* (when one is in *T* and the other in *A*), etc. The arrest of a thief by the first detective can be represented as an uncertain variable X_1 which can take the value 1 (thief arrested) or the value 0 (thief not arrested). If the arrest of a thief by the second detective is similarly represented as X_2, we shall accept the assumption that the two variables are independent.

We have:

p_{TA} = probability of arrest by the first detective + probability of arrest by the second if the first has not captured the thief

$$= p_T + p_A(1 - p_T)$$

$$= p_T + p_A - p_T p_A$$

$$= 0.3 + 0.4 - (0.3)(0.4) = 0.58; \qquad (24.3)$$

TABLE 24.10

		Thief		
		A	B	
	TT	0.51	0.75	x_1
	AA	0.64	0.36	x_2
	BB	0.19	0.91	x_3
Two detectives	TA	0.58	0.60	x_4
	TB	0.37	0.85	x_5
	AB	0.46	0.76	x_6
		y_1	y_2	

in the same way:

$$p_{TT} = p_T + p_T(1 - p_T)$$

$$= p_T + p_T - p_T p_T$$

$$= 0.3 + 0.3 - (0.3)(0.3) = 0.51. \qquad (24.4)$$

In this way we obtain Table 24.10, which constitutes a rectangular game of strategy between the detectives and the eventual thief; the former have at their disposal strategies with 6 components, the latter strategies with 2 components. The equations for the detectives are

$$0.51x_1 + 0.64x_2 + 0.19x_3 + 0.58x_4 + 0.37x_5 + 0.46x_6 \geqslant v,$$

$$0.75x_1 + 0.36x_2 + 0.91x_3 + 0.60x_4 + 0.85x_5 + 0.76x_6 \geqslant v, \qquad (24.5)$$

$$x_1 + x_2 + x_3 + x_4 + x_5 + x_6 = 1, \qquad x_i \geqslant 0, \qquad i = 1, 2,..., 6,....$$

Those of the thief are

$$
\begin{align}
(1) && 0.51y_1 + 0.75y_2 &\leqslant v, \\
(2) && 0.64y_1 + 0.36y_2 &\leqslant v, \\
(3) && 0.19y_1 + 0.91y_2 &\leqslant v, \\
(4) && 0.58y_1 + 0.60y_2 &\leqslant v, && (24.6) \\
(5) && 0.37y_1 + 0.85y_2 &\leqslant v, \\
(6) && 0.46y_1 + 0.76y_2 &\leqslant v, \\
&& y_1 + y_2 = 1, \qquad x_1, x_2 &\geqslant 0.
\end{align}
$$

Since the thief's strategies contain two components, it is possible to calculate the optimal strategy or strategies by the graphic method explained in Section 22, and by means of which we obtain Fig. 24.1. It follows that

$$y_1 = 0.8, \qquad y_2 = 0.2, \qquad v = 0.584. \qquad (24.7)$$

These values are obtained from the intersections of straight lines (2) and (4); hence, inequalities (1), (3), (5), and (6) in (24.6) are strict

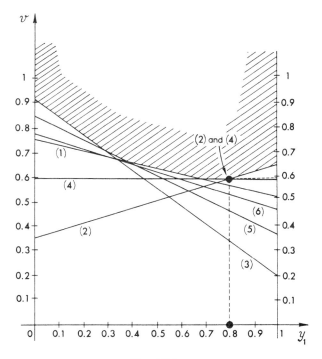

FIG. 24.1.

inequalities for the optimal solution, and from what we learned in Section 22, the variables x_1, x_3, x_5, and x_6 are zero in the optimal strategy of the detectives. The equations are therefore reduced to

$$0.64x_2 + 0.58x_4 \geqslant v,$$
$$0.36x_2 + 0.60x_4 \geqslant v, \tag{24.8}$$
$$x_2 + x_4 = 1, \qquad x_2, x_4 \geqslant 0.$$

By means of the graphic method it is easily seen that

$$x_2 = 1/15, \qquad x_4 = 14/15. \tag{24.9}$$

The following conclusions will therefore be reached.

If the detectives spend 1/15 of their time together in A and 14/15 of their time separately, with one in T and the other in A, they will have a probability of making an arrest equal to 0.584, whatever strategy may be chosen by the thief (or thieves, if they do not operate simultaneously). The thief (and, indeed, the set of thieves) would be led to follow the mixed strategy of stealing 4 out of 5 times in A and once out of 5 times in B, with the result that the probability can never exceed 0.584, whatever mixed strategy the detectives may adopt.

If the television set is eliminated from the problem, the probability of arrest given by the optimal strategy will only be slightly reduced. The use of television sets probably deters a number of potential thieves from entering the store, but has little influence on the probability of an arrest.

It would be interesting to expand this problem to cases where there is the frequent possibility that several thieves may be in the store at the same time, and even that they may be collaborating. This would make the problem a much more complex one, but it could still be treated by means of the theory of games of strategy, though the calculations would present considerable difficulty.

25. Use in Cases of a "Struggle against Nature"

A. NATURE AS AN ADVERSARY

It is possible to consider nature as an adversary whose intentions are entirely unknown to us, and the aim of a statistic analysis is to uncover these intentions (as far as that can be done), and to formulate laws of probability which will be more or less acceptable according to our hypotheses. Very frequently, however, a statistical analysis of the facts is very difficult (if not impossible) to carry out, and on other occasions, the conclusion is reached that a knowledge of the past is of little help in foretelling the future. If such is the case, how are we to act? If future states which are denumerable and can be given a function of value, do not reveal any law of probability, we have to admit (where we possess the power of decision) that we are involved in "a game with nature," and that we can, with certain reservations, treat it by the methods already explained for dealing with an intelligent opponent. Unfortunately, the "minimax" criterion used in the theory of games in the sense given to it by Borel and von Neumann, is open to various critical objections which will be discussed in Section 26. For the moment, we shall accept this criterion and shall attempt to discover how it can be used in a contest against nature.

B. Example 1: Simplified Investment Problem[1]

An investor plans to invest $10,000 during a period of international uncertainty whether there will be peace, a continuation of the cold war, or an actual war. He has a choice between government bonds, armament shares, or industrials for his investment.

The rates of interest which we have assumed are given in Table 25.1, where the reader must not be surprised at finding a negative value.

We shall consider this game as a struggle between a financier (who has so many adversaries that he cannot enumerate them) and nature.

To calculate the investor's optimal strategy, we see that line (3) dominates (but not strictly) line (1), and shall try to discover the optimal values of x_2 and x_3 and the value v of the game (see Tables 25.1 and 25.2).

	TABLE 25.1					TABLE 25.2			
	War	Cold war	Peace				(1)	(2)	(3)	
	(1)	(2)	(3)							
Government bonds (1)	2	3	3.2	x_1	(2)	18	6	-2	x_2
Armament shares (2)	18	6	-2	x_2	(3)	2	7	12	x_3
Industrials (3)	2	7	12	x_3					

[1] This example is mentioned in *La Stratégie dans les actions humaines* by J. D. Williams, Dunod, Paris, 1956 [J23].

The equations which control the game of Table 25.2 are

$$\text{(1)}\qquad 18x_2 + 2x_3 \geqslant v,$$

$$\text{(2)}\qquad 6x_2 + 7x_3 \geqslant v, \qquad x_2 + x_3 = 1. \qquad \text{(25.1)}$$

$$\text{(3)}\qquad -2x_2 + 12x_3 \geqslant v,$$

By the use of the graphic method, we find

$$x_2 = 5/17; \qquad x_3 = 12/17; \qquad v = 114/17 = 6.70. \qquad \text{(25.2)}$$

In this example the strategy $\{x_1, x_2, x_3\}$, is unique, as we shall find if we substitute the values of (25.2) in the system of equations for the complete game (Table 25.1).

Hence, the financier's optimal strategy will be

$$x_1 = 0, \qquad x_2 = 5/17, \qquad x_3 = 12/17. \qquad \text{(25.3)}$$

With this strategy he can be certain of a return of at least 6.70, whatever the international situation may be.

In the event of war, the return would be

$$(18)(5/17) + (2)(12/17) = 114/17 = 6.70; \qquad \text{(25.4)}$$

in the event of a cold war, it would be

$$(6)(5/17) + (7)(12/17) = 114/17 = 6.70;$$

in the event of peace, it would be

$$(-2)(5/17) + (12)(12/17) = 134/17 = 7.88.$$

By investing $10{,}000 \times 5/17 = \$2941$ in armament shares and $10{,}000 \times 12/17 = \$7058$ in industrials, the financier will be assured of an income of at least \$670.

C. EXAMPLE 2: USE OF ANTIBIOTICS[1]

Three antibiotics A_1, A_2, and A_3 and five types of bacilli M_1, M_2, M_3, M_4, and M_5 are involved in this problem, with A_1 having a probability 0.3 of destroying M_1, 0.4 of destroying M_2, etc. These probabilities are given in Table 25.3.

Without knowing the proportion in which these germs are distributed during an epidemic, in what ratio should the antibiotics be mixed to have a limit lower than the greatest probability of being effective?

[1] Suggested by an example given in Ref. [J23].

	TABLE 25.3						TABLE 25.4				TABLE 25.5			
	Type of germ													
	(1)	(2)	(3)	(4)	(5)		(1)	(4)	(5)		(1)	(4)	(5)	
(1)	0.3	0.4	0.5	1	0	x_1 (1)	0.3	1	0	x_1 (1)	0.3	1	0	x_1
(2)	0.2	0 3	0.6	0	1	x_2 (2)	0.2	0	1	x_2 (2)	0.2	0	1	x_2
(3)	0.1	0.5	0.3	0.1	0	x_3 (3)	0.1	0.1	0	x_3				

(Antibiotic — row labels)

It is observed that columns (2) and (3) dominate column (1), so that these two columns will be eliminated. The game is now reduced to the one in Table 25.4 in which line (3) is dominated by line (1) and is therefore removed, and we are left with the game shown in Table 25.5.

By the use of graphic means, we find

$$x_1 = 8/11, \qquad x_2 = 3/11; \qquad v = 3/11. \tag{25.5}$$

and thence:

$$x_1 = 8/11, \qquad x_2 = 3/11, \qquad x_3 = 0; \qquad v = 3/11. \tag{25.6}$$

By preparing an antibiotic containing $3/11$ of A_1 and $8/11$ of A_2, the probability of effective action would be at least equal to $3/11$:

if M_1 is alone, the probability will be $(0.3)(8/11) + (0.2)(3/11) = 3/11$,

if M_2 is alone, the probability will be $(0.4)(8/11) + (0.3)(3/11) = 41/11$,

if M_3 is alone, the probability will be $(0.5)(8/11) + (0.6)(3/11) = 5.8/11$,

if M_4 is alone, the probability will be $(1)(8/11) = 8/11$,

if M_5 is alone, the probability will be $(1)(3/11) = 3/11$. (25.7)

We can conclude that, in whatever proportions the different types are distributed, the probability must be greater than $3/11$.

D. EXAMPLE 3: A GAME AGAINST NATURE WITH INCOMPLETE INFORMATION[1]

In order to provide heating for a house, four tons of coal are needed if the winter is a mild one, five tons if it is normal, and six tons if it is

[1] In using an example of somewhat the same kind as the one referred to in [J23], we intend to show the influence of further information on the value of a game and the optimal strategies.

a severe one. If the supply of coal is exhausted, the cost of replenishing it is 200, 220, and 240 monetary units a ton for the respective types of winter. What decision should be taken?

TABLE 25.6

	Mild	Average	Severe	
4 Tons	−800	−1020	−1280	x_1
5 Tons	−1000	−1000	−1240	x_2
6 Tons	−1200	−1200	**−1200**	x_3

After buying four tons in the summer, the extra cost will be

$$800 \text{ units} \qquad \text{if the winter is mild;}$$
$$800 + 200 \text{ units} = 1000 \text{ units} \qquad \text{in an average winter;}$$
$$800 + 2 \times 240 = 1280 \text{ units} \qquad \text{in a severe one.}$$

If five tons have been stored in the summer, the further cost will be

$$1000 \text{ units} \qquad \text{if the winter is mild or average;}$$
$$1000 + 240 = 1240 \text{ units} \qquad \text{if the winter proves a severe one.}$$

When six tons are bought during the summer the cost will be 1200 units, whatever the winter is like.

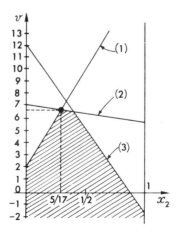

FIG. 25.1.

The problem is a game against nature (Table 25.6) which has a point of equilibrium, the element at the intersection of the third line and third column. Hence, the optimal strategy will be:

$$x_1 = 0, \qquad x_2 = 0, \qquad x_3 = 1, \qquad v = -1200.$$

On the basis of the minimax criterion, six tons should be purchased so that the expenditure is limited to 1200 units.

We shall now assume that the meteorological office has forecast that the coming winter will not be severe, and will decide on an optimal strategy.

To do this, we eliminate column (3) and also line (3), the new game being shown in Table 25.7. Like the previous one, it has a point of equilibrium, and the optimal strategy is

$$x_1 = 0, \qquad x_2 = 1, \qquad\qquad (25.8)$$

TABLE 25.7

	Winter		
	Mild	Average	
4 Tons	−800	−1020	x_1
5 Tons	−1000	**−1000**	x_2

and the value of the game is $v = -1000$.

As a result of this information the expense is reduced from 1200 units to 1000 for the optimal strategy, an intuitive result which leads us to an important and very simple generalization:

"If the minimizing player B has partial or complete information as to A's intentions, he can decide that a fresh game J' has been started in which the new value v' cannot exceed the value v of the previous one.

"If the maximizing player A has partial or complete information about B's intentions, he can decide that a fresh game J' has begun of which the new value cannot be less than the value v of the previous one."

In the case of a game against nature, this property may be expressed thus:

"Any statistical information about the state of nature, partial or total (it is total if the state or frequency of the states of nature are fully known), can never diminish the value of a game when the player is maximizing, or increase it if he is minimizing."

E. EXAMPLE 4: THE MINIMAX CRITERION CAN BE UNACCEPTABLE

Let us show by means of an example how this criterion can lead to unacceptable conclusions in certain cases of a struggle against nature.

A news agent sells a weekly magazine; he orders a certain number every Tuesday, and the unsold copies are returned the following Tuesday.

On each copy sold he makes a profit of fifty cents, and on each unsold copy (which he can return at a loss) he loses thirty cents. The only information he has about his customers is that his sales will never exceed fifty.

To simplify the table, we will assume that both the states of nature (sales) and the amount of the order increase by units of 10. Table 25.8 shows the news agent's profit for each state of nature and each decision in this game against nature.

TABLE 25.8

		Sales					
		0	10	20	30	40	50
	0	**0**	0	0	0	0	0
	10	−3	5	5	5	5	5
	20	6	2	10	10	10	10
Order	30	9	−1	7	15	15	15
	40	−12	−4	4	12	20	20
	50	−15	−7	1	9	17	25

This game has a point of equilibrium which represents the decision not to order any copies; but a news agent has to earn a living, so he will obviously discard this solution.

What he should do, in fact, is to obtain statistics for the sales, and he may then find that they always exceed twenty copies. A new game (Table 25.9) will then be formed in which the point of equilibrium corresponds to an order for thirty, an intuitive decision (considering the extremely prudent character of the minimax criterion), the use of which is open to the same objections as before.

TABLE 25.9

	30	40	50
30	**15**	15	15
40	12	20	20
50	9	17	25

The news agent should attempt to improve his information and thereby obtain an increasing knowledge of the frequency of the states of nature. As he does so, the problem will pass from the field of uncertainty to that of probabilities, and can then be treated by the method given in Volume I (p. 177)[1] for various problems of stocks.

26. Choice of a Criterion

We shall now make a brief study of various other criteria which can be used in cases of a struggle against nature.

A. LAPLACE'S CRITERION

If the probabilities of the different states of nature are unknown, they are assumed to be equal. Hence, if the player chooses the first line, his profit is found from the average

$$\frac{1}{n}(a_{i1} + a_{i2} + \cdots + a_{in}),\qquad(26.1)$$

and the criterion is to choose the line in which he finds

$$\underset{i}{\text{MAX}}\left[\frac{1}{n}(a_{i1} + a_{i2} + \cdots + a_{in})\right].\qquad(26.2)$$

What he does, in effect, is to choose the line with the highest average.

A criterion of this kind appears invalid if the elements of the rectangular game are widely dispersed. It is well known how frequently errors in evaluation and faulty decisions are caused by using a nonweighted average in summary estimates.

[1] Volume I refers to A. Kaufmann, *Methods and Models of Operations Research*. Prentice-Hall, Englewood, New Jersey, 1963.

B. WALD'S CRITERION[1]

This criterion is that of the minimax applied to a game against nature; the statistician and economist Wald developed von Neumann's ideas with *a theory of estimates* based on the minimax.

Let us recall the criterion:

(a)   Choose the line which contains

$$\underset{i}{\mathrm{MAX}}[\underset{j}{\mathrm{MIN}}\ a_{ij}]$$

if the game possesses a point of equilibrium.

(b)   If there is no point of equilibrium, choose the mixed strategy $\{x_1, x_2, ..., x_m\}$, for which

$$\underset{j}{\mathrm{MIN}}[a_{1j}x_1 + a_{2j}x_2 + \cdots + a_{mj}x_m] \quad \text{is maximal.} \tag{26.3}$$

In the previous section we examined the use of this criterion and showed, in the example of the news agent, the drawbacks it presents.

C. HURWICZ'S CRITERION[2]

The example given in Table 26.1 again shows the very dubious result obtained by the minimax criterion. This game has a point of equilibrium, and in accordance with this criterion the player should choose line (1) in order to win at least 2.

TABLE 26.1

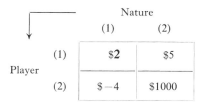

		Nature	
		(1)	(2)
Player	(1)	$2	$5
	(2)	$ −4	$1000

Would it not, however, be better to take the risk of losing $4 in the hope of winning $1000? Of course, if the player did not possess $4, or if its loss would be very inconvenient for him, he would perhaps be

---

[1] See references [J4] and [J15].

[2] L.Hurwicz, "Criterion for Decision-Making under Uncertainty," Cowles Commission Discussion Paper, *Statistics*, No. 355.

wise to choose line (1), but many people would take the risk of choosing line (2). In replacing the concept of value by that of "utility," which consists of weighting the values attached to the choices from a subjective standpoint, or of giving them an order, various difficulties can arise. As an indirect introduction to this concept of utility, Hurwicz has suggested the following criterion:

The optimism of the player is arbitrarily defined by a number $0 \leqslant \alpha \leqslant 1$. In each line (or weighting of lines) take the smallest element a_i, then the largest element A_i, and form the equation:

$$h_i = \alpha A_i + (1 - \alpha)a_i ; \qquad (26.4)$$

finally, choose line i corresponding to

$$\operatorname*{MAX}_i h_i .$$

For example, in Table 26.1 it is sufficient to take $\alpha = 1/100$, to have

line (1): $\quad h_1 = \dfrac{99}{100} \times (2) + \dfrac{1}{100} \times (5) = 2.03,$

line (2): $\quad h_2 = \dfrac{99}{100}(-4) + \dfrac{1}{100} \times (1000) = 5.94.$

Hence it is enough to take $\alpha = 1/100$ to obtain a decided preference for line (2).

To find the least coefficient of optimism which will make line (2) preferable to line (1), all that we need do is to solve the equation:

$$(1 - \alpha) \cdot 2 + \alpha \cdot 5 = (1 - \alpha) \cdot (-4) + \alpha \cdot 1000.$$

We find

$$\alpha = 6/1001 \approx 0.006,$$

which gives a very small lower bound for this optimism, so that very many people would consider it reasonable to bet on line (2).

For a further example, we will choose the one in Table 26.2 which has a point of equilibrium, and will provide a useful comparison between Hurwicz's criterion and that of the minimax (which corresponds to the case where $\alpha = 0$).

Taking the very optimistic value $\alpha = \frac{4}{5}$, it follows that

$$\begin{aligned}
h_1 &= (\tfrac{4}{5})(14) + (\tfrac{1}{5})(-4) = \tfrac{52}{5} \leftarrow \text{MAX}, \\
h_2 &= (\tfrac{4}{5})(11) + (\tfrac{1}{5})(-1) = \tfrac{43}{5} , \\
h_3 &= (\tfrac{4}{5})(7) \;\; + (\tfrac{1}{5})(2) = \tfrac{30}{5} .
\end{aligned} \qquad (26.5)$$

TABLE 26.2

	Nature		
	(1)	(2)	(3)

	(1)	(2)	(3)		
(1)	14	−6	−4	$A = 14,$	$a = -4$
(2)	2	11	−1	$A = 11,$	$a = -1$
(3)	4	7	2	$A = 7,$	$a = 2$

Now let us find the lowest value of α for which one line is preferable to the others. Line (1) will be preferable to line (2) if

$$14\alpha - 4(1 - \alpha) > 11\alpha - (1 - \alpha), \qquad a > \tfrac{1}{2}. \qquad (26.6)$$

Line (2) is preferable to line (3) if

$$11\alpha - (1 - \alpha) > 7\alpha + 2(1 - \alpha), \qquad \alpha > \tfrac{3}{7}. \qquad (26.7)$$

Line (3) is preferable to line (1) if

$$7\alpha + 2(1 - \alpha) > 14\alpha - 4(1 - \alpha), \qquad \alpha < \tfrac{6}{13}. \qquad (26.8)$$

After expressing the fractions with a common denominator, we find that it is possible to arrange the different values of α in order of preference.

Value of α	Preferences			Order of preferences
$0 \leqslant \alpha < \tfrac{3}{7}$	(2) preferred to (1) (3) (2) (3) (1)			(3), (2), (1)
$\tfrac{3}{7} < \alpha < \tfrac{6}{13}$	(2) preferred to (1) (2) (3) (3) (1)			(2), (3), (1)
$\tfrac{6}{13} < \alpha < \tfrac{1}{2}$	(2) preferred to (1) (2) (3) (1) (3)			(2), (1), (3)
$\tfrac{1}{2} < \alpha \leqslant 1$	(1) preferred to (2) (2) (3) (1) (3)			(1), (2), (3)

The values $\alpha = \tfrac{3}{7}$, $\tfrac{6}{13}$, and $\tfrac{1}{2}$ correspond to the indifference between the respective lines (2) and (3); (3) and (1); (1) and (2).

D. SAVAGE'S CRITERION[1]

Let us examine Savage's comparison between a decision made in ignorance of the state of nature and the one which would have been made if it had been known. Expressed differently, this means that we shall evaluate *the regret* or *failure to gain* resulting from this ignorance for each decision and state of nature. In this way we shall obtain a new game, which will be treated by the minimax criterion. Such is Savage's criterion, which can be expressed mathematically in Table 26.3.

TABLE 26.3

| | Nature | | | | Table of Regrets | | | |
	(1)	(2)	(3)		(1)	(2)	(3)	
(1)	4	0	2	(1)	−3	−1	−4	x_1
(2)	2	1	1	(2)	−5	0	−5	x_2
(3)	7	−1	6	(3)	0	−2	0	x_3
		(a)				(b)		

Taking a_{ij} as the elements of a given table, calculate a new table, the elements of which are

$$r_{ij} = a_{ij} - \underset{k}{\text{MAX}}\, a_{kj} \qquad (r_{ij} \leqslant 0), \tag{26.9}$$

choose line i, in such a way that we have

$$\underset{i}{\text{MAX}}(\underset{j}{\text{MIN}}\, r_{ij}). \tag{26.10}$$

If table r_{ij} does not have a point of equilibrium, choose

$$\{x_1, x_2, ..., x_n\}$$

in such a way that

$$\underset{j}{\text{MIN}}(r_{1j}x_1 + r_{2j}x_2 + \cdots + r_{mj}x_m) \quad \text{is maximal.} \tag{26.11}$$

Taking Table 26.3 as our example, we find the highest number in each column of Table 26.3a and subtract it from each element in its own column, which gives us Table 26.3b. Looking for an optimal strategy, we find that there is no point of equilibrium,[2] and we therefore seek

---

[1] L. J. Savage, "The Theory of Statistical Decision," *J. Amer. Statistical Assoc.* 238–248, 1947.

[2] For a point of equilibrium to be present, a line of zeros would be required.

a mixed optimal strategy. With this object, we first eliminate column (1) which dominates column (3), and it is now easy to find the solution of the remainder of the table. Our final solution is

$$x_1 = \tfrac{2}{5}, \qquad x_2 = 0, \qquad x_3 = \tfrac{3}{5}; \qquad v = -\tfrac{8}{5}. \tag{26.12}$$

By choosing this optimal strategy we can be certain that our regret will not exceed 8/5.

OTHER CRITERIA. It is possible to think of an infinitude of other criteria, but they would have to be mathematically and economically coherent. In this connection [J15] should be consulted.

27. The Theory of Statistical Decision[1]

A. THE THEORY OF STATISTICAL DECISION INTRODUCED INTO THE THEORY OF GAMES

Many problems of decision take the form of a game against nature in which the player is a statistician seeking information as to its behavior. The following cases are problems of this character:

(a)   Whether to accept or refuse a set of parts which may or may not be satisfactory in the knowledge of the detriment of accepting a bad or refusing a good one, as well as the cost of the parts control.

(b)   The study of a physical phenomenon to discover whether it follows a law L_1 or another L_2, when the cost of false conclusions is known, and also that of the various experiments required to establish the fact, as well as the order of magnitude of the experimental errors.

(c)   The evaluation of an unknown quantity m, when we know the cost of a false decision (e.g., the proportional cost of the square which separates the real quantity from the quantity chosen), as well as the degree of accuracy of the means of measurement.

(d)   To make experimental drillings with a view to deciding whether or not to invest capital in the hope of striking oil.

A mathematical theory for solving such problems was published some 20 years ago by Professor Abraham Wald, and a simplified resume of it is given here. Since the publication of his work, most of the theory of

---

[1] This subsection is a slightly modified text of part of "Aspects de la théorie statistique des décisions" by G. d'Herbemont in *Informations Scientifiques BULL*, and we wish to thank the author for permission to include it.

traditional statistics, applicable to tests of assumptions and to the estimation and evaluation of intervals of confidence, has been employed in accordance with his concept. Indeed, in a final analysis, all such problems can be regarded as defining the behavior of the player (the statistician) when confronted by nature.

We shall begin with a simple example.

Let us assume that the preferences of a group of customers can be shown in either of two ways:

(1)   A preference for the product P_1 (state E_1),

(2)   A preference for the product P_2 (state E_2).

Let us further assume that the manufacturer of this product has the choice between three possible decisions:

(1)   To continue to produce P_1,

(2)   To launch a new product P_2,

(3)   To cease production of both P_1 and P_2,

and that it is possible to evaluate the losses which would result from each of the decisions, though the behavior of nature (the customers) is unknown. By representing nature as B and "the decision maker" as A, we obtain the rectangular game shown in Table 27.1. In our initial explanation of Wald's theory we shall not give a precise definition of the criterion, leaving this important question until a little later.

TABLE 27.1

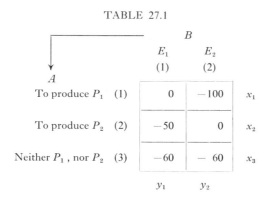

The customers have a choice between two pure strategies and the manufacturer has a choice of three such strategies; nevertheless, states E_1 and E_2 cannot be considered as conscious strategies on the part of the customers, and the latter can neither make a profit nor suffer a loss as a result of the manufacturer's decisions.

B. A Game against Nature with Espionage Introduced

Let us suppose that the manufacturer, whom we shall now call "the statistician," is able to gauge the state of nature by taking "samplings" and to use these in making his decisions. The game is now transformed into one with *unilateral spying*, and the pure strategies of the statistician are formed from the set of ways in which the results of his samplings can be associated with the final decisions from which he must choose, the cost of these samplings being taken into account when he weights the losses of these new strategies.

So as to make it simpler to understand the manner in which the statistician obtains his pure strategies, we shall represent P_1 by the color red (R) and P_2 by yellow (Y), while the state of nature will be represented by an urn containing:

for state E_1,   2 red balls and 1 yellow,
for state E_2,   1 red ball and 2 yellow.

The statistician is ignorant of the contents of the urn, which is placed beside him, but he can, if he wishes, obtain information about it by removing balls and noting their color before making his decisions.[1] At the same time, studies of the market and of future behavior, which are symbolized by these draws, entail expenditure, so that the removal of a ball requires a payment, which will be shown as a unit of money.

For his strategy, the statistician might decide to draw one ball, and if it is red, decide on E_1; if it is yellow on E_2. What average losses would such a strategy entail?

If the state is E_1, there is a $\frac{2}{3}$ probability of drawing a red ball and choosing P_1 (which is correct), the loss being confined to the cost of the draw (1). There is, however, a $\frac{1}{3}$ probability of drawing a yellow ball and of choosing P_2 (which is false), thereby losing $50 + 1 = 51$.

Faced by state E_1, this fourth pure strategy therefore results in an average loss:

$$(1)(\tfrac{2}{3}) + (51)(\tfrac{1}{3}) = 17.6 \tag{27.1}$$

In the same way, if the state is E_2, the average loss is:

$$(1)(\tfrac{2}{3}) + (101)(\tfrac{1}{3}) = 34.3. \tag{27.2}$$

Table 27.2 shows the new rectangular game obtained by adding this new pure strategy.

---

[1] The draw is assumed to be nonexhaustive, that is to say, that a ball which is taken out is immediately replaced, and the contents of the urn are then well shaken.

TABLE 27.2

	E_1 (1)	E_2 (2)
(1)	0	-100
(2)	-50	0
(3)	-60	-60
(4)	-17.6	-34.3

C. Multistage Strategy

A more elaborate pure strategy, a fifth for example, could be based on the principle of not making a decision until at least two balls have been drawn. If they are both red, P_1 will be chosen; if they are both yellow, P_2. There are obviously two further cases when one ball of each color is drawn (RY or YR), and since the statistician is then no wiser than before, it seems only logical that he should make further draws and only take a decision when both are the same color. A strategy of this kind, where the number of draws is not fixed in advance, is called "a multistage strategy," and the best possible strategy is often of this type.

Let us find the losses if the pure strategy (5) is adopted.

If the state is E_1, the probability of drawing RR is

$$\tfrac{2}{3} \cdot \tfrac{2}{3} = \tfrac{4}{9}$$

that of drawing YY is

$$\tfrac{1}{3} \cdot \tfrac{1}{3} = \tfrac{1}{9}$$

and that of drawing RY or YR is

$$\tfrac{2}{3} \cdot \tfrac{1}{3} + \tfrac{1}{3} \cdot \tfrac{2}{3} = \tfrac{4}{9}.$$

Hence, the probability of deciding P_1 at the first draw is then $\tfrac{4}{9}$, of deciding P_2 at the first draw is $\tfrac{1}{9}$, while that of a further draw[1] is $\tfrac{4}{9}$.

---

[1] We must remember that a ball which is drawn is immediately replaced in the urn. A draw consists of shaking the urn, removing a ball, and then replacing it.

These probabilities are for E_1, and we can verify that all the cases E_1 have been considered, since $\frac{4}{9} + \frac{1}{9} + \frac{4}{9} = 1$.

The fraction $\frac{4}{9}$ of the undefined cases can again be subdivided into three eventualities:

(a) To decide P_1 after the second draw; probability: $\frac{4}{9} \times \frac{4}{9}$

(b) To decide P_2 after the second draw; probability: $\frac{4}{9} \times \frac{1}{9}$

(c) Impossibility of a decision after the second draw;
     probability: $\frac{4}{9} \times \frac{4}{9}$

$$\text{Total:} \quad \frac{4}{9}$$

Progressively, it would be found in the same way if the state is E_1:

(a) Probability of choosing P_1 with the nth draw: $(\frac{4}{9})^n$

(b) Probability of choosing P_2 with the nth draw: $(\frac{4}{9})^{n-1} \cdot (\frac{1}{9})$

(e) Probability of no decision after the nth draw: $(\frac{4}{9})^{n-1}$

Total probability of no decision after $(n-1)$ draws: $(\frac{4}{9})^{n-1}$

The probability of deciding P_1 at any given moment, if the state is E_1, is therefore the sum of the geometrical progression:

$$\frac{4}{9} + (\tfrac{4}{9})^2 + \cdots + (\tfrac{4}{9})^n + \cdots = \frac{\frac{4}{9}}{1 - \frac{4}{9}} = \frac{4}{5} = 0.8. \qquad (27.3)$$

In the same way, it can be seen that the probability of choosing P_2 if the state is E_1 is 0.2. The probability of there being no decision is the limit of $(\frac{4}{9})^n$ for $n \to \infty$, that is, 0. In practice, therefore, the decision would be made after a finite number of draws.

The probability of there not being a decision after the 10th draw is only $(\frac{4}{9})^{10} \approx 0.0003$, which means that it would only occur on average 3 times out of 10,000.

Finally, if a decision (P_1 or P_2) is made after n draws, the samplings cost $2n$ with a probability

$$(\tfrac{4}{9})^n + (\tfrac{4}{9})^{n-1} \times \tfrac{1}{9} = \tfrac{5}{4}(\tfrac{2}{3})^{2n}. \qquad (27.4)$$

The average loss due to the samplings is therefore:

$$\tfrac{5}{4}[2 \times (\tfrac{2}{3})^2 + 4 \times (\tfrac{2}{3})^4 + 6(\tfrac{2}{3})^6 + \cdots] = 3.6. \qquad (27.5)$$

For the final decision (0 with probability 0.8 and 50 with probability 0.2), the loss is

$$(0.8)(0) + (0.2)(50) = 10. \qquad (27.6)$$

Hence, we find that the average loss for strategy (5), if the state is E_1, has the value:

$$w_{E_1} = 3.6 + 10 = 13.6. \tag{27.7}$$

In the same way, if the state is E_2, we should find:

$$w_{E_2} = 23.6.$$

We then obtain the rectangular game of Table 27.3.

TABLE 27.3

	E_1 (1)	E_2 (2)
(1)	0	100
(2)	−50	0
(3)	−60	−60
(4)	−17.6	−34.3
(5)	−13.6	−23.6

Using the same method, it would be possible for any strategy S of the statistician, to evaluate the losses $w_{E_1}(S)$ and $w_{E_2}(S)$.

D. ADMISSIBLE STRATEGIES

The task of enumerating all the pure strategies, before deciding between them, may well seem a hopeless one, since they are infinite in number. Hence, it is essential to eliminate as many of them as possible in order to reduce the dimensions of the problem.

We shall apply the term *admissible pure strategy* to one which is not dominated by another (see Section 23). Hence, strategy (3) is not acceptable because it is dominated by strategy (2).

$$[-50 \quad 0] > [-60 \quad -60];$$

in the same way (5) dominates (4):

$$[-13.6 \quad -23.6] > [-17.6 \quad -34.3].$$

There are, however, two games which do not have any admissible strategy. Hence, in our example, there would be no acceptable strategies if the draw were free: there would always be a better strategy as the draws became more numerous.

A class of strategies is said to be *complete* if every strategy outside the class can be opposed by at least one strategy of the class which dominates it. There is always one complete class, the one comprising all the strategies, and the allowable strategies must necessarily be included in the complete class.

After the elimination of strategies (3) and (4) the rectangular game is reduced to that of Table 27.4. The game does not have a point of

TABLE 27.4

	E_1 (1)	E_2 (2)	
(1)	0	−100	x_1
(2)	−50	0	x_2
(5)	−13.6	−23.6	x_3
	y_1	y_2	

equilibrium, and to study the kind of mixed strategies which will be of value to us, we shall make use of the representation of the problem shown in Fig. 27.1.

Each of the N states of nature is represented by a rectilinear coordinate axis which can be orthogonal or otherwise. Each strategy of the statisti-

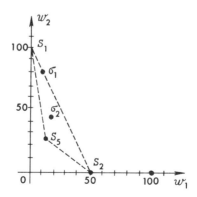

FIG. 27.1.

cian is then shown by a point, the coordinate of which on the i axis is the loss w_i corresponding to the state of nature i. As $N = 2$ in the present problem, there will be two axes, Ow_1 and Ow_2. Hence, strategy (1) (which we shall call S_1) of the game of Table 27.1 has the coordinates (0, 100). For strategy (2) or (S_2) they are (50, 0), and for S_5 (13.6, 23.6).

Let us now consider a mixed strategy of the statistician: $x_1 = \frac{4}{5}$, $x_2 = \frac{1}{5}$, $x_3 = 0$, which has corresponding values $w_1 = 10$, $w_2 = 80$. The corresponding point σ_1 is situated on the straight line connecting (50, 0) and (0, 100), and is the center of gravity of the masses $\frac{4}{5}$ and $\frac{1}{5}$ situated, respectively, at S_1 and S_2. If we take another mixed strategy $x_1 = \frac{2}{6}$, $x_2 = \frac{1}{6}$, and $x_3 = \frac{3}{6}$ with corresponding values $w_1 = 15.1$ and $w_2 = 45.1$, the corresponding point is inside the polygon $S_1S_2S_5S_1$, and at the center of gravity of the masses $\frac{2}{6}$, $\frac{1}{6}$, $\frac{3}{6}$ situated respectively at S_1, S_2, and S_5. This property is a general one: for every mixed strategy containing pure or mixed strategies there is a corresponding point in the *referential*[1] w_1Ow_2, which is inside or on the perimeter of the polygon of sustentation[2] formed by the pure strategies which are being considered in the game by the adversary of nature.

Still referring to the present example, we at once notice that all the strategies situated in the hatched portion of the first quadrant of Fig. 27.2 are unacceptable. Indeed, every mixed strategy[3] such as σ is dominated by at least one strategy σ' situated on the broken line $S_1S_5S_2$ (a strategy σ' dominates a strategy σ, if the segment from σ' towards σ will cut the first quadrant of the axes, with understood limits; that is to say, in this example, when the components of the directed segment are both positive).

The line $S_1S_5S_2$ does not alone constitute a complete class, but it will do so if the strategies (as yet unknown) situated to the side nearer the point of origin are added to it (Fig. 27.2).

The statistician, spy, and adversary of nature, may equally employ mixed strategies (it rains p days out of q, 3% of the parts inspected by the parts control are bad, etc.).

In our example, nature may sometimes reveal state E_1 and sometimes E_2 with respective probabilities of p_{E_1} and p_{E_2}:

$$p_{E_1} + p_{E_2} = 1.$$

[1] The w_1 values are losses shown *negatively* on the matrix of the rectangular game (27.1 to 27.4), and *positively* on the referentials w_1Ow_2 (Figs. 27.1 and 27.2).

[2] In the case of a space with n dimensions (n possible states of nature) this property is enunciated as follows: "Every mixed strategy is situated inside or on the perimeter of the convex polyhedron which contains the set of strategies."

[3] In what follows, σ will stand for a *pure or mixed strategy*, S for a *pure strategy*.

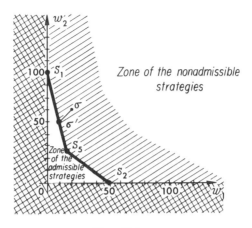

FIG. 27.2.

Whether or not it is realistic to make such an assumption about the preferences and changing taste of our customers, is of small importance, as long as we give p_{E_1} an arbitrary value. Indeed we can always choose p_{E_1} as a parameter which enables us to represent state E_1 ($p_{E_1} = 1$ being the certitude of E_1) or state E_2 ($p_{E_1} = 0$ being the certitude of E_2), as well as intermediate cases.

A mixed strategy of this kind can be shown shown graphically by a straight line D passing through O and with a slope:

$$r = p_{E_2}/p_{E_1} . \tag{27.8}$$

This straight line may coincide with one of the axes and will then represent a pure state E_1 or E_2.

The game is then transformed in the following way. Acting independently of each other, nature chooses a straight line D and the statistician a point from among the figurative set of his pure or mixed strategies (Fig. 27.3).

It can then be shown that the loss resulting from the pair (D, σ) [in other words, from the choice of D by nature and of σ by the statistician, which will be called $w_D(\sigma)$] can be obtained by constructing the perpendicular from σ to D, and finding its intersection with the bisector Od. By so doing, we determine point 1, the coordinates of which will be called w. Let us prove that

$$w_D(\sigma) = w. \tag{27.9}$$

This rule will obviously apply in particular when D coincides with one of the axes.

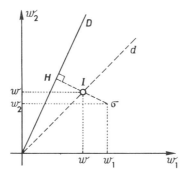

FIG. 27.3.

Indeed, if σ has w_1 (loss against state E_1) and w_2 (loss against state E_2) as its coordinates, the average loss against D is obtained by weighting these two losses by their probabilities

$$w_D = p_{E_1} w_1 + p_{E_2} w_2 . \qquad (27.10)$$

This expression is the *scalar product* of the vector $\mathbf{O\sigma}$ by the vector $\mathbf{p}$ of coordinates p_{E_1} and p_{E_2} and which is carried by D. So

$$w_D = \mathbf{O\sigma \cdot p} = \mathbf{OH \cdot p} = \mathbf{OI \cdot p}$$
$$= w \cdot p_{E_1} + w \cdot p_{E_2} = w(p_{E_1} + p_{E_2}) = w. \qquad (27.11)$$

This construction shows that all strategies σ situated on the straight line σH have an equal loss against D. This loss is the weaker because point H is nearer to the origin.

E. BAYES'S STRATEGY

Thomas Bayes was an English clergyman of the 18th century whose work *An Essay Toward Solving a Problem in the Doctrine of Chance* posthumously published in 1763, introduced for the first time the concept of *a priori* and *a posteriori* probabilities.

At that period, of course, nothing was known about the theory of games of strategy nor of the theory of statistics, but, as we shall discover, the use of Bayes's name is fully justified for describing a whole class of strategies which are of basic importance.

Let us assume the problem solved, and consider the set of strategies (pure or mixed) of the statistician. The representative points of these strategies form a "convex set," that is to say, a set which, if it contains two points and σ_1 and σ_2, also contains all the points of the segment (σ_1 , σ_2) (Figs. 27.4 and 27.5).

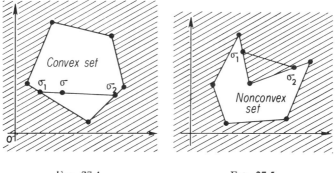

FIG. 27.4.   FIG. 27.5.

Every point in this segment can, indeed, be found by weighting strategies σ_1 and σ_2, as we discovered earlier.

We shall agree that this convex set contains its boundary. Let us now consider that portion of this boundary formed by those of its points which are lowest for every value of w_1 compatible with the points of the set or those furthest to the left for each value of w_2 compatible with the points of this set (the part $S_a S_b S_c S_e$ shown as a heavy line in Fig. 27.6). We can then conclude that every strategy σ situated outside this line can be opposed by a strategy σ' which dominates it. Elsewhere there is no strategy dominating $S_a S_b S_c S_e$ and this line therefore represents a complete class of admissible strategies.

Let there now be a straight line D representing a mixed strategy of nature (y_1, y_2), and let us now find those among the statistician's

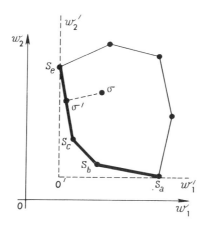

FIG. 27.6.

strategies σ, which result in a minimal loss against D. For such cases, H must be as near as possible to O, as we observed earlier. It will be seen in Fig. 27.7 that at least one solution must be obtained if the convex set includes its boundary.

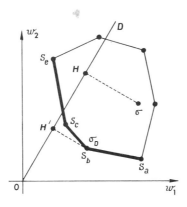

FIG. 27.7.

The strategies such as σD which produce a minimal loss against D are called *Bayes's strategies against D*.

It will be observed that by varying the straight line D we can obtain all the acceptable strategies (line $S_a S_b S_c S_e$), and reciprocally. Some important special cases may arise if D is parallel to an axis or if some of the segments of the perimeter are parallel to one or both of the axes of reference.

The problem of finding the admissible strategies is therefore a matter of discovering Bayes's strategies against the various *a priori* probabilities which can be given to the states of nature.

In some problems, nature may be safely assumed to behave according to a probable "scheme" and Bayes's corresponding strategy can then be applied. In others, the concept of *a priori* probability is only used to describe the set of acceptable strategies.

We will now return to our example and will show the form taken by Bayes's strategies. The type of reasoning which we shall use could be adapted to most multistage statistical games containing two states of nature, two final decisions, and a cost for sampling proportional to the number of draws.

It must be stated at the outset that the strategies S_1 and S_2 of the statistician are necessarily admissible. In this connection, it seems desirable to draw attention to the fact that when a term (in this case admissible) has been given an exact mathematical connotation it can be disastrous to interpret it by its popular meaning: what inspection department would regard it as admissible to make its decisions without any examination of the items which it is offered?

Nevertheless, each draw, whatever its result, increases all the elements of the matrix by unity. It follows that the loss against E_1, as well as against E_2, can never be less than 1. None of the strategies such as $w_1 \geqslant 1$ and $w_2 \geqslant 1$ can dominate either S_1 or S_2, since the strategies which can dominate any strategy σ must be contained in the rectangle formed by the straight lines $w_1 = 0$, $w_1 = w_{1\sigma}$; $w_2 = 0$, $w_2 = w_{2\sigma}$, where $w_{1\sigma}$ and $w_{2\sigma}$ are the projections of point σ on Ow_1 and Ow_2. S_1 and S_2 are, therefore, acceptable in a mathematical sense.

Let us now consider the convex set of strategies *with a draw*. This set contains S_6, which is on the side nearer the origin in relation to $S_1 S_2$. Hence, there are two straight lines $S_1\sigma'$ and $S_2\sigma''$ which touch the outline of the strategies containing a draw at σ' and σ''.

Let us now draw the perpendiculars D' and D'' from O to these two straight lines. The gradient of D' is r' and that of D'' is r''.

Given a straight line D of slope $r > r''$, it can be seen that Bayes's strategy is S_1; for a straight line D with gradient $r < r'$, Bayes's strategy is S_2. Lastly, Fig. 27.8 shows that it is only when

$$r' < r < r'' \qquad (27.12)$$

that at least one draw is required. In the restricted cases such as $r = r'$, we may use either S_1 or σ', or their combination, since the loss against D will be the same.

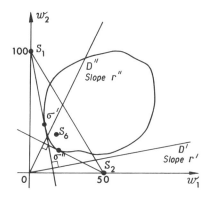

FIG. 27.8.

Hence, the rule to be followed is

$$\text{if } r \leqslant r' \text{ or } r \geqslant r'' \text{ decide without a draw in favor of}$$
$$S_1 \text{ or } S_2, \text{ according to the case;} \qquad (27.13)$$

$$\text{if } r' < r < r'' \text{ make at least one draw.} \qquad (27.14)$$

All we now have left is to decide how to proceed in the last case. By making one draw, the color X of the ball (where X may be either red or yellow) gives us information about the state of nature. Let $\pi_{E_1}(X)$ represent the probability of drawing X when the state is E_1, and $\pi_{E_2}(X)$ the probability when it is E_2.

In this case, we have

$$\pi_{E_1}(X)\,(\text{red}) = \tfrac{2}{3}, \qquad \pi_{E_1}(X)\,(\text{yellow}) = \tfrac{1}{3};$$
$$\pi_{E_2}(X)\,(\text{red}) = \tfrac{1}{3}, \qquad \pi_{E_2}(X)\,(\text{yellow}) = \tfrac{2}{3}. \qquad (27.15)$$

The probability p_{E_1} of state E_1 is modified by the knowledge of X and becomes p_{E_1}/X, and similarly p_{E_2} becomes p_{E_2}/X. By calculating the probabilities, it is found that

$$p_{E_1/X} = \frac{p_{E_1} \cdot \pi_{E_1}(X)}{p_{E_1} \cdot \pi_{E_1}(X) + p_{E_2} \cdot \pi_{E_2}(X)}, \qquad (27.16)$$

$$p_{E_2/X} = \frac{p_{E_2} \cdot \pi_{E_2}(X)}{p_{E_1} \cdot \pi_{E_1}(X) + p_{E_2} \cdot \pi_{E_2}(X)}. \qquad (27.17)$$

Hence, the knowledge of X leaves us in the same position as before the draw, except that the quotient of probability of the red state has become

$$r_1 = \frac{p_{E_2/X}}{p_{E_1/X}} = r\,\frac{\pi_{E_2}(X)}{\pi_{E_1}(X)} = rY_1 \qquad (27.18)$$

by assuming

$$Y_1 = \frac{\pi_{E_2}(X)}{\pi_{E_1}(X)}. \qquad (27.19)$$

From this we can deduce that Bayes's strategy consists in applying the rule already laid down but replacing r by r_1 (r' and r'' remain unchanged):

$$\text{if } r_1 \leqslant r' \qquad \text{decide } S_1;$$
$$\text{if } r_1 \geqslant r'' \qquad \text{decide } S_2; \qquad (27.20)$$
$$\text{if } r' < r < r'' \quad \text{make at least one draw.}$$

By progressive stages it will be found that Bayes's strategy against D with slope r is: if n draws have been made without a decision, so that

$$r' < r_n < r'', \qquad (27.21)$$

draw again and observe X_{n+1}. Calculate

$$Y_{n+1} = \frac{\pi_{E_2}(X_{n+1})}{\pi_{E_1}(X_{n+1})} \qquad (27.22)$$

and

$$r_{n+1} = r_n Y_{n+1} ; \qquad (27.23)$$

if

$$
\begin{aligned}
&r_{n+1} \leqslant r' && \text{decide } P_1 , \\
&r_{n+1} \geqslant r'' && \text{decide } P_2 , \\
&r' < r_{n+1} < r'' && \text{do not decide and continue the process.} \quad (27.24)
\end{aligned}
$$

Hence, it can be seen that all Bayes's strategies can be enunciated as soon as r' and r'' are found, so that the problem is reduced to finding these two unknowns.

In the general case, the latter depend on the matrix of loss and on the form of the functions $\pi_{E_1}(X)$ and $\pi_{E_2}(X)$, the values of X not being limited to two. In addition, there are cases when the convex set of the strategies with a draw is situated above and to the left of the straight line $S_2\sigma$, and no draw is then admissible; whatever the value of r, the decision must be made without a draw.

In Part II, Section 73, readers with a more advanced knowledge of mathematics will find various methods for calculating r' and r''. For the moment, let us discover whether the statistician would find it useful to employ a strategy founded on the minimax criterion.

F. USING THE MINIMAX CRITERION

In the above example we have discovered a complete class of acceptable strategies, and it would obviously be absurd to choose a strategy outside it, since there would always be a dominating one inside. But none of the admissible strategies dominates another, and the question arises as to the criterion which should be used in making a choice.

If it is possible to attribute an *a priori* probability to the states of nature the answer is simple: we should apply the strategy (or equivalent strategies) of Bayes against D which gives the least loss. But to know D, that is to say, r, amounts to knowing the mixed strategy of the opponent (in this case nature), and while this may be possible when the opponent is an indifferent player, we must be prepared to face a situation where r is unknown.

According to the theory of games of strategy our course would be to use the minimax criterion, but according to the statistical theory of decision this would provide a dubious result: there is generally no *a priori* reason why nature should adopt the same behavior as an intelligent player who is the statistician's opponent.

If σ represents a strategy of the statistician, and w_1 and w_2 his respective losses against E_1 and E_2, the greatest loss against any possible strategy (pure or mixed) of nature is

$$w_M = \max(w_1, w_2). \tag{27.25}$$

As we know, the use of a minimax strategy results in a minimal loss for the player against any possible strategy of his opponent, whether or not the latter is intelligent. If the statistician has found the class containing all his admissible strategies it will then be easy by the use of graphic methods to discover the minimax strategy or strategies.

Figure 27.9 shows that the lines $\max(w_1, w_2) = Cte$ consist of two straight lines which meet at right angles at the point M on the bisector. Hence, the minimax strategies will be those when M has the lowest possible position. In Fig. 27.10, the point which represents the minimax strategy is σ_{mM}, the intersection of the bisector with the line of acceptable strategies. This point can represent several equivalent strategies, and in certain cases there may even be minimax strategies which are not on the bisector.

In Part II we have shown how the minimax strategy in this example can be calculated. These calculations are not simple, and they would be

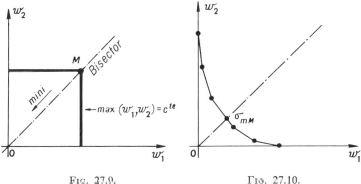

Fig. 27.9.   Fig. 27.10.

still more complicated if more than two states of nature were taken into account. But on the basis of this theory we are able to enunciate rules of rational behavior for decision making, whenever statistical multistage "spying" on nature is possible.

28. Multistage Games with Two Players

A. THE CASE WHERE NEITHER ADVERSARY HAS ANY INFORMATION

The game to be considered has two players and three stages.[1] Each of the two players, A and B, has the choice of a color: red (R) or green (G) in each of the three stages.

Stage 1: A chooses R or G;
Stage 2: B chooses R or G;
Stage 3: A chooses R or G.

At the end of Stage 3, B pays A a loss corresponding to the function $M(x, y, z)$ where x is A's choice in Stage 1, y is B's choice in 2, and z is A's choice in 3.

$$M(R, R, R) = -2, \qquad M(G, R, R) = 5,$$
$$M(R, R, G) = -1, \qquad M(G, R, G) = 2,$$
$$M(R, G, R) = 3, \qquad M(G, G, R) = 2, \tag{28.1}$$
$$M(R, G, G) = -4, \qquad M(G, G, G) = 6.$$

Let us assume that B has no knowledge of A's choice, either in Stage 1 or 3, and that A is ignorant of B's choice in Stage 2. This is the same as

TABLE 28.1

		B		
		R	G	
	RR	−2	3	x_1
	RG	−1	−4	x_2
A	GR	5	2	x_3
	GG	2	6	x_4
		y_1	y_2	

[1] We have used an example from Reference [J16].

if A and B play simultaneously, and we can now obtain the rectangular game of Table 28.1 which is without a point of equilibrium, so that the solution of the game is

$$x_1 = 0, \qquad x_2 = 0, \qquad x_3 = \tfrac{4}{7}, \qquad x_4 = \tfrac{3}{7};$$
$$y_1 = \tfrac{4}{7}, \qquad y_2 = \tfrac{3}{7}; \qquad\qquad\qquad (28.2)$$
$$g = \tfrac{26}{7}.$$

B. CASE WHERE PARTIAL INFORMATION IS AVAILABLE

We will now modify the rules of the game in the following manner: A plays first (Stage 1), then B plays (Stage 2) after noting A's choice; finally, A plays again (Stage 3) after noting B's choice in the previous stage, but without taking account of his own play in Stage 1. (Let us imagine that A is a team of two players A_1 and A_2 who play in turn and are associated financially.)

We will represent B's strategies as follows:

f_{11}, if A has chosen R then B will choose R
f_{11}, if A has chosen G then B will choose R

f_{12}, if A has chosen R then B will choose R
f_{12}, if A has chosen G then B will choose G

(28.3)

f_{21}, if A has chosen R then B will choose G
f_{21}, if A has chosen G then B will choose R

f_{22}, if A has chosen R then B will choose G
f_{22}, if A has chosen G then B will choose G

A's strategies will be represented with a different notation:

[1, 11] means: A chooses R in Stage 1; and in Stage 3, A chooses R if B has chosen R, and R if B has chosen G.

[1, 12] means: A chooses R in Stage 1; and in Stage 3, A chooses R if B has chosen R, and G if B has chosen G.

(28.4)

[1, 21] means: A chooses R in Stage 1; and in Stage 3, A chooses G
   ⋮   if B has chosen R, and R if B has chosen G.

[2, 12] means: A chooses G in Stage 1; and in Stage 3, A chooses R if B has chosen R, and G if B has chosen G.

$[i_0, i_1, i_2]$

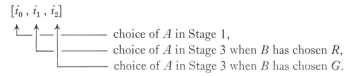

choice of A in Stage 1,
choice of A in Stage 3 when B has chosen R,
choice of A in Stage 3 when B has chosen G.

TABLE 28.2

		f_{11}	f_{12}	f_{21}	f_{22}
				B	
	[1, 11]	-2	-2	3	3
	[1, 12]	-2	-2	-4	-4
	[1, 21]	-1	-1	3	3
A	[1, 22]	-1	-1	-4	-4
	[2, 11]	5	2	5	2
	[2, 12]	5	6	5	6
	[2, 21]	2	2	2	2
	[2, 22]	2	6	2	6

How are we to establish the table for this new game (Table 28.2)? As a beginning, what number should we place at the intersection of line 1 and column 1? The symbol [1, 11] shows that A has chosen R in Stage 1, f_{11} that B has chosen R, and [1, 11] that A has chosen R in Stage 3. Hence, we have

$$a_{11} = M(R, R, R) = -2. \tag{28.5}$$

Turning to a_{21}, the symbol [1, 12] indicates that A has chosen R in Stage 1, f_{11} that B has chosen R, and [1, 12] that A has chosen R in Stage 3. We find that

$$a_{21} = M(R, R, R) = -2. \tag{28.6}$$

For a_{31}, the symbol [1, 21] indicates that A has chosen R in Stage 1, f_{11} that B has chosen R, and [1, 21] that A has chosen G in Stage 3. Hence, we find

$$a_{31} = M(R, R, G) = -1.$$

For a_{53}, [2, 11] shows that A has chosen G in Stage 1, f_{21} that B has chosen R, and [2, 11] that A has chosen R in Stage 3. It follows that

$$a_{53} = M(G, R, R) = 5.$$

In this way we have obtained the game of Table 28.2, which contains two *saddle points* $a_{61} = 5$ and $a_{63} = 5$. Hence, the optimal strategy for A is [2, 12], and for B it is f_{11} or f_{21} or any mixed strategy formed from these two pure strategies.

In multistage games, the way in which information is defined is obviously very important, in view of the role which it assumes, and we shall now examine this question with the help of a graphic representation.

C. REPRESENTATION OF A MULTISTAGE GAME AS AN ARBORESCENCE

The graph in Fig. (28.1) represents the previous multistage game, and the successive choices are shown as the arcs of a tree. A "history" of a game is represented by a "path" leading from Stage 1 to the payment. Thus, in the game we are considering, (R, R, G), (G, G, R), and (G, G, G) are histories whose paths can be followed in the tree of Fig. 28.1. To each history a value is assigned, as in (28.1).

It is hardly necessary to point out that the graphic representation of a multistage game may become very complicated if a number of stages or choices are involved, so that in such cases the representation is of

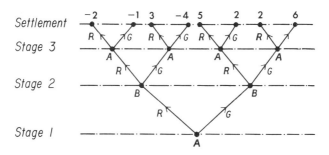

FIG. 28.1.

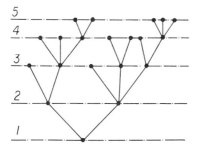

FIG. 28.2.

pedagogic interest only. Its main value, as we shall see, is in bringing to our notice various properties of the information which the players are allowed to obtain, depending on the rules of the game.

It should be noted that certain games are represented by trees, the extremities of which do not necessarily end at the last stage. This means that the number of stages in the game can vary according to the decisions which are taken (see Fig. 28.2).

D. THE TYPE OF INFORMATION AVAILABLE IN A MULTISTAGE GAME

To show the information which the players possess, a closed contour is drawn containing all those vertices (representing the position of the players) between which an ignorance of position exists. We shall now show how this representation may be utilized, with the aid of our previous example.

Example 1 (Fig. 28.3). In Stage 2, B knows A's choice; in Stage 3, A is unaware of B's choice and has forgotten his own choice[1] in Stage 1. The multistage game of Fig. 28.3 is defined by the rectangular game of Table 28.2.

Example 2 (Fig. 28.4). In Stage 2, B knows A's choice; in Stage 3, A is unaware of B's choice but remembers his own choice in Stage 1.

Example 3 (Fig. 28.5). In Stage 2, B is ignorant of A's choice; in Stage 3, A knows B's choice and recalls his own choice in Stage 1.

Example 4 (Fig. 28.6). In Stage 2, B does not know A's choice; in Stage 3, A knows B's choice, but no longer recalls his choice in Stage 1.

Example 5 (Fig. 28.7). In Stage 2, B is aware of A's choice; in Stage 3, A knows B's choice, but has forgotten his choice in Stage 1.

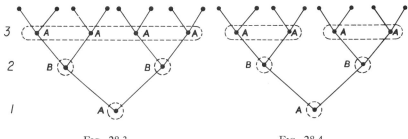

FIG. 28.3.   FIG. 28.4.

[1] Player A would in practice consist of a team (A_1, A_2) in which A_2 would be ignorant of his partner's choice, as stated for the game in Table 28.2.

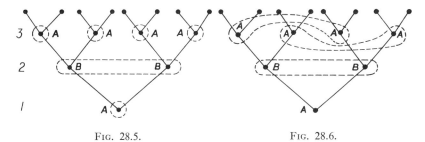

FIG. 28.5.   FIG. 28.6.

Example 6 (Fig. 28.8).   In Stage 2, *B* knows *A*'s choice; in Stage 3, *A* is aware of *B*'s choice and remembers his own in Stage 1 (a game with complete information).

Example 7 (Fig. 28.9).   Here we are concerned with a different type of game. In Stage 1, *A* chooses *R* or *G*. He is opposed by two partners B_1 and B_2 who play in turn, but are ignorant of the number of their stage.

The last of these examples might take the form of three players *A*, B_1, and B_2, isolated from each other in different rooms, with B_1 and B_2 as partners who constitute player *B*. In Stage 1, an usher goes to *A*'s room and asks him to choose red or green (*R* or *G*). If *A* chooses *R*, the usher goes to B_1 and asks him to choose *R* or *G*; but if *A* has chosen *G*, the usher visits B_2 and asks for his choice. When this stage is completed,

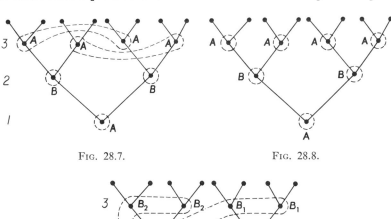

FIG. 28.7.   FIG. 28.8.

FIG. 28.9.

the usher goes to the player of the B team who has not yet made a choice and asks him which color he will choose. When the three choices have been made, the game is settled by a function such as that of (28.1).

Nearly all games played for recreation are multistage: chess, checkers (games into which probability does not enter), backgammon, game of goose, card games (where probability enters). To define an overall strategy for games of this kind is almost impossible, for it would require tables with an astronomical number of lines and/or columns. Thus, in a game of chess it would be virtually impossible to describe strategies for more than three moves at the end of the game.

E. ENTRANCE OF PROBABILITY. VARIOUS EXAMPLES

It is easy to think of multistage games, such as bridge and poker (with two players) into which probability enters: indeed it might be considered as a third player whose intentions are known in terms of probability. Probability can intervene in many different ways and in particular:

(1)   By affecting the function representing the settlement;
(2)   By affecting the number and/or nature of the elements of the sets from which the choices are made;
(3)   By deciding the order in which the choices are made;
(4)   In a complex manner by acting separately on the preceding elements.

Example 1. Probability affects the function representing the settlement.

We will use the game shown in the graph of Fig. 28.3, but will now assume that the payments at the end of the game are decided by the use of an urn containing, for example, 1 black and 3 white balls. If a black ball is drawn, payment will be made according to the function $M_1(x, y, z)$; if a white ball is drawn, according to the function $M_2(x, y, z)$. Let us assume that the functions are the following, where x, y, z represent the decisions of the players who have a choice of red (R) or green (G):

$$
\begin{array}{lll}
M_1(R, R, R) = 2, & M_1(R, R, G) = -3, & M_1(R, G, R) = 0, \\
 & M_1(R, G, G) = -1, & \\
M_1(G, R, R) = -3, & M_1(G, R, G) = 2, & M_1(G, G, R) = 2, \\
 & M_1(G, G, G) = 4, & \\
 & & (28.7) \\
M_2(R, R, R) = 1, & M_2(R, R, G) = -2, & M_2(R, G, R) = 1, \\
 & M_2(R, G, G) = 4, & \\
M_2(G, R, R) = -2, & M_2(G, R, G) = 2, & M_2(R, G, R) = 2, \\
 & M_2(G, G, G) = -3. &
\end{array}
$$

TABLE 28.3

Payment According to M_1

	$f^{(1)}_{22}$	$f^{(1)}_{21}$	$f^{(1)}_{12}$	$f^{(1)}_{11}$
$[1, 11]^{(1)}$	2	2	0	0
$[1, 12]^{(1)}$	2	2	−1	−1
$[1, 21]^{(1)}$	−3	−3	0	0
$[1, 22]^{(1)}$	−3	−3	−1	−1
$[2, 11]^{(1)}$	−3	2	−3	2
$[2, 12]^{(1)}$	−3	4	−3	4
$[2, 21]^{(1)}$	2	2	2	2
$[2, 22]^{(1)}$	2	4	2	4

TABLE 28.4

Payment According to M_2

	$f^{(2)}_{11}$	$f^{(2)}_{12}$	$f^{(2)}_{21}$	$f^{(2)}_{22}$
$[1, 11]^{(2)}$	1	1	1	1
$[1, 12]^{(2)}$	1	1	4	3
$[1, 21]^{(2)}$	−2	−2	1	1
$[1, 22]^{(2)}$	−2	−2	4	4
$[2, 11]^{(2)}$	−2	2	−2	2
$[2, 12]^{(2)}$	−2	−3	−2	−3
$[2, 21]^{(2)}$	2	2	2	2
$[2, 22]^{(2)}$	2	−3	2	−3

The symbols used to represent the strategies of A and B will be those previously employed in (28.3) and (28.4). We obtain the two rectangular games shown in Tables 28.3 and 28.4.

In a game where probability enters in this way, both players can agree that the expected value of the payments will produce the new game shown in Table 28.5, which is obtained by considering the function of payment:

$$M(x, y, z) = \tfrac{1}{4}M_1(x, y, z) + \tfrac{3}{4}M_2(x, y, z), \qquad (28.8)$$

which also gives a game with two points of equilibrium.

More often, if there are m functions $M_i(x, y, z, t,...)$, where $i = 1, 2,..., m$, in a multistage game, and if every function M_i has a probability r_i of being drawn, the game to be considered will have a function of value:

$$M(x, y, z, t,...) = E(M_i) = \sum_{i=1}^{m} M_i(x, y, z, t,...)r_i . \qquad (28.9)$$

Example 2. Probability affects the nature of the choices.

TABLE 28.5

Payment According to $M = \frac{1}{4} M_1 + \frac{3}{4} M_2$

		B			
		f_{11}	f_{12}	f_{21}	f_{22}
	[1, 11]	$\frac{5}{4}$	$\frac{5}{4}$	$\frac{3}{4}$	$\frac{3}{4}$
	[1, 12]	$\frac{5}{4}$	$\frac{5}{4}$	$\frac{11}{4}$	$\frac{11}{4}$
	[1, 21]	$-\frac{9}{4}$	$-\frac{9}{4}$	$\frac{3}{4}$	$\frac{3}{4}$
	[1, 22]	$-\frac{9}{4}$	$-\frac{9}{4}$	$\frac{11}{4}$	$\frac{11}{4}$
A	[2, 11]	$-\frac{9}{4}$	2	$-\frac{9}{4}$	2
	[2, 12]	$-\frac{9}{4}$	$-\frac{13}{4}$	$-\frac{9}{4}$	$-\frac{13}{4}$
	[2, 21]	2	2	2	2
	[2, 22]	2	$-\frac{13}{4}$	2	$-\frac{13}{4}$

Let us consider the multistage game of Fig. 28.10 where probability
enters in Stage 1 in the following manner: four balls are placed in an urn,
two marked with the number 1, one marked with 2, and one marked with
3; this stage is completed by removing a ball. In Stage 2, player A,
who has been told the result of the first draw, makes a choice of a color,
red (R) or green (G). In Stage 3, B, who has been told the result of A's

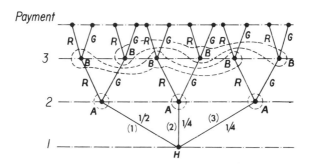

FIG. 28.10.

draw, but not of the initial one, chooses R or G. We shall assume the function of payment as:

$$M(1, R, R) = \quad 3, \qquad M(1, R, G) = \quad 2, \qquad M(1, G, R) = -1,$$
$$M(1, G, G) = \quad 4,$$

$$M(2, R, R) = \quad 5, \qquad M(2, R, G) = -3, \qquad M(2, G, R) = \quad 3, \quad (28.10)$$
$$M(2, G, G) = -2,$$

$$M(3, R, R) = -1, \qquad M(3, R, G) = \quad 2, \qquad M(3, G, R) = -3,$$
$$M(3, G, G) = \quad 5.$$

A strategy of A will be defined by the function f_{ijk} $(i, j, k = 1, 2)$:

f_{111}: A chooses R if the draw gives 1,
  A chooses R if the draw gives 2,
  A chooses R if the draw gives 3;

f_{112}: A chooses R if the draw gives 1,
  A chooses R if the draw gives 2,
  A chooses G if the draw gives 3;

f_{121}: A chooses R if the draw gives 1,
  A chooses G if the draw gives 2,
  A chooses R if the draw gives 3;

$$\vdots$$

f_{222}: A chooses G if the draw gives 1,
  A chooses G if the draw gives 2,
  A chooses G if the draw gives 3.

A strategy for B, will be defined by the function g_{ij} , $i, j = 1, 2$:

g_{11}: B chooses R if A has chosen R, B chooses R if A has chosen G,

g_{12}: B chooses R if A has chosen R, B chooses G if A has chosen G,

g_{21}: B chooses G if A has chosen R, B chooses R if A has chosen G,

g_{22}: B chooses G if A has chosen R, B chooses G if A has chosen G.

We will now assume that A uses strategy f_{111} and B strategy g_{11} . If the draw gives 1, the payment will be $M(1, R, R) = 3$; if the draw gives 2, the payment will be $M(2, R, R) = 5$; if it gives 3, the payment will be $M(3, R, R) = -1$. The expected value is

$$\tfrac{1}{2}M(1, R, R) + \tfrac{1}{4}M(2, R, R) + \tfrac{1}{4}M(3, R, R)$$

$$= (\tfrac{1}{2})(3) + (\tfrac{1}{4})(5) + (\tfrac{1}{4})(-1) = \tfrac{10}{4}. \quad (28.11)$$

TABLE 28.6

B

		g_{11}	g_{12}	g_{21}	g_{22}
	f_{111}	$\frac{10}{4}$	$\frac{10}{4}$	$\frac{3}{4}$	$\frac{3}{4}$
	f_{112}	$\frac{8}{4}$	$\frac{6}{4}$	$-\frac{2}{4}$	$\frac{6}{4}$
	f_{121}	$\frac{10}{4}$	$\frac{3}{4}$	$\frac{9}{4}$	$\frac{4}{4}$
	f_{122}	$\frac{6}{4}$	$\frac{9}{4}$	$\frac{4}{4}$	$\frac{7}{4}$
A	f_{211}	$\frac{2}{4}$	$\frac{12}{4}$	$-\frac{3}{4}$	$\frac{7}{4}$
	f_{212}	0	$\frac{8}{4}$	$-\frac{8}{4}$	$\frac{10}{4}$
	f_{221}	0	$\frac{5}{4}$	$\frac{3}{4}$	$\frac{8}{4}$
	f_{222}	$-\frac{2}{4}$	$\frac{11}{4}$	$-\frac{2}{4}$	$\frac{11}{4}$

Let us assume that A uses f_{112} and B uses g_{11}. If the draw produces 1, the payment will be $M(1, R, R) = 3$; if the draw gives 2, it will be $M(2, R, R) = 5$; if the draw gives 3, it will be $M(3, G, R) = -3$; the expected value will be $\tfrac{8}{4}$.

Assuming that A uses f_{121} and B uses g_{21}, the drawing of 1 will result in a payment of $M(1, R, G) = 2$; drawing 2 in a payment of $M(2, G, R) = 3$; drawing 3 in one of $M(3, R, G) = 2$; the expected value will be $\tfrac{9}{4}$.

It is by these means that Table 28.6 is obtained. It will be observed that the dimensions of the game increase very rapidly as new probabilities and choices are offered.

Example 3. Probability affects the order in which the players are called upon to choose.

In the following multistage game (Fig. 28.13) the first stage consists of drawing one of the numbers 1 or 2, each bearing a probability of $\frac{1}{2}$. If 1 is drawn, A will start and choose either red (R), green (G), or black (N); B will then choose from R or G. It is assumed that the information is complete and the payments are made by the function $M(u, x, y)$ such that

$$
\begin{array}{ll}
M(1, R, R) = 2, & M(2, R, R) = 4, \\
M(1, R, G) = -1, & M(2, R, G) = 5, \\
M(1, G, R) = 3, & M(2, G, R) = -3, \\
M(1, G, G) = -2, & M(2, G, G) = 1, \\
M(1, N, R) = 0, & M(2, N, R) = 0, \\
M(1, N, G) = 2, & M(2, N, G) = -3.
\end{array}
\tag{28.12}
$$

Figures 28.11–28.13 show how the elements of the arborescence are separated.

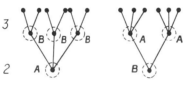

FIG. 28.11.   FIG. 28.12.

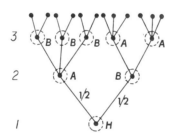

FIG. 28.13.

There are 27 strategies for A and 16 for B, and they will now be defined. Using f_{ijk} $(i, j, k = 1, 2, 3)$, A's strategies are defined as follows:

$$
\begin{array}{llllll}
f_{111}: & \text{if } u = 1, & x = R; & \text{if } u = 2, & \text{if } y = R, & x = R; \\
 & & & \text{if } u = 2, & \text{if } y = G, & x = R; \\
f_{112}: & \text{if } u = 1, & x = R; & \text{if } u = 2, & \text{if } y = R, & x = R; \\
 & & & \text{if } u = 2, & \text{if } y = G, & x = G; \\
f_{121}: & \text{if } u = 1, & x = R; & \text{if } u = 2, & \text{if } y = R, & x = G; \\
\vdots & & & \text{if } u = 2, & \text{if } y = G, & x = R;
\end{array}
\tag{28.13}
$$

$$f_{131}: \text{ if } u = 1, \quad x = R; \qquad \text{if } u = 2; \qquad \text{if } y = R, \quad x = N;$$
$$\text{if } u = 2; \qquad \text{if } y = G, \quad x = R;$$

$$\vdots$$

$$f_{211}: \text{ if } u = 1, \quad x = G; \qquad \text{if } u = 2; \qquad \text{if } y = R, \quad x = R;$$
$$\text{if } u = 2; \qquad \text{if } y = G, \quad x = R;$$

$$\vdots \tag{28.13}$$

$$f_{323}: \text{ if } u = 1, \quad x = N; \qquad \text{if } u = 2; \qquad \text{if } y = R, \quad x = G;$$
$$\text{if } u = 2; \qquad \text{if } y = G, \quad x = N;$$

$$f_{333}: \text{ if } u = 1, \quad x = N; \qquad \text{if } u = 2; \qquad \text{if } y = R, \quad x = N;$$
$$\text{if } u = 2; \qquad \text{if } y = G, \quad x = N.$$

We shall take g_{ijk} $(i, j, k, l - 1, 2)$, for B's strategies, defined as follows:

$$g_{1111}: \text{ if } u = 1, \qquad \text{if } x = R, \quad y = R; \qquad \text{if } u = 1, \qquad \text{if } x = G, \qquad y = R;$$
$$\text{if } u = 1, \qquad \text{if } x = N, \quad y = R; \qquad \text{if } u = 2, \qquad\qquad y = R,$$

$$g_{1112}: \text{ if } u = 1, \qquad \text{if } x = R, \quad y = R; \qquad \text{if } u = 1, \qquad \text{if } x = G, \qquad y = R;$$
$$\text{if } u = 1, \qquad \text{if } x = N, \quad y = R; \qquad \text{if } u = 2, \qquad\qquad y = G,$$

$$g_{1121}: \text{ if } u = 1, \qquad \text{if } x = R, \quad y = R; \qquad \text{if } u = 1, \qquad\quad x = G, \qquad y = R;$$
$$\text{if } u = 1, \qquad \text{if } x = N, \quad y = G; \qquad \text{if } u = 2, \qquad\qquad y = R,$$

$$\vdots$$

$$g_{1221}: \text{ if } u = 1, \qquad \text{if } x = R, \quad y = R; \qquad \text{if } u = 1, \qquad\quad x = G, \qquad y = G;$$
$$\text{if } u = 1, \qquad \text{if } x = N, \quad y = G; \qquad \text{if } u = 2, \qquad\qquad y = R,$$

$$\vdots \tag{28.14}$$

$$g_{2122}: \text{ if } u = 1, \qquad \text{if } x = R, \quad y = G; \qquad \text{if } u = 1, \qquad\quad x = G, \qquad y = R;$$
$$\text{if } u = 1, \qquad \text{if } x = N, \quad y = G; \qquad \text{if } u = 2, \qquad\qquad y = G,$$

$$g_{2222}: \text{ if } u = 1, \qquad \text{if } x = R, \quad y = R; \qquad \text{if } u = 1, \qquad\qquad y = G, \qquad y = G;$$
$$\text{if } u = 1, \qquad \text{if } x = N, \quad y = G; \qquad \text{if } u = 2, \qquad\qquad y = G.$$

The matrix of the rectangular game can now be constructed. Assuming, for example, that A uses strategy f_{323} and B strategy g_{1221}, the first hypothesis is $u = 1$; then with strategy f_{323}, $x = N$, and with strategy g_{1221}, $y = G$. For the second hypothesis, $u = 1$, and using strategy g_{1221}, $y = R$, and with strategy f_{323}, $x = G$. Hence, the expected value of the settlement will be

$$\tfrac{1}{2}M(1, N, G) + \tfrac{1}{2}M(2, R, G) = \tfrac{1}{2}(2) + \tfrac{1}{2}(5) = \tfrac{7}{2},$$

and the 27×16 elements of the table will be found in this manner.

F. NORMAL FORM OF A GAME

Every game with two players and a zero value in which the number of separate situations is finite can be shown as a rectangular game which is called *the normal form of the game.*

Hence any problem of real competition or combat between two firms or people can be shown as a rectangular game, provided it contains the above conditions (the hypothesis of a game with a zero value leads to the study of games with three people); but this theoretical conclusion is often invalid in practice, for the problems encountered are highly combinatorial and would require thousands of lines and/or columns to be given their normal form. Nor must it be forgotten that a strategy must take into account every probability and every decision of the opponent. The examples given in this subsection have shown how complicated it becomes when we attempt to define the strategies of an adversary.

G. GAMES WHERE THE INFORMATION IS COMPLETE

A game is said to provide *complete information* when every player is aware, at each point which represents him on the arborescence, not only of the choices already made but of the subsequent results which will be caused by the intervention of probability. Hence, the games of Figs. 28.8 and 28.13 have complete information.

H. FUNDAMENTAL THEOREM ON GAMES WITH COMPLETE INFORMATION[1]

Every multistage game with two players and with complete information has a normal form which contains a point of equilibrium.

---

[1] The proof of this proposition is given in [J16].

PART II

MATHEMATICAL DEVELOPMENTS

Symbols for the Theory of Sets Used in Part II

The developments in Part II presuppose an elementary knowledge of the theory of sets on the part of the reader.

Symbol	Meaning	Remarks
$a \in \mathbf{A}$	The element a "belongs" to set $\mathbf{A}$	Written "a belongs to $\mathbf{A}$"
$a \notin \mathbf{A}$	The element a "does not belong" to $\mathbf{A}$	"a does not belong to $\mathbf{A}$"
$\varnothing$	The set is empty	Written "empty" or "zero"
$\mathbf{A} = \mathbf{B}$	Set $\mathbf{A}$ is "identical" to set $\mathbf{B}$	"$\mathbf{A}$ identical to $\mathbf{B}$" or "$\mathbf{A}$ equals $\mathbf{B}$"
$\mathbf{A} \neq \mathbf{B}$	Set $\mathbf{A}$ "is not identical" to set $\mathbf{B}$	Written "$\mathbf{A}$ different from $\mathbf{B}$"
$\mathbf{B} \subset \mathbf{A}$	Set $\mathbf{B}$ is "included" in set $\mathbf{A}$	"$\mathbf{B}$ is included in $\mathbf{A}$"
$\mathbf{B} \subsetneqq \mathbf{A}$	Set $\mathbf{B}$ is "strictly included" in set $\mathbf{A}$	Means: $\mathbf{A} \subset \mathbf{B}$ and $\mathbf{A} \neq \mathbf{B}$; Written "$\mathbf{B}$ is strictly included in $\mathbf{A}$"
$\mathbf{A} \cup \mathbf{B}$	Union of two sets	"$\mathbf{A}$ union $\mathbf{B}$"
$\mathbf{A} \cap \mathbf{B}$	Intersection of two sets	"$\mathbf{A}$ intersects $\mathbf{B}$"
$\complement_\mathbf{A} \mathbf{B}$	Complementary of $\mathbf{B}$ in relation to $\mathbf{A}$, where $\mathbf{B} \subset \mathbf{A}$	Called "complementary of $\mathbf{B}$ in relation to $\mathbf{A}$"
$\mathbf{A} - \mathbf{B}$	Difference $\mathbf{A} - \mathbf{B} = \mathbf{A} - \mathbf{A} \cap \mathbf{B}$	$\mathbf{A}$ minus $\mathbf{B}$
$\complement \mathbf{B}$	Complementary of $\mathbf{B}$ in relation to the referential	"Complement of $\mathbf{B}$"
$\mathbf{A} \times \mathbf{B}$	Cartesian product of sets $\mathbf{A}$ and $\mathbf{B}$	
$\mid \mathbf{A} \mid$	Number of elements in set $\mathbf{A}$	
$P_1 \Rightarrow P_2$	The property P_1 implies the property P_2	This is the logical inference. Written "P_1 involves P_2 or implies P_2"
$P_1 \Leftrightarrow P_2$	The property P_1 is "equivalent" to property P_2	The logical equivalence. Written "P_1 equivalent to P_2"
$x \equiv y$	For $x \in \mathbf{X}$ and $y \in \mathbf{Y}$; x has equivalence with y, in respect to a certain property	Written: "x equivalent to y"
$\forall x$	"Whatever the value of x"	Universal quantifier
$\exists x$	"There is at least one x"	Existential quantifier
$y \in \Gamma x$	For $x \in \mathbf{X}$ and $y \in \mathbf{Y}$, mapping of $\mathbf{X}$ into $\mathbf{Y}$	
$x \in \Gamma^{-1} y$	For $x \in \mathbf{X}$ and $y \in \mathbf{Y}$, reciprocal mapping of $\mathbf{Y}$ into $\mathbf{X}$, let $$\Gamma^{-1} y = \{x \mid y \in \Gamma x\}$$	
$\Gamma^n x$ $n = 1, 2, 3...$	n successive mappings Γ of $\mathbf{X}$ into $\mathbf{X}$. For $x \in \mathbf{X}$: $$\Gamma^n x = \Gamma(\Gamma^{n-1} x) = \cdots = \Gamma^{n-1}(\Gamma x)$$	
$\hat{\Gamma} x$	Transitive closure of $x \in \mathbf{X}$ (mapping $\hat{\Gamma}$ of $\mathbf{X}$ into $\mathbf{X}$ such that $$\hat{\Gamma} x = x \cup \Gamma x \cup \Gamma^2 x \cap \Gamma^3 x \cup \cdots)$$	
$\Gamma_1(\Gamma_2 x)$	Product of composition where $$z \in \Gamma_1 y, \quad y \in \Gamma_2 x$$	

Symbols for the Theory of Sets (*cont.*)

Symbol	Meaning	Remarks
N	Set of natural integers	
Z	Set of related integers	
Q	Set of rational numbers	
R	Set of real numbers	
R$^+$	Set of positive real numbers	
C	Set of complex numbers	
R^n or **R**$_n$	Set of points in real space with n dimensions	One point is an "n-tuple"
E^n or **E**$_n$	Set of points in an Euclidean space with n dimensions	Real space into which the Euclidean metric has been introduced
]ab[	Open interval in **R**	
[ab]	Closed interval in **R**	

CHAPTER IV

THE PRINCIPAL PROPERTIES OF GRAPHS

29. Introduction

The aim of this chapter is to explain to the reader the theoretical aspects of the concepts which were used in the Part I. We shall therefore go back to the simplified definitions and the properties given therein, and express them in the symbols of modern algebra. New concepts will be introduced, as well as theorems of general interest. The fundamental work of Professor Berge [H2], the chief authority in this field of mathematics, has been taken as our volume of reference, and the reader will frequently be referred to it.

The theory of graphs is deserving of special study, and it is safe to prophesy that it will be of fundamental importance in operations research. Certain passages in the present chapter may present some difficulties to those who are not familiar with modern algebra, but we have tried to make them easier to understand by examples to illustrate the definitions and theorems, and we have also recapitulated a number of basic ideas.

Among the concepts introduced by the theory of graphs, the reader is advised to pay particular attention to the sections dealing with the shortest path, the optimal flow, Hamiltonian paths, trees, and arborescences, and the mapping of a graph into a graph; these valuable concepts are currently employed for numerous problems of "business economy."[1]

30. Definition and Figure

A. GRAPH OF ORDER n

If we consider a denumerable set $\mathbf{X} = \{X_1, X_2, ..., X_n\}$ and a multivocal[2] mapping Γ of $\mathbf{X}$ into $\mathbf{X}$, the pair $G = (\mathbf{X}, \Gamma)$ constitutes a *graph of order n.*

[1] We have borrowed this term from Brunet, who also used it as the title of the series in which the present work is included.

[2] A correspondence that is not always "one-to-one."

(a)   Each element of **X** is represented on the paper by a point called a *vertex* of the graph, which is marked as the element to which it corresponds.

(b)   Two vertices X_i, $X_j \in$ **X** are connected by an arrow pointing from X_i towards X_j if $X_j \in \Gamma X_i$. This arrow produces what is called an *arc* of the graph.

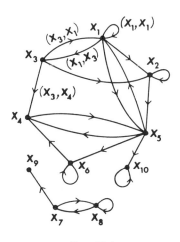

FIG. 30.1.

B. REPRESENTATION BY A FIGURE FORMED BY VERTICES AND ARCS

Figure 30.1 is an example of a representation of this kind. We have

$$\Gamma(X_1) = \{X_1, X_2, X_3, X_5\},$$

$$(30.1)$$

$$\Gamma(X_6) = \{X_4, X_6\}, \qquad \Gamma\{X_{10}\} = \{X_{10}\}, \qquad \Gamma(X_9) = \varnothing, \qquad \Gamma(X_4) = \{X_5\}.$$

The set of arcs will be called **U** and a graph can be referred to as either the pair (**X**, Γ) or the pair (**X**, **U**).

31. Oriented Concepts

A. EXTREMITIES

Given an arc $u = (X_i, X_j)$, then X_i is called the *initial extremity* and X_j the *terminal extremity*.

B. ADJACENT ARCS

Arcs $u = (X_i, X_j)$ where $X_i \neq X_j$ which have a common extremity and are separate are *adjacent*. In Fig. 30.1, (X_3, X_4), (X_1, X_3), and (X_3, X_1) are adjacent.

C. ADJACENT VERTICES

Two vertices, X_i and X_j, are adjacent if they are separate and if there is an arc $u = (X_i, X_j)$ or $v = (X_j, X_i)$. For example, in Fig. 30.1, X_4 and X_6 are adjacent, as well as X_1 and X_5, but X_3 and X_5 are not adjacent.

D. ARC CONNECTED TO A VERTEX. HALF-DEGREE

An arc u is *initially connected to a vertex* X_i if X_i is the initial extremity of u and if the terminal extremity of u is other than X_i. If $u = (X_i, X_j)$, with $X_i \neq X_j$, u satisfies the definition.

An arc v is *terminally connected to a vertex* X_i if X_i is the terminal extremity of v and if the initial extremity of v is other than X_i. If $v = (X_k, X_i)$, with $X_i \neq X_k$, v satisfies the definition.

The *internal half-degree* of a vertex is the number of terminally connected arcs at this vertex. The *external half-degree* of a vertex is the number of initially connected arcs at this vertex.

E. ARC WHICH IS CONNECTED TO A SET OF VERTICES

If **A** is a set of vertices, then an arc u is *initially connected to* **A**, if

$$u = (A, X_i), \qquad A \in \mathbf{A}, \qquad X_i \notin \mathbf{A}. \tag{31.1}$$

Example (Fig. 30.1). If $\mathbf{A} = \{X_1, X_2, X_3\}$, then (X_3, X_4) is initially connected to **A**.

An arc v is *terminally connected to* **A**, if

$$v = (X_i, A), \qquad A \in \mathbf{A}, \qquad X_i \notin \mathbf{A}. \tag{31.2}$$

Example (Fig. 30.1). If $\mathbf{A} = \{x_4, x_5, x_6\}$, then (X_2, X_5) is terminally connected to **A**.

The set of arcs terminally connected to **A** is represented as $\mathbf{U}_A{}^+$, that of the arcs initially connected as $\mathbf{U}_A{}^-$. The set of arcs initially and terminally connected is known as $\mathbf{U}_A$. By definition:

$$\mathbf{U}_A = \mathbf{U}_A{}^+ \cup \mathbf{U}_A{}^-. \tag{31.3}$$

F. PARTIAL GRAPH OF A GRAPH

A *partial graph* of G is, by definition, a graph $(\mathbf{X}, \Delta)$, where $(\forall X_i)$,

$$\Delta X_i \subset \Gamma X_i \,.$$

Expressed differently, there are the same vertices, but at least one arc is missing. Thus, the graph of Fig. 31.2 is a partial graph of that in Fig. 31.1.

G. SUBGRAPH OF A GRAPH

A *subgraph* of G is by definition a graph $(\mathbf{A}, \Gamma_A)$, where

$$\mathbf{A} \subset \mathbf{X} \quad \text{and} \quad \forall X_i \in \mathbf{A}, \qquad \Gamma_A X_i = (\Gamma X_i) \cap \mathbf{A}.$$

To obtain a subgraph all that is needed is to remove a certain nonzero number of vertices and their adjacent arcs: the graph thus obtained is a subgraph of the given graph.

Thus Fig. 31.3 shows a subgraph of the graph in Fig. 31.1.

A *partial subgraph of a graph* can also be defined by combining the two definitions given above (see, for example, Fig. 31.4).

H. COMPLETE GRAPH

A graph is *complete* if

$$\forall X_i, \forall X_j (i \neq j): \quad (X_i, X_j) \notin \mathbf{U} \Rightarrow (X_j, X_i) \in \mathbf{U}. \tag{31.4}$$

In other words, when each pair of vertices is connected in at least one of two directions. For example, Fig. 31.5 shows a complete graph, but this is not the case in Fig. 31.1.

I. PATH, SIMPLE PATH, AND ELEMENTARY PATH

A *path* is a sequence $(u_1, u_2, \ldots)$ of arcs such that the terminal extremity of each arc coincides with the initial extremity of the following one. A path can be *finite* or *infinite*.

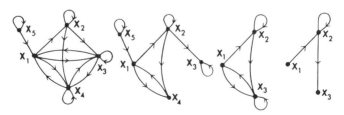

FIG. 31.1 –4.

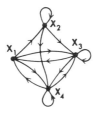

FIG. 31.5.

Example (Fig. 31.6).

$$(a, c, m, i), \qquad (f, m, d, b, a), \qquad (g, h, n, j, h, q)$$

are paths.

A path may also be represented by the vertices which it contains:

$$(X_3, X_1, X_2, X_5, X_4), \qquad (X_3, X_2, X_5, X_1, X_3, X_1),$$

$$(X_3, X_4, X_5, X_6, X_4, X_5, X_{10}).$$

Simple Path.   A path is *simple* when it does not make use of the same arc twice. If it does, it is *composite*.

Example (Fig. 31.6).

$$(a, c, m, i) \qquad \text{is a simple path;}$$

$$(g, h, n, j, h, q) \qquad \text{is a composite path.}$$

Elementary Path.   A path is *elementary* if it does not make use of the same vertex twice. If it does, it is called *nonelementary*.

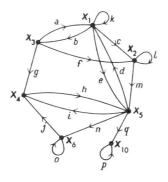

FIG. 31.6.

Example (Fig. 31.6).

(a, c, m, i) is elementary (and simple);

(h, d, c, m, q) is nonelementary (but simple);

(g, h, n, j, h, q) is composite and nonelementary.

J. CIRCUIT, ELEMENTARY CIRCUIT, LOOP

A *circuit* is a finite path $(u_1, u_2, ..., u_k)$ in which the initial vertex, u_1, coincides with the terminal vertex, u_k. A circuit can be represented by the arcs or vertices contained in it.

Example (Fig. 31.6).

(n, j, h) or (X_5, X_6, X_4, X_5) is a circuit;

(b, g, h, n, j, h, d) or $(X_1, X_3, X_4, X_5, X_6, X_4, X_5, X_1)$ also;

(d, e) or (X_5, X_1, X_5), as well.

Elementary Circuit. A circuit is called *elementary* if all the vertices through which it passes are separate except for the initial and terminal vertices which coincide.

Example (Fig. 31.6).

(d, b, g, h) or $(X_5, X_1, X_3, X_4, X_5)$ is an elementary circuit.

Loop. This is a circuit composed of a single arc and a single vertex. Thus, in Fig. 31.6, k, l, o, and p are loops.

K. LENGTH OF A PATH

The *length* of a path $(u_1, u_2, ..., u_r)$ is the number of arcs contained in the sequence. Assuming

$$\mu = (u_1, u_2, ..., u_s),$$

the length will be represented by $l(\mu)$. Hence, $l(\mu) = r$; if the path is infinite, $l(\mu) = \infty$.

Example (Fig. 31.6).

$$\mu = (a, c, m, i) : \quad l(\mu) = 4,$$
$$\mu = (g, h, n, j, h, q) : \quad l(\mu) = 6.$$

L. SYMMETRICAL AND ANTISYMMETRICAL GRAPH

A graph $G = (\mathbf{X}, \mathbf{U})$ is *symmetrical* if

$$\forall X_i, \forall X_j: \quad (X_i, X_j) \in \mathbf{U} \Rightarrow (X_j, X_i) \in \mathbf{U}. \tag{31.5}$$

Example (Fig. 31.7).   This graph is symmetrical.

In certain cases it is convenient to represent a symmetrical graph with the symmetrical area replaced by nonarrowed lines (Fig. 31.8), but care must be exercised over the use of this representation, especially when there are loops.

Antisymmetrical Graph.   A graph $G = (\mathbf{X}, \mathbf{U})$ is *antisymmetrical* when

$$\forall X_i, \forall X_j: \quad (X_i, X_j) \in \mathbf{U} \Rightarrow (X_j, X_i) \notin \mathbf{U}. \tag{31.6}$$

An antisymmetrical graph cannot contain a loop and there cannot be more than one arc between any pair of vertices.

Example (Fig. 31.9).   This graph is antisymmetrical.

M. STRONGLY CONNECTED GRAPH AND TRANSITIVE CLOSURE

A graph is *strongly connected* if $\forall X_i$ and $\forall X_j (X_i \neq X_j)$ there exists a path from X_i towards X_j.

Example.   The graph in Fig. 31.10 is strongly connected, but the graph in Fig. 31.11 is not, since we cannot join X_1 with X_2 or X_4, nor X_2 with X_4, and reciprocally.

Transitive Closure.   Given a finite graph $G = (\mathbf{X}, \Gamma)$ the mappings $\Gamma^2, \Gamma^3, \ldots$ mean:

$$\Gamma^2(X_i) = \Gamma(\Gamma X_i) \tag{31.7}$$

$$\Gamma^3(X_i) = \Gamma(\Gamma^2(X_i)) = \Gamma(\Gamma(\Gamma(X_i))). \tag{31.8}$$

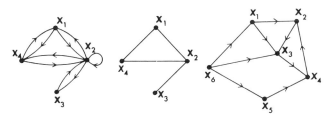

FIG. 31.7–9.

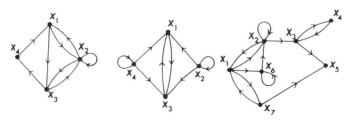

FIG. 31.10 –12.

The *transitive closure* of Γ is a mapping $\hat{\Gamma}$ of **X** in **X**, defined by

$$\hat{\Gamma}(X_i) = \{X_i\} \cup \Gamma(X_i) \cup \Gamma^2(X_i) \cup \Gamma^3(X_i) \cup \cdots. \tag{31.9}$$

Example (Fig. 31.12).   The reader is advised to follow this example.

$$\Gamma(X_1) = \{X_2, X_6, X_7\},$$
$$\Gamma(X_2) = \{X_1, X_2, X_3\},$$
$$\Gamma(X_3) = \{X_4, X_5\},$$
$$\Gamma(X_4) = \{X_3\},$$
$$\Gamma(X_5) = \varnothing,$$
$$\Gamma(X_6) = \{X_2, X_6\},$$
$$\Gamma(X_7) = \{X_1, X_5\},$$

$$\Gamma^2(X_1) = \Gamma(X_2, X_6, X_7) = \{X_1, X_2, X_3, X_5, X_6\},$$
$$\Gamma^2(X_2) = \Gamma\{X_1, X_2, X_3\} = \{X_1, X_2, X_3, X_4, X_5, X_6, X_7\},$$
$$\Gamma^2(X_3) = \Gamma\{X_4, X_5\} = \{X_3\},$$
$$\Gamma^2(X_4) = \Gamma\{X_3\} = \{X_4, X_5\},$$
$$\Gamma^2(X_5) = \varnothing,$$
$$\Gamma^2(X_6) = \Gamma\{X_2, X_6\} = \{X_1, X_2, X_3, X_6\},$$
$$\Gamma^2(X_7) = \Gamma\{X_1, X_5\} = \{X_2, X_6, X_7\},$$

$$\Gamma^3(X_1) = \Gamma\{X_1, X_2, X_3, X_5, X_6\}$$
$$\qquad = \{X_1, X_2, X_3, X_4, X_5, X_6, X_7\} = \mathbf{X},$$
$$\Gamma^3(X_2) = \Gamma\{X_1, X_2, X_3, X_4, X_5, X_6, X_7\}$$
$$\qquad = \{X_1, X_2, X_3, X_4, X_5, X_6, X_7\} = \mathbf{X},$$
$$\Gamma^3(X_3) = \Gamma\{X_3\} = \{X_4, X_5\},$$
$$\Gamma^3(X_4) = \Gamma\{X_4, X_5\} = \{X_3\},$$
$$\Gamma^3(X_5) = \varnothing,$$
$$\Gamma^3\{X_6\} = \Gamma\{X_1, X_2, X_3, X_6\} = \{X_1, X_2, X_3, X_4, X_5, X_6, X_7\},$$

$$\Gamma^3(X_7) = \Gamma\{X_2\,,\,X_6\,,\,X_7\} = \{X_1\,,\,X_2\,,\,X_3\,,\,X_5\,,\,X_6\},$$
$$\Gamma^4(X_1) = \mathbf{X},$$
$$\Gamma^4(X_2) = \mathbf{X},$$
$$\Gamma^4(X_3) = \Gamma\{X_4\,,\,X_5\} = \{X_3\},$$
$$\Gamma^4(X_4) = \Gamma\{X_3\} = \{X_4\,,\,X_5\},$$
$$\Gamma^4(X_5) = \varnothing,$$
$$\Gamma^4(X_6) = \mathbf{X},$$
$$\Gamma^4(X_7) = \Gamma\{X_1\,,\,X_2\,,\,X_3\,,\,X_5\,,\,X_6\,,\,X_7\} = \mathbf{X},$$

whence:

$$\hat{\Gamma}(X_1) = \mathbf{X},$$
$$\hat{\Gamma}(X_2) = \mathbf{X},$$
$$\hat{\Gamma}(X_3) = \{X_3\} \cup \{X_4\,,\,X_5\} \cup \{X_4\,,\,X_5\} \cup \{X_3\} \cdots$$
$$\qquad\quad = \{X_3\,,\,X_4\,,\,X_5\},$$
$$\hat{\Gamma}(X_4) = \{X_4\} \cup \{X_3\} \cup \{X_4\,,\,X_5\} \cup \{X_3\} \cup \cdots$$
$$\qquad\quad = \{X_3\,,\,X_4\,,\,X_5\},$$
$$\hat{\Gamma}(X_5) = \{X_5\},$$
$$\hat{\Gamma}(X_6) = \mathbf{X},$$
$$\hat{\Gamma}(X_7) = \mathbf{X}.$$

Another Example. Given a set of people $\mathbf{X}$, $X_i \in \mathbf{X}$, and representing the set of children of X_i by $\Gamma(X_i)$:

$\Gamma^2(X_i)$ is the set of the grandchildren of X_i ;

$\hat{\Gamma}(X_i)$ is the set of X_i and of all his descendants.

N. MEANING OF INVERSE MAPPING

The inverse of Γ is a mapping Γ^{-1} defined by

$$\Gamma^{-1}(X_j) = \{X_i \mid X_j \in \Gamma X_i\}. \tag{31.12}$$

In other words, it is the set of all the vertices X_i, such that X_j belongs to the mapping of X_i. If $\mathbf{A} \subset \mathbf{X}$:

$$\Gamma^{-1}\mathbf{A} = \{X_i \mid \Gamma(X_i) \cap \mathbf{A} \neq \varnothing\}. \tag{31.13}$$

Thus $\Gamma^{-1}(X_j)$ is the set of vertices connected to X_j by arcs connected terminally with X_j, and $\Gamma^{-1}\mathbf{A}$ is the set of vertices of X_i which can be joined to one of the vertices of **A** by an arc which is connected to it terminally.

Examples (Fig. 31.12).

$$\Gamma^{-1}\{X_1\} = \{X_2, X_7\}, \qquad \Gamma^{-1}\{X_6\} = \{X_1, X_6\},$$
(31.14)
$$\Gamma^{-1}\{X_5\} = \{X_3, X_7\}$$

Assuming

$$\mathbf{A} = \{X_1, X_2, X_6\},$$
(31.15)

it follows that

$$
\begin{aligned}
\Gamma(X_1) &= \{X_2, X_6, X_7\}, & \{X_2, X_6, X_7\} \cap \mathbf{A} &= \{X_2, X_6\} \neq \varnothing, \\
\Gamma(X_2) &= \{X_1, X_2, X_3\}, & \{X_1, X_2, X_3\} \cap \mathbf{A} &= \{X_1, X_2\} \neq \varnothing, \\
\Gamma(X_3) &= \{X_4, X_5\}, & \{X_4, X_5\} \cap \mathbf{A} &= \varnothing, \\
\Gamma(X_4) &= \{X_3\}, & \{X_3\} \cap \mathbf{A} &= \varnothing, \\
\Gamma(X_5) &= \varnothing, & & \\
\Gamma(X_6) &= \{X_2, X_6\}, & \{X_2, X_6\} \cap \mathbf{A} &= \{X_2, X_6\} \neq \varnothing, \\
\Gamma(X_7) &= \{X_1, X_5\}, & \{X_1, X_5\} \cap \mathbf{A} &= \{X_1\} \neq \varnothing.
\end{aligned}
$$
(31.16)

Hence:

$$\Gamma^{-1}\mathbf{A} = \{X_1, X_2, X_6, X_7\}.$$

O. PRE-ORDER ASSOCIATED WITH A GRAPH

A binary relation between two elements X and Y of a set **E** is a *pre-order* represented as $\leqslant$ if

(1)   It is *reflexive* : $X \leqslant X$;

(2)   It is *transitive* : $(X \leqslant Y$ and $Y \leqslant Z) \Rightarrow (X \leqslant Z)$.

The following is an example of a pre-order associated with a graph. Let us take the following binary relation:

$$X_i \leqslant X_j \qquad \text{if} \quad X_i = X_j$$
(31.17)

or there is a path from X_i to X_j.

Example (Fig. 31.12).   Using the above pre-order,

$$X_1 \leqslant X_1, X_2, X_3, X_4, X_5, X_6, X_7 ;$$
$$X_2 \leqslant X_1, X_2, X_3, X_4, X_5, X_6, X_7 ;$$
$$X_3 \leqslant X_3, X_4, X_5 ;$$
$$X_4 \leqslant X_3, X_4, X_5 ;$$ (31.18)
$$X_5 \leqslant X_5 ;$$
$$X_6 \leqslant X_1, X_2, X_3, X_4, X_5, X_6, X_7 ;$$
$$X_7 \leqslant X_1, X_2, X_3, X_4, X_5, X_6, X_7 .$$

P. Equivalence Relation Associated with a Graph

If $X \leqslant Y$ and $Y \leqslant X$ we write $X \equiv Y$ (X equivalent to Y); this is an *equivalence relation* which has the following properties:

(1)   It is *reflexive* : $X \equiv X$;

(2)   It is *transitive* : $(X \equiv Y, Y \equiv Z) \Rightarrow (X \equiv Z)$; (31.19)

(3)   It is *symmetrical* : $(X \equiv Y) \Rightarrow (Y \equiv X)$.

The following is an example of an equivalence relation associated with a graph. Let us take the binary relation:

$$X_i \equiv X_j \qquad \text{if} \quad X_i = X_j$$ (31.20)

or if there is a path from X_i to X_j and from X_j to X_i, then, in our example (Fig. 31.12),

$$X_1 \equiv X_2 \equiv X_6 \equiv X_7 ; \qquad X_3 \equiv X_4 ; \qquad X_5 \equiv X_5 .$$ (31.21)

The equivalence classes defined by this relation form a *partition* of **X** (each vertex in the graph belongs to one class and one only).

If we consider the equivalence classes obtained from the equivalence relation (31.20), we can deduce the following property:

Let **X**$_i$ be one of the classes

$$\mathbf{X}_i = \{X_{i_1}, X_{i_2}, ..., X_{i_p}\}.$$

The subgraph of G, $G_1 = (\mathbf{X}_i, \Gamma)$ is strongly connected, since there is a path from each X_{i_r} to each X_{i_s} belonging to **X**$_i$.

Q. RELATION OF STRICT ORDER

If $X \leqslant Y$, and if we do not have $X \equiv Y$, we write $X < Y$ (X is an ancestor of Y or Y is a descendant of X). This constitutes a relation of strict order.

If a graph is to include such a relation it is necessary and sufficient that no circuit is contained in it.

Let us consider, for example, the case of technological operations which have to be carried out in a certain chronological sequence, and take as a relation of strict order that operation X_i must take place before operation X_j. The graph of Fig. 31.13 gives an example of strict order. There is no circuit in the graph. We have, for example,

$$X_9 < X_1 < X_2 < X_3 < X_6 .$$

R. MAJORANT, MAXIMUM

Let $\mathbf{A} \subset \mathbf{X}$, $X_j \in \mathbf{X}$. If

$$X_j \geqslant A_r , \qquad \forall A_r \in \mathbf{A} \tag{31.22}$$

then X_j is a *majorant* of $\mathbf{A}$.

Expressed differently, the vertex X_j, which follows all the vertices of $\mathbf{A}$, is a majorant of $\mathbf{A}$.

If an element of $\mathbf{A}$ is a majorant of $\mathbf{A}$, this element is called the *maximum* of $\mathbf{A}$. If X_k and X_l are two maximums of $\mathbf{A}$ it follows $X_k \leqslant X_i$ and $X_l \leqslant X_k$; hence $X_k \equiv X_l$. All the maximums are equivalent.

Example (Fig. 31.13).
X_3, X_5, X_6, and X_7 are the majorants of

$$\mathbf{A} = \{X_1 , X_2 , X_3 , X_4 , X_8 , X_9 , X_{10}\}$$

X_3 is a maximum of

$$\mathbf{A} = \{X_1 , X_2 , X_3 , X_4 , X_8 , X_9 , X_{10}\}.$$

S. MINORANT, MINIMUM

Let $\mathbf{A} \subset \mathbf{X}$, $X_j \in \mathbf{X}$, if

$$X_j \leqslant A_r , \qquad \forall A_r \in \mathbf{A};$$

then X_j is a *minorant* of $\mathbf{A}$.

Otherwise expressed, the vertex X_j, which precedes all the vertices of $\mathbf{A}$, is a minorant of $\mathbf{A}$.

If an element of **A** is a minorant of **A**, this element is called the *minimum* of **A**. When there are two minimums they are *equivalent*.

Example (Fig. 31.13).
X_1, X_8, X_9, X_{10} are the minorants of

$$\mathbf{A} = \{X_1, X_2, X_3, X_4, X_5, X_6, X_7\}$$

X_1 is a minimum of

$$\mathbf{A} = \{X_1, X_2, X_3, X_4, X_5, X_6, X_7\}.$$

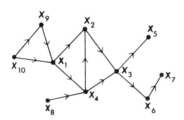

FIG. 31.13.

32. Nonoriented Concepts

A. LINK

The name *link* in a graph $G = (\mathbf{X}, \mathbf{U})$ is given to a set of two elements X_i and X_j such that

$$(X_i, X_j) \in \mathbf{U} \quad \text{and/or} \quad (X_j, X_i) \in \mathbf{U}, \qquad X_i \neq X_j. \qquad (32.1)$$

In other words, a link is a pair of vertices joined by one arc in either direction or by two arcs in opposite directions.
The graph in Fig. 32.1 has 21 arcs and 12 links.

A link is usually denoted by a small letter under a bar:

$$\bar{u} = \overline{(X_i, X_j)}$$

and the set of links is called $\overline{\mathbf{U}}$.

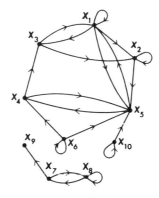

FIG. 32.1.

B. CHAIN, SIMPLE CHAIN, ELEMENTARY CHAIN

A *chain* is a sequence of links $(\bar{u}_1, \bar{u}_2, ...)$ in which each link $\bar{u}_k$ is connected to $\bar{u}_{k-1}$ by one of its extremities and to $\bar{u}_{k+1}$ by the other. A chain is usually referred to by the vertices which it contains. It may be finite or infinite.

Example (Fig. 32.1).

$$(X_4, X_3, X_1, X_5, X_2), \qquad (X_3, X_4, X_6, X_5, X_4, X_6, X_5, X_{10})$$

are chains.

Simple Chain. A chain is *simple* if all the links used are different; in the opposite case, it is *composite*.

Example (Fig. 32.1).

$$(X_4, X_3, X_1, X_5, X_2), \qquad (X_2, X_5, X_1, X_3, X_4, X_5, X_{10})$$

are simple chains;

$$(X_3, X_4, X_6, X_5, X_4, X_6, X_5, X_{10})$$

is a composite chain.

Elementary Chain. A chain is *elementary* if it does not meet the same vertex twice.

Example. $(X_1, X_2, X_5, X_6, X_4)$ is an elementary chain.

C. CYCLE, ELEMENTARY CYCLE

A *cycle* is a finite chain which leaves a vertex X_i and ends at the same vertex.

Example (Fig. 32.1).

$$(X_4, X_6, X_5, X_4), \qquad (X_3, X_1, X_2, X_5, X_1, X_2, X_5, X_6, X_4, X_3)$$

are cycles.

Elementary Cycle. A cycle is *elementary* if no vertex in the cycle is encountered more than once, except for the vertex which is both the initial and terminal one.

Examples. The cycle (X_4, X_6, X_5, X_4) is elementary. The cycle $(X_1, X_2, X_5, X_6, X_4, X_5, X_1)$ is not elementary.

D. Connected Graph, Connected Components of a Graph

A graph is said to be *connected* if $\forall X_i$ and $\forall X_j(X_i \neq X_j)$, there is a chain going from X_i towards X_j.

Example. The graph of Fig. 31.12 is connected (though not strongly connected); that of Fig. (32.1) is not.

A strongly connected graph is connected.

Connected Components of a Graph. Given a vertex X_i and $\mathbf{C}_{X_i}$, the set of vertices which can be connected to X_i by a chain to which X_i itself is added, the term *connected component* of the graph is applied to the subgraph formed by a set of the form $\mathbf{C}_{X_i}$.

Example (Fig. 32.1). The subgraphs formed by the vertices X_1, X_2, X_3, X_4, X_5, X_6, X_{10}, on the one hand, and by X_7, X_8, X_9, on the other, are the connected components of the whole graph.

The different connected components of the graph $G = (\mathbf{X}, \Gamma)$ constitute a partition of $\mathbf{X}$; in other words, if $X_i \in \mathbf{X}$:

$$(1) \quad \mathbf{C}_{X_i} \neq \varnothing,$$
$$(2) \quad \mathbf{C}_{X_i} \neq \mathbf{C}_{X_j} \Rightarrow \mathbf{C}_{X_i} \cap \mathbf{C}_{X_j} = \varnothing, \qquad (32.2)$$
$$(3) \quad \bigcup_{X_i \in \mathbf{X}} \mathbf{C}_{X_i} = \mathbf{X}.$$

A graph is connected, if and only if, it does not possess more than one component.

E. DEGREE OF A VERTEX; REGULAR SUBGRAPH WITH DEGREE d

The *degree* of a vertex X_i is the number of links which have one extremity at X_i, the other extremity not being X_i. The degree of X_i is written as $d(X_i)$.

Example (Fig. 32.1).

$$d(X_1) = 3, \qquad d(X_5) = 5, \qquad d(X_8) = 1, \qquad d(X_{10}) = 1. \qquad (32.3)$$

Regular Subgraph with Degree d. We say that a subgraph $G(\mathbf{A}, \Gamma_A)$ where $\mathbf{A} \subset \mathbf{X}$ is *regular with degree d* if the degree of all the vertices is the same and equal to d.

Example (Fig. 32.1). The subgraph formed with the vertices X_4, X_5, X_6 is regular with degree 2.

33. Product and Sum of Graphs

A. PRODUCT OF GRAPHS

Let there be n graphs

$$G^{(1)} = (\mathbf{X}^{(1)}, \Gamma^{(1)}), \qquad G^{(2)} = (\mathbf{X}^{(2)}, \Gamma^{(2)}),..., G^{(n)} = (\mathbf{X}^{(n)}, \Gamma^{(n)})$$

and let us consider the graph $G = (\mathbf{X}, \Gamma)$, defined by the following *cartesian product*:

$$\mathbf{X} = \mathbf{X}^{(1)} \times \mathbf{X}^{(2)} \times \cdots \times \mathbf{X}^{(n)}; \qquad (33.1)$$

that is to say, the elements of $\mathbf{X}$ are formed by the elements

$$X_{ij...r} = X_i^{(1)} \cdot X_j^{(2)} \cdot \cdots \cdot X_r^{(n)}, \qquad (33.2)$$

where

$$X_i^{(1)} \in \mathbf{X}^{(1)}, \qquad X_j^{(2)} \in \mathbf{X}^{(2)},..., X_r^{(n)} \in \mathbf{X}^{(n)}.$$

On the other hand,

$$\Gamma(X_{ij...r}) = (\Gamma^{(1)} X_i^{(1)}) \times (\Gamma^{(2)} X_j^{(2)}) \times \cdots \times (\Gamma^{(n)} X_r^{(n)}). \qquad (33.3)$$

This graph is denoted by

$$G = G^{(1)} \times G^{(2)} \times \cdots \times G^{(n)}$$

and is called the *product of the graphs* $G^{(i)}$.

B. EXAMPLE

Let us consider two machines $M^{(1)}$ and $M^{(2)}$. Machine $M^{(1)}$ can be in a certain number of states $\mathbf{X} = \{X_1, X_2, X_3\}$, and the graph in Fig. 33.1 shows which state can follow any of the states X_1, X_2, X_3. Machine $M^{(2)}$ can be in either of the two states of $\mathbf{Y} = \{Y_1, Y_2\}$, and Fig. 33.2 shows the transitions.

FIG. 33.1. $G^{(1)} = (\mathbf{X}, \Gamma)$.   FIG. 33.2. $G^{(2)} = (\mathbf{Y}, \Delta)$.

Assuming that a mechanic can operate the two machines, each situation is defined by an oriented pair (X_i, Y_j), and the set of situations is the cartesian product:

$$\mathbf{X} \times \mathbf{Y} = \{(X_1, Y_1), (X_1, Y_2), (X_2, Y_1), (X_2, Y_2), (X_3, Y_1), (X_3, Y_2)\}.$$
(33.4)

If the mechanic can operate the two machines *simultaneously*, situation (X_i, X_j) may be followed by one of the situations $\Gamma(X_i) \times \Delta(Y_j)$. We obtain the graph of Fig. 33.3 in which the arcs are calculated in the following manner:

$$\Gamma(X_1) = \{X_2\}, \qquad \Gamma(X_2) = \{X_1, X_3\}, \qquad \Gamma(X_3) = \{X_1\}$$
$$\Delta(Y_1) = \{Y_2\}, \qquad \Delta(Y_2) = \{Y_1, Y_2\}.$$
(33.5)

$$\Gamma(X_1) \times \Delta(Y_1) = \{(X_2, Y_2)\},$$
$$\Gamma(X_1) \times \Delta(Y_2) = \{(X_2, Y_1), (X_2, Y_2)\}$$
$$\Gamma(X_2) \times \Delta(Y_1) = \{(X_1, Y_2), (X_3, Y_2)\},$$
$$\Gamma(X_2) \times \Delta(Y_2) = \{(X_1, Y_1), (X_1, Y_2), (X_3, Y_1), (X_3, Y_2)\}$$
$$\Gamma(X_3) \times \Delta(Y_1) = \{(X_1, Y_2)\},$$
$$\Gamma(X_3) \times \Delta(Y_2) = \{(X_1, Y_1), (X_1, Y_2)\}.$$
(33.6)

C. SUM OF GRAPHS

With the same notation as above, the *sum of the graphs* $G^{(i)}$, which is denoted by
$$G = G^{(1)} + G^{(2)} + \cdots + G^{(n)},$$
is defined by
$$\mathbf{X} = \mathbf{X}^{(1)} \times \mathbf{X}^{(2)} \times \cdots \times \mathbf{X}^{(n)}$$
(33.7)

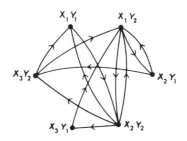

FIG. 33.3. $G^{(1)} \times G^{(2)}$.

and

$$\Gamma(X_{ij...r}) = [\{\Gamma^{(1)} X_i^{(1)}\} \times \{X_j^{(2)}\} \times \cdots \times \{X_r^{(n)}\}]$$

$$\cup \; [\{X_i^{(1)}\} \times \{\Gamma^{(2)} X_j^{(2)}\} \times \cdots \times \{X_r^{(n)}\}] \cup \cdots$$

$$\cup \; [\{X_i^{(1)}\} \times \{X_j^{(2)}\} \times \cdots \times \{\Gamma^{(n)} X_r^{(n)}\}]. \qquad (33.8)$$

D. EXAMPLE

Let us return to the previous example and assume that the mechanic can only operate *one machine at a time*. Situation (X_i, Y_j) may now be followed by one of the following situations:

(a) If he operates $M^{(1)}$, we obtain the set $\{\Gamma(X_i)\} \times \{Y_j\}$, of the situations described by the oriented pairs (X_i', Y_j), where X_i' is one of the situations of $\Gamma(X_i)$, and where Y_j remains fixed;

(b) If he operates $M^{(2)}$, we obtain the set $\{X_i\} \times \{\Delta(Y_j)\}$ of the situations described by the oriented pairs (X_i, Y_j'), where Y_j' is one of the situations of $\Delta(Y_j)$, where X_i remains fixed.

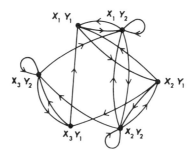

FIG. 33.4. $G^{(1)} + G^{(2)}$.

We obtain the graph of Fig. 33.4 in which the arcs have been calculated as follows, with γ the mapping of the sum of the graph into itself:

$$
\begin{aligned}
\gamma(X_1\,,\,Y_1) &= \{\Gamma(X_1)\} \times \{Y_1\} \cup \{X_1\} \times \{\varDelta(Y_1)\} \\
&= \{(X_2\,,\,Y_1),(X_1\,,\,Y_2)\}, \\
\gamma(X_1\,,\,Y_2) &= \{\Gamma(X_1)\} \times \{Y_2\} \cup \{X_1\} \times \{\varDelta(Y_2)\} \\
&= \{(X_2\,,\,Y_2),(X_1\,,\,Y_1),(X_1\,,\,Y_2)\}, \\
\gamma(X_2\,,\,Y_1) &= \{(\Gamma(X_2)\} \times \{Y_1\} \cup \{X_2\} \times \varDelta(Y_1)\} \\
&= \{(X_1\,,\,Y_1),(X_3\,,\,Y_1),(X_2\,,\,Y_2)\}, \\
\gamma(X_2\,,\,Y_2) &= \{\Gamma(X_2)\} \times \{Y_2\} \cup \{X_2\} \times \{\varDelta(Y_2)\} \\
&= \{(X_1\,,\,Y_2),(X_3\,,\,Y_2),(X_2\,,\,Y_1),(X_2\,,\,Y_2)\}, \\
\gamma(X_3\,,\,Y_2) &- \{\Gamma(X_3)\} \times \{Y_2\} \cup \{X_3\} \times \{\varDelta(Y_2)\} \\
&= \{(X_1\,,\,Y_2),(X_3\,,\,Y_1),(X_3\,,\,Y_2)\}, \\
\gamma(X_3\,,\,Y_1) &= \{\Gamma(X_3)\} \times \{Y_1\} \cup \{X_3\} \times \{\varDelta(Y_1)\} \\
&= \{(X_1\,,\,Y_1),(X_3\,,\,Y_2)\}.
\end{aligned}
\tag{33.9}
$$

34. Various Generalizations

Given p-graphs $G_1 = (\mathbf{X},\,\Gamma_1)$, $G_2 = (\mathbf{X},\,\Gamma_2),...,$ $G_p = (\mathbf{X},\,\Gamma_p)$, let us unite in the same figure the arcs of these graphs, showing them in different colors representing the graph to which they belong. By doing this, we have constructed what is called a "p-colored graph," which is only a graph according to the terminology we have adopted if $p = 1$ is a superposition of graphs. Figure 34.1 gives an example of a three-colored graph.

Let us now consider another concept. Let us unite the elements of a denumerable set by arcs in such a way that several arcs with the same

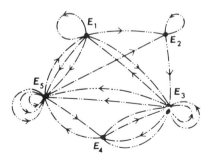

FIG. 34.1.

direction can connect an oriented pair of vertices X_i and X_j. If the greatest number of arcs in the same oriented pair is p, we have formed a "p-applied graph." In the three-colored graph of Fig. 34.1, if we eliminate the differences of color we obtain a three-applied graph, but it would be easy to show other examples where a p-colored graph would, if treated in this way, give a q-applied graph where $q \leqslant p$.

Another generalization deals with the nonoriented concept. Given a finite set **X**, if certain vertices X_i are joined by links to certain other vertices X_j in such a way that there can be more than one link between X_i and X_j and if the greatest number of links between this pair X_i, X_j is p, we say that we have a p-graph or multigraph. A p-graph is not strictly a graph because it is a nonoriented concept.

We must point out that the definitions which we have just given for p-colored graphs, p-applied graphs, and p-graphs will often be found to differ from those given in the various works dealing with graphs and their generalizations.

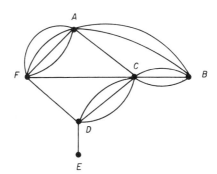

FIG. 34.2.   Representation of a 4-graph.

35. Typical Numbers in a p-Graph

A. CYCLOMATIC NUMBER

Let G be a p-graph with n vertices, m links, and r connected components.

We can assume that

$$\rho(G) = n - r, \tag{35.1}$$

$$\nu(G) = m - \rho(G) = m - n + r. \tag{35.2}$$

The number ν is called the *cyclomatic number* of the p-graph.

In certain p-graphs such as those of electrical networks (Fig. 35.1a), $\nu(G)$, and $\rho(G)$ have a physical significance: they represent, respectively, the number of cycles which can constitute basic mesh currents in independent linear links and the number of pairs of vertices (pairs of nodes) which can constitute an arrangement of differences of potential between pairs of nodes. Hence, in the electrical network shown in Fig. 35.1a it is possible to evaluate all the currents circulating in the links

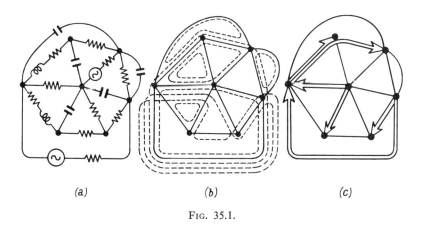

(a)   (b)   (c)

FIG. 35.1.

(branches) by using *currents of independent linear links* of the number ν (Fig. 35.1b), and in the same way we can evaluate the differences of potential between the extremities of the links by using the *independent linear differences of potential* of number ρ (Fig. 35.1c).

In the given example (Fig. 35.1), we have

$$n = 7, \qquad m = 15, \qquad r = 1, \tag{35.3}$$

whence:

$$\rho = n - r = 7 - 1 = 6 \tag{35.4}$$

and

$$\nu = m - \rho = 15 - 6 = 9. \tag{35.5}$$

Hence, a linking of cables requires nine cables and a linkage of pairs of nodes requires six pairs of nodes. The choice of these linkings is clearly arbitrary and we can state the following properties (bearing in mind the concept of a tree as defined in Section 44):

(a)   In order to arrange a linking between pairs of nodes, we need only construct a tree in each connected component and define the ρ differences of separate potential in it.

(b)   In order to construct basic mesh currents, we need only construct a tree in each connected component and to pass one mesh, and one only, through each link of the circuit which does not belong to the tree. It is in this way that the tree of Fig. 35.1c has been used to construct the basic mesh currents of Fig. 35.1b.

B. Chromatic Number

We say that a 1-graph is *r-chromatic*[1] if it is possible to mark the vertices with r distinct colors without any two adjacent vertices having the same color. The smallest number r, which makes the graph r-chromatic, is called the *chromatic number* of the graph and is represented by $\gamma(G)$.

If we look at the map of ten counties shown in Fig. 35.2 we see that it can be shaded with four different colors represented by B, J, V, and R.[2] We can verify that four is the least number of colors which can be used if no two adjacent counties are to have the same color. The graph of Fig. 35.2 is four-chromatic: $\gamma(G) = 4$.

König has proved (see [H2], p. 31) that a graph is two-chromatic if, and only if, it does not contain any cycles of uneven length.

To determine the chromatic number it is necessary to solve a linear program with integral values (see [H2]); and this reference also gives a useful empirical method.

C. Chromatic Class

This is an integer q with the following properties:

(1)   It is possible with q separate colors to color each link of the graph in such a way that no two adjacent links are marked with the same color.
(2)   This is not possible with $q - 1$.

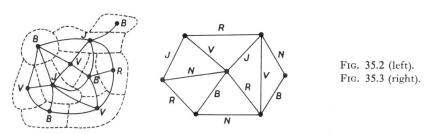

Fig. 35.2 (left).
Fig. 35.3 (right).

[1] A "*p*-colored" graph (Sect. 34) must not be confused with a "*p*-chromatic 1-graph."
[2] *Note*: B, J, V, R represent colors used in original French edition: Bleu (Blue), Jaune (Yellow), Vert (Green), and Rouge (Red).

Expressed differently, q is the minimal number of colors that can be used to mark the links of a graph in such a way that no two adjacent links will have the same color.

An example is given in Fig. 35.3 where the number of the chromatic class is five, so that at least five colors must be used (B, N, V, J, and R).[1]

To find the chromatic class of a graph it is necessary to consider that of another graph, which is obtained from the original graph in the following manner:

The chromatic class[2] of graph $G = (\mathbf{X}, \mathbf{\bar{U}})$ is the chromatic number of graph $G^* = (\mathbf{\bar{U}}, \varGamma^*)$ which is defined thus: the vertices of G^* are the links of G; $\bar{u}' \in \varGamma^* \bar{u}$ if the links $\bar{u}$ and $\bar{u}'$ are adjacent in the initial graph.

We shall now show how to obtain G^* from G by means of an example (Fig. 35.4).

Let

$$G = (\mathbf{X}, \mathbf{\bar{U}}),$$

where

$$\mathbf{X} = \{X_1, X_2, X_3, X_4, X_5, X_6\}, \tag{35.6}$$

$$\mathbf{\bar{U}} = \{\bar{u}_1, \bar{u}_2, \bar{u}_3, \bar{u}_4, \bar{u}_5, \bar{u}_6, \bar{u}_7, \bar{u}_8\}. \tag{35.7}$$

The graph is shown by heavy points and dark lines (see Fig. 35.4).

Each link $\bar{u}_i$ being considered as a vertex, $\bar{u}_i$ will be joined to $\bar{u}_j$ by a link, if, in graph G, the links $\bar{u}_i$ and $\bar{u}_j$ are adjacent. Thus $\bar{u}_i$ will be joined to $\bar{u}_2$, since $\bar{u}_1$ and $\bar{u}_2$ are adjacent in G, and also in $\bar{u}_2$ and $\bar{u}_5$ or $\bar{u}_6$ and $\bar{u}_7$, etc. In this way we obtain graph G^*, the vertices of which are shown by light points and dotted lines.

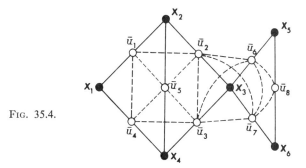

FIG. 35.4.

---

[1] N represents Noir (Black).

[2] When writing $G = (\mathbf{X}, \mathbf{\bar{U}})$ we are considering the graph obtained with the vertices belonging to $\mathbf{X}$ and the links belonging to the set $\mathbf{\bar{U}}$ of links. (See p. 230 where we explained that we should take $\mathbf{U}$ to represent the set of arcs and, p. 241, $\mathbf{\bar{U}}$ the set of links).

It is then clear that the chromatic number of G^* is equal to the chromatic class of G. In addition, by coloring the vertices of G^* we obtain the colors for the links in G.

D. SET OF INTERNALLY STABLE VERTICES

Any subset $\mathbf{S} \subset \mathbf{X}$, such that *no two of its vertices are adjacent*, is said to be *internally stable*. We must therefore have

$$\mathbf{S} \cap \Gamma\mathbf{S} = \varnothing. \tag{35.8}$$

Example (Fig. 35.5).   The subsets

$$\mathbf{S}_1 = \{A, D, G\}, \qquad \mathbf{S}_2 = \{B, G\},$$
$$\mathbf{S}_3 = \{A, C, D, G\} \tag{35.9}$$

are internally stable. Indeed, for $\mathbf{S}_1$, for example,

$$\Gamma(A) = \{B, E, F\}, \qquad \Gamma(D) = \{E\}, \qquad \Gamma(G) = \{H\}, \tag{35.10}$$

whence

$$\Gamma\mathbf{S}_1 = \{B, E, F, H\} \tag{35.11}$$

and

$$\{A, D, G\} \cap \{B, E, F, H\} = \varnothing. \tag{35.12}$$

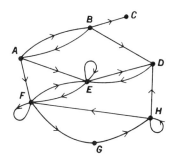

FIG. 35.5.

E. NUMBER OF INTERNAL STABILITY OF A GRAPH

This is the number of elements of the largest internally stable subset or subsets, and is represented as $\alpha(G)$:

$$\alpha(G) = \underset{\mathbf{S}\in\Sigma}{\text{MAX}} \mid \mathbf{S} \mid, \tag{35.13}$$

where Σ is the family of internally stable sets.

In the given example (Fig. 35.5), $\alpha(G) = 4$. The largest internally stable subset is $\mathbf{S} = \{C, A, D, G\}$. It should be understood that there may be several such subsets with the same number of elements α in a graph.

Between $\alpha(G)$ and $\gamma(G)$ there is a useful relation:

$$\alpha(G) \cdot \gamma(G) \geqslant |\mathbf{X}|. \tag{35.14}$$

The concept of internal stability is useful in various fields such as in the theory of information. The following example is given in Berge's work ([H2], p. 37).

Let us assume that a transmitter can give out five signals a, b, c, d, e, which are received in such a way that each signal gives rise to two interpretations: a can be p or q; b can be q or r; c can be r or s; d can be s or t; and e can be t or p (Fig. 35.6). What is the maximal number of signals that must be used to avoid any possible confusion? The problem can be treated as that of finding an internally stable and maximal set of a graph G, in which two vertices are adjacent if the signals they represent can be confused. From the graph shown in Fig. 35.6 we shall construct that of Fig. 35.7, in which $\alpha(G) = 2$, and shall choose one of the following sets:

$$\{a, c\}, \qquad \{b, d\}, \qquad \{c, e\}, \qquad \{d, a\}, \qquad \{e, b\}.$$

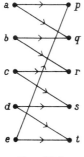

FIG. 35.6.

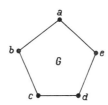

FIG. 35.7.

F. S??   ?ABLE VERTICES

?h that each vertex $X_i \notin \mathbf{T}$ is connected to at ?rc with its origin at X_i , is said to be *externally* ? say, $\mathbf{T}$ is externally stable if

$$\forall X_i \notin \mathbf{T}, \qquad \mathbf{T} \cap \Gamma X_i \neq \varnothing.$$

Example (Fig. 35.5). The subset $\{C, D, E, F, H\}$ is externally stable, as we can verify, since all the other vertices A, B, and G are such that

$$\Gamma(A) = \{B, E, F\}, \qquad \Gamma(B) = \{A, C, D\}, \qquad \text{and} \qquad \Gamma(G) = \{H\}$$

each contains at least one vertix of $\{C, D, E, F, H\}$.

G. NUMBER OF EXTERNAL STABILITY OF A GRAPH

This is the number of elements in the smallest externally stable subset or subsets

$$\beta(G) = \operatorname*{MIN}_{\mathbf{T} \in \mathbf{T}} |\mathbf{T}|, \tag{35.15}$$

where $\mathbf{T}$ is the family of externally stable sets.

In Fig. 35.5, $\beta(G) = 3$, and the smallest externally stable subset is $\mathbf{T} = \{C, E, H\}$. It should be understood that in other cases there may be more than one such minimal subset.

H. NUCLEI OF A GRAPH

Given a graph $G = (\mathbf{X}, \Gamma)$, which is finite or infinite, the subset $\mathbf{S}_0 \subset \mathbf{X}$ is a nucleus of G if $\mathbf{S}_0$ is internally and externally stable, that is to say, if

$$\forall X_i \in \mathbf{S}_0: \qquad \mathbf{S}_0 \cap \Gamma X_i = \varnothing, \tag{35.16}$$

$$\forall X_j \notin \mathbf{S}_0: \qquad \mathbf{S}_0 \cap \Gamma X_j = \varnothing. \tag{35.17}$$

From these conditions it follows that a nucleus cannot contain a vertex with a loop, that it must include every vertex X_i such that $\Gamma X_i = \varnothing$, and finally, that $\varnothing$ is not a nucleus.

Example. In Fig. 35.5 $\{A, C, D, G\}$ is a nucleus, since no two of these vertices are adjacent and every other vertex B, E, F, or H is joined to at least one of the vertices $\{A, C, D, G\}$ by an arc with its initial extremity at B, E, F, or H. This subset is the only nucleus in the graph.

I. PROPERTIES RELATING TO THE NUCLEUS OF A GRAPH

(a) A graph does not necessarily have a nucleus;

(b) A graph may have several nuclei;

(c)   If a graph has a nucleus $\mathbf{S}_0$, then with $|\mathbf{S}_0|$ as the number of vertices of this nucleus:

$$\alpha(G) \geqslant |\mathbf{S}_0| \geqslant \beta(G); \tag{35.18}$$

(d)   If $\mathbf{S}_0$ is a nucleus of G, the set $\mathbf{S}_0$ is a maximal set of the family Σ of the internally stable sets;

(e)   In a symmetrical graph without loops, every maximal set of the family Σ of the internally stable sets, is a nucleus;

(f)   A symmetrical graph without a loop always has a nucleus;

(g)   If a finite graph does not have a circuit of uneven length, it contains a nucleus.

The concept of a nucleus plays an important part in the theory of games of strategy (see [H2]).

36. Separation or Shortest Path in a Graph

A. SEPARATION BETWEEN TWO VERTICES

Given two vertices X_i and X_j of a graph $G = (\mathbf{X}, \Gamma)$, the *separation* between X_i and X_j is the number of arcs in the path of minimal length connecting the vertices, and is shown as $d(X_i, X_j)$. Thus, in Fig. 36.1, $d(A, B) = 4$, $d(B, A) = 7$. When there is no route from X_i to X_j, we assume $d(X_i, X_j) = \infty$.

To find the path or paths corresponding to the separation between two vertices the following simple algorithm is used:

(a)   Every vertex X_i is progressively given a number m equal to the length of the shortest route from X_0 (the origin) to X_i, up to X_n (the destination).

The proof is immediate and is derived from the principle of dynamic

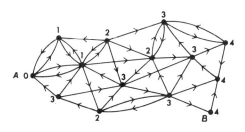

FIG. 36.1.

programming itself: "A path of minimal length can only be formed from partial paths which are minimal."

We give the value 0 to X_0, then 1 to all the vertices of ΓX_0, then 2 to all those of $\Gamma^2 X_0$ (unless already marked), proceeding to $\Gamma^3 X_0$, etc., and stopping when X_n has been reached. If r is the number of X_n, then $d(X_0, X_n) = r$. The optimal route (or routes) are found by returning from a higher number to the one below it.

A simple example is given in Fig. 36.1, where the minimal path from A to B can be found with little difficulty.

B. PATH OF MINIMAL VALUE (WEIGHTED LENGTH)

In a graph $G = (\mathbf{X}, \mathbf{U})$, every arc $u \in \mathbf{U}$ is assigned a number $l(u) \geqslant 0$ called the *value*[1] of u; we wish to find a path μ from a vertex $A \in \mathbf{X}$ to a vertex $B \in \mathbf{X}$ such that the total value

$$l(\mu) = \sum_{u \in \mu} l(u) \tag{36.1}$$

is minimal. Several algorithms can be used to solve this problem (see [H2] and [H9], for example), and we shall explain two of them.

C. FORD'S ALGORITHM

(1)   Mark every vertex X_i with an index λ_i; we begin by taking $\lambda_0 = 0$ and

$$\lambda_i = +\infty \qquad \text{if} \quad i \neq 0;$$

(2)   We look for an arc (X_i, X_j) such that

$$\lambda_j - \lambda_i > l(X_i, X_j);$$

We then replace λ_j by $\lambda_j' = \lambda_i + l(X_i, X_j) < \lambda_j$, observing that $\lambda_j' > 0$ if $j \neq 0$. The process is continued until there is no arc left which makes it possible to reduce the λ_i terms;

(3)   There is a vertex X_{p_1} such that

$$\lambda_n - \lambda_{p_1} = l(X_{p_1}, X_n);$$

λ_n has diminished monotonically in the course of the process and X_{p_1} is the last vertex used to reduce λ_n. In the same way, give X_{p_2} such that

$$\lambda_{p_1} - \lambda_{p_2} = l(X_{p_2} - X_{p_1}), ..., \text{etc.....} . \tag{36.2}$$

---

[1] Different authors call $l(u)$ "the generalized length," "the weighted length," or simply "the length."

The sequence λ_n, λ_{p_1}, λ_{p_2} ,... being strictly diminishing, there will be a certain moment

$$X_{p_{k+1}} = X_0 .$$

Then λ_n is the value of the minimal path from X_0 to X_n and

$$\mu = [X_0 , X_{p_k} , X_{p_{k-1}} ,..., X_{p_1} , X_n]$$

is this path.

Figure 36.2 provides an example for the reader to study.

D. The Bellman–Kalaba Algorithm [H9]

This is a variant of the previous one, shown with the help of dynamic programming.[1] It depends on the following obvious property: "Every minimal route which does not contain more than r arcs is formed by partial paths which do not contain more than k arcs ($k \leqslant r$) which are themselves minimal."

For this algorithm we shall use a different notation. Taking $c_{ij} \geqslant 0$ as the value of arc (X_i , X_j), we assume that, for every $(X_i , X_j) \notin \mathbf{U}$, $c_{ij} = \infty$, and that for every (X_i , X_i), $c_{ii} = 0$. It is required to find a path:

$$\mu = [X_0 , X_{i_1} , X_{i_2} ,..., X_{i_k} , X_n] \tag{36.3}$$

such that

$$c_{0i_1} + c_{i_1 i_2} + c_{i_2 i_3} + \cdots + c_{i_k n} \tag{36.4}$$

is minimal.

All we need do is to solve the equations:

$$v_i = \underset{j \neq i}{\text{MIN}}(v_j + c_{ij}), \qquad i = 0, 1, 2,..., n - 1, \tag{36.5}$$

$$v_n = 0, \tag{36.6}$$

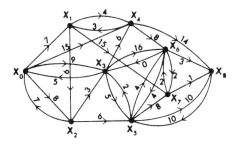

FIG. 36.2.

[1] The method explained in Part I, Section 11 is simply a special and more elementary case of this.

where $v_i (i = 0, 1, 2,..., n - 1)$ represents the value of the optimal path from vertex i to the terminal vertex. The $n + 1$ vertices are assumed as numbered from 0 to n, from the origin to the final vertex.

We proceed as follows. Let

$$v_i^{(0)} = c_{in}, \qquad i = 0, 1, 2,..., n - 1, \tag{36.7}$$

$$v_n^{(0)} = 0. \tag{36.8}$$

We next calculate

$$v_i^{(1)} = \underset{j \neq i}{\mathrm{MIN}}(v_j^{(0)} + c_{ij}) \qquad \begin{cases} i = 0, 1, 2,..., n - 1, \\ j = 0, 1, 2,..., n, \end{cases} \tag{36.9}$$

$$v_n^{(1)} = 0; \tag{36.10}$$

then, in succession,

$$v_i^{(k)} = \underset{j \neq i}{\mathrm{MIN}}(v_j^{(k-1)} + c_{ij}) \qquad \begin{cases} i = 0, 1, 2,..., n - 1, \\ j = 0, 1, 2,..., n, \end{cases} \tag{36.11}$$

$$v_n^{(k)} = 0, \tag{36.12}$$

for all the values of $k = 1, 2, 3,...$, stopping when

$$v_i^{(k)} = v_i^{(k-1)}, \qquad i = 0, 1, 2,..., n. \tag{36.13}$$

Thus $v_0^{(k)}$ expresses the value of the optimal route between the vertices X_0 and X_n. It has been proved that if the graph has $n + 1$ vertices, $n - 1$ iterations are sufficient to reach the optimum.

In various problems, the c_{ij} terms represent times of transit t_{ij} (examples of this would be the minimal time required for the transit of a message in a communications network or for a shipment of goods to get through a network of roads or streets); in other problems these c_{ij} terms are transport costs or operational costs in a technological process.

E. EXAMPLE (Fig. 36.2)

The matrix of the costs is given in Table 36.1.
We first have

$$v_0^{(0)} = c_{08} = \infty, \qquad v_1^{(0)} = c_{18} = \infty,$$

$$v_2^{(0)} = c_{28} = \infty, \qquad v_3^{(0)} = c_{38} = \infty,$$

$$v_4^{(0)} = c_{48} = 14, \qquad v_5^{(0)} = c_{58} = 10, \tag{36.14}$$

$$v_6^{(0)} = c_{68} = 3, \qquad v_7^{(0)} = c_{78} = 1,$$

$$v_8^{(0)} = c_{88} = 0.$$

TABLE 36.1

	(0)	(1)	(2)	(3)	(4)	(5)	(6)	(7)	(8)
(0)	0	7	8	9	15	∞	∞	∞	∞
(1)	∞	0	6	∞	4	∞	∞	15	∞
(2)	7	∞	0	3	∞	6	∞	∞	∞
(3)	5	∞	∞	0	6	5	16	∞	∞
$c_{ij} =$ (4)	∞	3	∞	∞	0	∞	8	∞	14
(5)	∞	∞	∞	2	∞	0	4	8	10
(6)	∞	∞	∞	0	∞	4	0	2	3
(7)	∞	∞	∞	∞	∞	∞	2	0	1
(8)	∞	∞	∞	∞	∞	10	∞	∞	0

$$v_0^{(1)} = \underset{j \neq 0}{\text{MIN}}(v_j^{(0)} + c_{0j}) = \min[v_1^{(0)} + c_{01}, v_2^{(0)} + c_{02}, ..., v_8^{(0)} + c_{08}]$$
$$= \min[\infty + 7, \infty + 8, \infty + 9, 14 + 15, 10 + \infty, 3 + \infty,$$
$$1 + \infty, 0 + \infty] = 29,$$

$$v_1^{(1)} = \underset{j \neq 1}{\text{MIN}}(v_j^{(0)} + c_{1j}) = \min[v_0^{(0)} + c_{10}, v_2^{(0)} + c_{12}, ..., v_8^{(0)} + c_{18}]$$
$$= \min[\infty + \infty, \infty + 6, \infty + \infty, 14 + 4, 10 + \infty,$$
$$3 + \infty, 1 + 15, 0 + \infty] = 16,$$

$$v_2^{(1)} = \underset{j \neq 2}{\text{MIN}}(v_j^{(0)} + c_{2j}) = \min[v_0^{(0)} + c_{20}, v_0^{(0)} + c_{21}, ..., v_8^{(0)} + c_{28}]$$
$$= \min[\infty + 7, \infty + \infty, \infty + 3, 14 + \infty, 10 + 6, 3 + \infty,$$
$$1 + \infty, 0 + \infty] = 16,$$ (36.15)

$$v_3^{(1)} - \underset{j \neq 3}{\text{MIN}}(v_j^{(0)} + c_{3j}) = \min[v_0^{(0)} + c_{30}, v_1^{(0)} + c_{31}, .., v_8^{(0)} + c_{38}]$$
$$= \min[\infty + 5, \infty + \infty, \infty + \infty, 14 + 6, 10 + 5, 3 + 16,$$
$$1 + \infty, 0 + \infty] = 15,$$

and so on. We shall now form Table 36.2, finding that with $k = 4$, we have

$$\forall i \quad v_i^{(4)} = v_i^{(3)},$$

hence $v_0^{(4)} = 21$ is the optimal value. To find the optimal path or paths, all we need do is to note on the return path from $v_i^{(k-1)}$ to $v_i^{(0)}$ those optimal portions of path which form the optimal path or paths. There are four optimal solutions:

$$\mu_1 = [X_0, X_2, X_5, X_6, X_8],$$
$$\mu_2 = [X_0, X_2, X_5, X_6, X_7, X_8],$$
$$\mu_3 = [X_0, X_3, X_5, X_6, X_8],$$
$$\mu_4 = [X_0, X_3, X_5, X_6, X_7, X_8].$$

(36.16)

TABLE 36.2

$v_i^{(k)} = $ Value of optimal route from X_i to X_n if it is $\leqslant k$	$v_0^{(k)}$	$v_1^{(k)}$	$v_2^{(k)}$	$v_3^{(k)}$	$v_4^{(k)}$	$v_5^{(k)}$	$v_6^{(k)}$	$v_7^{(k)}$	$v_8^{(k)}$
$k = 0$	∞	∞	∞	∞	14	10	3	1	0
$k = 1$	29	16	16	15	11	7	3	1	0
$k = 2$	23	15	13	12	11	7	3	1	0
$k = 3$	21	15	13	12	11	7	3	1	0
$k = 4$	21	15	13	12	11	7	3	1	0

F. FINDING A ROUTE OF MAXIMAL VALUE

The algorithm of Bellman and Kalaba can be used to find a route of maximal value, but there is one obvious restriction: the graph cannot contain a circuit or loop, for if it did, the length of a route (the number of adjacent arcs) could be nonfinite and the value of the route would be without bounds. The search for a path of maximal value will be carried out by using Eqs. (36.5)–(36.12) in which $\text{MIN}_{j \neq 1}$ will be replaced by $\text{MAX}_{j \neq 1}$. In addition $c_{ij} = -\infty$ will be substituted in Table 36.2 of c_{ij} terms if X_i is not joined to X_j by an arc.

Example (Fig. 36.3). The circled numbers show the values of the paths of maximal value from each point X_i to the terminal point X_n. Another example was given in Section D.

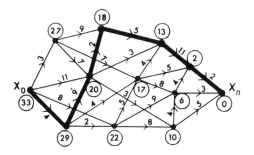

FIG. 36.3.

37. Transport Network

A. DEFINITION

The concepts explained in Part I (Section 6) will now be considered in a more theoretical form.

We use the term *transport network* for a *finite* graph without loops in which each arc u is assigned a number $c(u) \geqslant 0$, known as the *capacity of the arc,* and in which

(1) There is one and only one vertex X_0 such that $\Gamma^{-1}X_0 = \phi$; this vertex is called *the entry to the network*;

(2) There is one and only one vertex X_n such that $\Gamma X_n = \phi$; this vertex is called *the exit from the network*.

B. FLOW IN A TRANSPORT NETWORK

Given $\mathbf{U}_{X_i}^{-}$ the set of arcs connected terminally to X_i, and $\mathbf{U}_{X_i}^{+}$ that of the arcs connected initially to X_i, we say that a function $\phi(u)$ defined on $\mathbf{U}$ and with *integral values* is a *flow* in a transport network if we have

$$\phi(u) \geqslant 0, \qquad \forall u \in \mathbf{U}, \tag{37.1}$$

$$\sum_{u \in \mathbf{U}_{X_i}^{-}} \phi(u) = \sum_{u \in \mathbf{U}_{X_i}^{+}} \phi(u) \qquad (X_i \neq X_0, X_i \neq X_n), \tag{37.2}^1$$

$$\phi(u) \leqslant c(u), \qquad \forall u \in \mathbf{U}. \tag{37.3}$$

$\phi(u)$ can be used to represent a quantity of matter which enters arc (u) and which must never exceed the capacity $c(u)$ of this arc.

---

[1] This property is the one which is known in electricity as Kirchhoff's *law of nodes*; it is thus verified except for X_0 and X_n.

From (37.2) we obtain

$$\sum_{u \in \mathbf{U}^+_{X_0}} \phi(u) = \sum_{u \in \mathbf{U}^-_{X_n}} \phi(u) = \phi_0 . \tag{37.4}$$

The number ϕ_0 expresses the quantity of matter which leaves X_0 and reaches X_n through the network.

C. CUT

If $\mathbf{A}$ is a set of the vertices of the network which include X_n but not X_0, the set $\mathbf{U}_A^-$ of arcs connected terminally to $\mathbf{A}$ is a *cut* of the network.

Example (Fig. 37.1).   Let

$$\mathbf{A} = \{X_n, X_2, X_3, X_6, X_9\}. \tag{37.5}$$

$$\mathbf{U}_A^- = \{(X_8, X_n), (X_{10}, X_n), (X_0, X_2), (X_1, X_2), (X_5, X_2),$$
$$(X_0, X_3), (X_5, X_9), (X_{10}, X_9)\} \tag{37.6}$$

is the cut corresponding to $\mathbf{A}$ and is shown by a broken line in the graph.[1]

Since $\mathbf{A}$ includes the exit, any unit of matter passing from X_0 to X_n borrows at least one arc from $\mathbf{U}_A^-$; hence, whatever a flow ϕ and a cut $\mathbf{U}_A^-$ may be, we have

$$\phi(X_n) \leqslant c(\mathbf{U}_A^-). \tag{37.7}$$

If there is a flow $\phi_0(X_n)$ and a cut $\mathbf{V}$ such that

$$\phi_0(X_n) = c(\mathbf{V}), \tag{37.8}$$

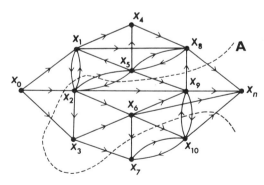

FIG. 37.1.

[1] The reason for drawing a cut on the graph is merely to show the terminally connected arcs which are encountered, and it may also form nonconnected open or closed lines, depending on the choice of $\mathbf{A}$.

the flow $\phi_0(X_n)$ therefore has a maximal value and cut **V** a minimal capacity; hence the following theorem.

D. THE FORD–FULKERSON THEOREM (MIN-CUT, MAX-FLOW)

In a given transport network, the maximal value of a flow is equal to the minimal capacity of a cut.

$$\text{MAX}\,\phi_0 = \underset{\substack{X_0 \in \mathbf{A} \\ X_n \in \mathbf{A}}}{\text{MIN}}\ c(\mathbf{U}_A^{-}). \tag{37.9}$$

The Ford–Fulkerson Algorithm. This algorithm was given in Part I, Section 6, and the method is justified by the above theorem (Eq. 37.9).

E. FINDING THE MINIMAL FLOW WHICH SATISFIES $\phi(u) \geqslant c(u)$, $\forall u$

First Method. The Ford–Fulkerson algorithm can be used with suitable modifications, but we must first change our definition of a complete flow. We shall state that a flow φ is complete if it is impossible to find a path from the entry X_0 to the exit X_n which does not include an arc u such that $\varphi(u) > c(u)$. The algorithm now becomes

(a) Pass an arbitrary flow φ, such that $\varphi(u) \geqslant c(u)$, through the network;

(b) Find a complete flow by reducing the flow in the arcs which constitute the paths from X_0 to X_n ;

(c) Find the minimal flow as follows:

(i) Mark X_0 with a $+$;

(ii) After marking a vertex X_i , mark with $[+X_i]$ all the unmarked vertices X_j where there is an arc (X_i, X_j) such that $\varphi(X_i, X_j) > c(X_i, X_j)$, then mark with $[-X_i]$ each vertex X_j where there is an arc (X_j, X_i);

(iii) If this process enables us to mark X_n , we can diminish the flow along the chain from X_0 to X_n . We repeat process (iii) until it is no longer possible to mark X_n , and we now have the minimal flow.

Second Method, if the Graph is Antisymmetrical.[1]

(1) We try to find a flow $\phi_1(u) \geqslant c(u)$, $\forall u$ by sending a unit of matter towards the least saturated arcs;

(2) We assume

$$c_2(u) = \phi_1(u) - c(u), \quad \forall u$$

[1] In point of fact, this method is valid for any graph, but $\varphi = \varphi_1 - \varphi_2$ is not always the minimal flow; if the graph is antisymmetrical we can be sure that $\varphi_1 - \varphi_2$ is the minimal flow.

and we construct a maximal flow

$$\phi_2(u) \leqslant c_2(u), \quad \forall u$$

with the Ford–Fulkerson algorithm.

(3) We have

$$\phi_1(u) - \phi_2(u) \geqslant \phi_1(u) - c_2(u) = c(u),$$

where $\phi = \phi_1 - \phi_2$ is the minimal flow for which we have been looking.

38. Linking Two Disjoint Sets

A. DEFINITION

Given two finite disjoint sets **X** and **Y** such that **X** ∩ **Y**= ∅, and a multivocal mapping Γ of **X** into **Y**, we have a graph which we shall call $G = (\mathbf{X}, \mathbf{Y}, \Gamma)$. A *linking* is a set of arcs **W** which belong to G, and of which no two are adjacent.

In Fig. 38.1 a linking of **X** and **Y** is shown by heavy lines.

If a linking makes a set **A** ⊂ **X** correspond in a biunivocal manner with a set **B** ⊂ **Y**, it is said we "link **A** on **B**" or "**A** in **Y**."

A typical example of linking is given by the allocation of p workmen X_1, X_2,..., X_p to q vacancies Y_1, Y_2,..., Y_q, when each workman is fitted for one or more of the positions.

$$\Gamma X_1 = \{Y_1, Y_2, Y_4\}$$
$$\Gamma X_2 = \{Y_2, Y_4\}$$
$$\vdots$$

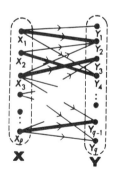

FIG. 38.1.

B. THE KÖNIG–HALL THEOREM

We can link **X** in **Y** if, ond only if

$$\forall \mathbf{A} \subset \mathbf{X}, \tag{38.1}$$

we have

$$| \Gamma \mathbf{A} | \geqslant | \mathbf{A} |.$$

In other terms, if every subset $\mathbf{A}$ of $\mathbf{X}$ with k vertices has at least k arcs, we can always find a linking.

C. MAXIMAL LINKING

A linking is *maximal* if it contains the maximal number of arcs.

To find one, we construct a transport network with an origin A and a destination B as shown in Fig. 38.2. We then reverse the direction of

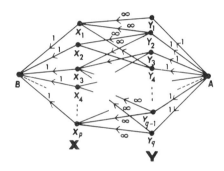

FIG. 38.2.

the arcs between $\mathbf{X}$ and $\mathbf{Y}$, giving a capacity of 1 to those leaving A or reaching B, and an infinite capacity to those joining $\mathbf{Y}$ to $\mathbf{X}$. We thus obtain a transport network in which the Ford–Fulkerson algorithm is used to find the maximal flow.

In [H2] numerous theorems will be found dealing with the concept of linking.

D. PROBLEM OF ALLOCATION. HUNGARIAN ALGORITHM[1]

Given p workmen X_1, X_2,..., X_p, and p vacancies Y_1, Y_2,..., Y_p, a value is given to each allocation (X_i, Y_j)

$$c_{ij} \geqslant 0, \qquad i,j = 1, 2,..., p.$$

---

[1] This method is the outcome of the successive studies of König, Ergévary, and Kuhn.

Some c_{ij} terms can be infinite, which means that the corresponding allocation is impossible. Our problem is to assign p workmen to p positions so that each workman has only one position and the total value of the allocations is minimal.

The problem can be solved by means of the Hungarian algorithm, and our explanations refer to Fig. 38.3 and Table 38.1. The matrix c_{ij} of the values given to the allocations is shown in the latter, and these values may represent costs, periods, etc., depending on the problem.

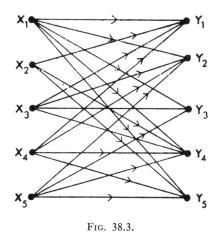

FIG. 38.3.

Before introducing the algorithm we must give an important lemma: increasing or decreasing each element of a line or column of the matrix c_{ij} by a quantity λ, in a problem of allocation, alters the total value but does not affect the optimal solutions.

TABLE 38.1

	Y_1	Y_2	Y_3	Y_4	Y_5
X_1	7	3	5	7	10
X_2	6	∞	∞	8	7
$c_{ij} = X_3$	6	5	1	5	∞
X_4	11	4	∞	11	15
X_5	∞	4	5	2	10

Stage 1. Obtaining the Zeros. Remove the smallest element of each column, then remove the smallest element in each line of the new matrix. By this process, we are sure of obtaining a matrix $c_{ij}^{(1)}$ with at least one zero in each line and column. In the matrix shown in Table 38.1 we have removed 6 from column (1), 3 from (2), etc., and have produced the matrix in Table 38.2.

		TABLE 38.2						TABLE 38.3		
	(1)	(2)	(3)	(4)	(5)		(1)	(2)	(3)	(4)	(5)
(1)	1	0	4	5	3	(1)	1	0	4	5	3
(2)	0	∞	∞	6	0	(2)	0	∞	∞	6	0
(3)	0	2	0	3	∞	(3)	0	2	0	3	∞
(4)	5	1	∞	9	8	(4)	4	0	∞	8	7
(5)	∞	1	4	0	3	(5)	∞	1	4	0	3

Stage 2. Seeking the Optimal Solution. Using the zeros of $c_{ij}^{(1)}$, we try to find a zero solution, in other words, an allocation in which all the $c_{ij}^{(1)}$ terms are zeros. If this is possible, we have found the optimal solution, and if not, we go on to Stage 3.

To find this zero solution, we first consider one of the lines which contain the fewest zeros; we enclose one of the zeros and then cross out the zeros in the same line or column as the enclosed zero. All the lines are then treated in the same manner.

In Table 38.4 we have encased $c_{12}^{(1)}$ and have crossed out $c_{42}^{(1)}$, enclosed $c_{54}^{(1)}$, etc. It can be seen that a zero solution was unobtainable.

Stage 3. Obtaining a Minimal Set of Lines and Columns Containing all the Zeros. Take the following steps:

(a) Ascertain whether the linking obtained in Stage 2 is maximal; if not, use the Ford–Fulkerson algorithm as shown earlier.

(b) Mark with a cross all the lines without an encased zero;

(c) Mark each column which contains one or more crossed out zeros;

(d) Mark each line with an enclosed zero which is in a marked column;

(e) Repeat (c) and (d) until no new marked lines or columns can be found.

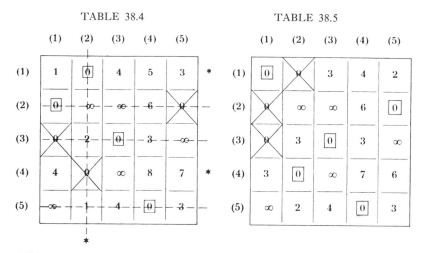

TABLE 38.4

	(1)	(2)	(3)	(4)	(5)
(1)	1	⬚0	4	5	3	*
(2)	⬚0	∞	∞	6	✗
(3)	✗	2	⬚0	3	∞
(4)	4	✗	∞	8	7	*
(5)	∞	1	4	⬚0	3

TABLE 38.5

	(1)	(2)	(3)	(4)	(5)
(1)	⬚0	✗	3	4	2
(2)	✗	∞	∞	6	⬚0
(3)	✗	3	⬚0	3	∞
(4)	3	⬚0	∞	7	6
(5)	∞	2	4	⬚0	3

This process enables us to obtain a minimal set of lines and/or columns containing all the zeros which are enclosed or crossed out (see König's theorem further on).

In our example we have marked line (4), then column (2) and line (1), but could not proceed further.

Stage 4. Draw a dashed line through each *unmarked line* and each *marked column.* This makes it possible to obtain the minimal set of lines and/or columns which contain all the zeros.

In Table 38.4 it will be seen that dashed lines have been drawn through lines (2), (3), (5), and column (2).

Stage 5. Eventual Displacement of Certain Zeros. Examine the partial matrix of elements which are not crossed by dashed lines, and find the smallest element in it; subtract this number from the columns without a dashed line and add it to the lines marked with a dashed line.[1]

In Table 38.4, 1 has been subtracted from columns (1), (3), (4), and (5), and added to lines (2), (3), and (5) to produce Table 38.5.

Stage 6. Obtaining an Optimal Solution or Establishing a New Cycle. Reproduce Stage 2 on the new table $c_{ij}^{(2)}$ obtained in Stage 5, stopping when an optimal solution is obtained, but otherwise continuing

---

[1] This is the same as subtracting the number from the part without dashed lines and adding it to the elements at the intersection of two dashed lines, while leaving all other elements unchanged.

until Stage 2 produces an optimal solution. There may be more then one optimal solution.

In Table 38.5 we have enclosed $c_{42}^{(2)}$ and have crossed out $c_{12}^{(2)}$, enclosed $c_{54}^{(2)}$, etc. An optimal solution has been found with

$$c_{11}^{(2)} = c_{25}^{(2)} = c_{33}^{(2)} = c_{42}^{(2)} = c_{54}^{(2)} = 0.$$

In the present problem this gives us

$$c_{11} + c_{25} + c_{33} + c_{42} + c_{54} = 7 + 4 + 1 + 2 + 7 = 21.$$

Figure 38.4 shows the linking with a minimal value. This result can again be obtained and thus checked. In Stage 1 we subtracted $6 + 3 + 1 + 2 + 7 = 19$ from the columns, and 1 from the lines, making a total of 20; in Stage 3 we subtracted 4 times 1 and added 3 times 1 for a sum of -1, giving us a total of 21 subtracted for all the stages.

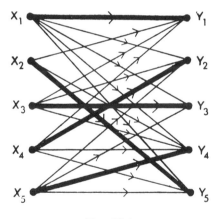

FIG. 38.4.

The Hungarian method may be applied with modifications to cases where the number of workmen differs from the number of vacancies. Reference [H2] should be consulted for such cases.

Finding the maximum: This was explained in Vol. I, Sect. 18. (English language edition: A. Kaufmann, *Methods and Models of Operations Research*. Prentice-Hall, Englewood Cliffs, New Jersey, 1963).

E. KÖNIG'S THEOREM

The minimal number of rows (lines and/or columns) containing all the zeros of a table is equal to the maximal number of zeros situated in separate lines and columns.

This theorem may be enunciated in a different way. We use the term *support of a graph* $G = (\mathbf{X}, \mathbf{Y}, \Gamma)$ constituting a linking, for a set $\mathbf{S}$ of points of $\mathbf{X}$ and/or $\mathbf{Y}$ such that any arc of the graph has a point of $\mathbf{S}$ as its extremity. König's theorem can now be expressed as follows:

In any 1-graph $G = (\mathbf{X}, \mathbf{Y}, \Gamma)$, the minimal number of points of support is equal to the maximal number of arcs in a linking:

$$\underset{\mathbf{S} \in \mathbf{S}^*}{\text{MIN}} \mid \mathbf{S} \mid \ = \underset{\mathbf{C} \in \mathbf{C}^*}{\text{MAX}} \mid \mathbf{C} \mid, \tag{38.2}$$

where $\mathbf{S}^*$ is the set of supports and $\mathbf{C}^*$ is the set of linkings.

The reader will find the proof of König's theorem in Reference [H2].[1]

39. Hamiltonian Path and Circuit

A. DEFINITIONS

In a graph $G = (\mathbf{X}, \Gamma)$, we say that a path

$$\mu = [X_{i_1}, X_{i_2}, ..., X_{i_n}]$$

is *Hamiltonian* when it passes once, and once only, through each vertex of the graph. If the graph is of order n, the length of a Hamiltonian path (if there is one) is $l(\mu) = n - 1$. Hence, Hamiltonian paths constitute the class of elementary paths of length $(n - 1)$, which may be empty.

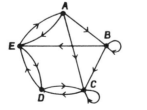

FIG. 39.1a.

---

[1] Or in *Invitation à la Recherche Opérationnelle*, A. Kaufmann and R. Faure, Dunod, Paris, 1962. (English language edition to be published in 1968 by Academic Press, New York under the title "Introduction to Operations Research.")

Example (Fig. 39.1a). $\mu = [B, E, A, C, D]$ is a Hamiltonian path. In a graph $G = (\mathbf{X}, \Gamma)$ we say that a circuit $\mu = [X_{i_1}, X_{i_2}, ..., X_{i_n}, X_{i_1}]$ is Hamiltonian if it passes once, and once only, through each vertex except X_{i_1}. If the graph is of order n, the length of the circuit (if there is one) is $l(\mu) = n$. Hence, Hamiltonian circuits form the class of elementary circuits of length n, a class which may be empty. An obvious condition for the existence of at least one Hamiltonian path is that the graph must be strongly connected. In Fig. 39.1a, $\mu = [A, B, C, D, E, A]$ is a Hamiltonian circuit.

Hamiltonian circuits obtained by circular permutations are called "equivalent." Hence the five circuits

$$[A, B, C, D, E, A], \quad [B, C, D, E, A, B], \quad [C, D, E, A, B, C],$$

$$[D, E, A, B, C, D], \quad [E, A, B, C, D, E]$$

are equivalent. To distinguish between equivalent Hamiltonian circuits, we refer to the initial and terminal vertex.

B. FINDING HAMILTONIAN PATHS. LATIN MULTIPLICATION[1]

There are various methods for finding Hamiltonian paths and circuits. The reader will find these in [H2] and [H56], but the simple method which will now be explained is due to the author of the present work.

Latin Multiplication. By means of a special type of matricial multiplication, it is successively possible to enumerate *without redundancy* all the elementary paths of length $1, 2, ..., n - 1$, stopping when the latter, which comprise all the Hamiltonian paths of an order n graph, have been found. If the graph does not have a Hamiltonian path, we can still find all the sets of elementary paths of length $1, 2, ..., k$ if the longest elementary path or paths have a length k.

The method of latin multiplication will be shown with the help of examples and will be found to be very simple.

Using the graph in Fig. 39.1a, we construct a *latin matrix*[2] as follows: if vertex i is joined to vertex j, we put ij in square (i, j); if i is not joined to j we put 0 in the same square. In addition, in every square (i, i) we write 0. Hence, the matrix for Fig. 39.1a becomes

[1] See A. Kaufmann and Y. Malgrange, *Revue Française de Recherche Opérationelle*, No. 26, Dunod, Paris, 1963.
[2] The epithet is used in the sense of "latin squares," a well-known term in statistics.

	A	B	C	D	E
A	0	AB	AC	0	AE
B	0	0	BC	0	BE
$[\mathcal{M}]^{(1)} = C$	0	0	0	CD	0
D	0	0	DC	0	DE
E	EA	0	0	ED	0

$$(39.1a)$$

From this latin matrix we now deduce another one by removing the first letter or *initial* in each square, and have

	A	B	C	D	E
A	0	B	C	0	E
B	0	0	C	0	E
$[\tilde{\mathcal{M}}]^{(1)} = C$	0	0	0	D	0
D	0	0	C	0	E
E	A	0	0	D	0

$$(39.1b)$$

We now "multiply" $[\mathcal{M}]^{(1)}$ by $[\tilde{\mathcal{M}}]^{(1)}$ to form $[\mathcal{M}]^{(2)}$. We shall follow the usual method of "line by column," but instead of obtaining products (which would be meaningless here), whenever a square (i, j) would be "multiplied" by a square (j, k), we put a zero in (i, k) of $[\mathcal{M}]^{(2)}$, if for every j, one or other, or both, of these squares contains a zero. A zero is also entered if, from all the "products," we cannot obtain a sequence in which no letter is repeated when a sequence of the square (j, k) of $[\tilde{\mathcal{M}}]^{(1)}$ is placed after a sequence of (i, j) of $[\mathcal{M}]^{(1)}$, for very j. (In this case, the sequences of $[\tilde{\mathscr{M}}]^{(1)}$ are reduced to one letter, but a generalization can be made, as we shall show later). Finally, when a sequence of (j, k) of $[\tilde{\mathscr{M}}]^{(1)}$ has been placed after a sequence of (i, j) of $[\mathscr{M}]^{(1)}$, and one or more sequences are found in which no letter is repeated, we write the sequence (or sequences) in square (i, k) of $[\mathscr{M}]^{(2)}$, and continue the same process for each j.

This gives

$[\mathscr{M}]^{(1)} \mathbf{L} [\tilde{\mathscr{M}}]^{(1)}$

	A	B	C	D	E			A	B	C	D	E
A	0	AB	AC	0	AE		A	0	B	C	0	E
B	0	0	BC	0	BE		B	0	0	C	0	E
= C	0	0	0	CD	0	**L**	C	0	0	0	D	0
D	0	0	DC	0	DE		D	0	0	C	0	E
E	EA	0	0	ED	0		E	A	0	0	D	0

(39.1c)

	A	B	C	D	E
A	0	0	ABC	ACD / AED	ABE
B	BEA	0	0	BCD / BED	0
= C	0	0	0	0	CDE
D	DEA	0	0	0	0
E	0	EAB	EAC / EDC	0	0

$= [\mathscr{M}]^{(2)}.$

Matrix $[\mathscr{M}]^{(1)}$ gives the list of elementary paths of length 1, and $[\mathscr{M}]^{(2)}$ those of length 2.

Let us proceed:

$[\mathcal{M}]^{(2)}$ L $[\mathcal{M}]^{(1)}$

	A	B	C	D	E
A	0	0	ABC	ACD AED	ABE
B	BEA	0	0	BCD BED	0
= C	0	0	0	0	CDE
D	DEA	0	0	0	0
E	0	EAB	EAC EDC	0	0

L

	A	B	C	D	E
A	0	B	C	0	E
B	0	0	C	0	E
C	0	0	0	D	0
D	0	0	C	0	E
E	A	0	0	D	0

(39.1d)

	A	B	C	D	E
A	0	0	AEDC	ABCD ABED	ACDE
B	0	0	BEAC BEDC	0	BCDE
= C	CDEA	0	0	0	0
D	0	DEAB	DEAC	0	0
E	0	0	EABC	EACD	0

$= [\mathcal{M}]^{(3)}$,

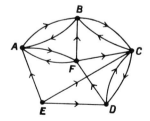

Fig. 39.1b.

which gives us the set of elementary paths of length 3. To continue:

$[\mathcal{M}]^{(3)} \ \mathbf{L} \ [\tilde{\mathcal{M}}]^{(1)}$

	A	B	C	D	E
A	0	0	AEDC	ABCD ABED	ACDE
B	0	0	BEAC BEDC	0	BCDE
= C	CDEA	0	0	0	0
D	0	DEAB	DEAC	0	0
E	0	0	EABC	EACD	0

L

	A	B	C	D	E
A	0	B	C	0	E
B	0	0	C	0	E
C	0	0	0	D	0
D	0	0	C	0	E
E	A	0	0	D	0

(39.1e)

	A	B	C	D	E
A	0	0	ABEDC	0	ABCDE
B	BCDEA	0	0	BEACD	0
= C	0	CDEAB	0	0	0
D	0	0	DEABC	0	0
E	0	0	0	EABCD	0

$= [\mathcal{M}]^{(4)}$

We stop at $[\mathcal{M}]^{(4)}$, for we have found all seven Hamiltonian paths.

[ABEDC],   [ABCDE],   [BCDEA],   [BEACD],

[CDEAB],   [DEABC],   [EABCD].

By shortening the method, latin multiplication can be extended to matrices $[\tilde{\mathscr{M}}]^{(s)}$, $s > 1$ defined below. Let

$$[\mathscr{M}]^{(r)}\mathbf{L}[\tilde{\mathscr{M}}]^{(s)} = [\mathscr{M}]^{(r+s)}, \qquad (39.1f)$$

where $[\tilde{\mathscr{M}}]^{(s)}$ is the matrix $[\mathscr{M}]^{(s)}$ from which, at each stage representing an elementary path, the first letter has been eliminated. If the enumeration of the elementary paths shorter than n is not necessary, all we need is to calculate progressively:

$$[\mathscr{M}]^{(2)}, [\mathscr{M}]^{(4)} = [\mathscr{M}]^{(2)}\mathbf{L}[\tilde{\mathscr{M}}]^{(2)}, [\mathscr{M}]^{(8)} = [\mathscr{M}]^{(4)}\mathbf{L}[\tilde{\mathscr{M}}]^{(4)},...,$$

$$[\mathscr{M}]^{(2^k)} = [\mathscr{M}]^{(2^{k-1})}\mathbf{L}[\tilde{\mathscr{M}}]^{(2^{k-1})} \qquad (39.1g)$$

where $2^k \leqslant n$.

The process is continued, using all the powers already obtained. To find $[\mathscr{M}]^{(19)}$, we should first calculate $[\mathscr{M}]^{(2)}$, $[\mathscr{M}]^{(4)}$, $[\mathscr{M}]^{(8)}$, $[\mathscr{M}]^{(16)}$ then

$$[\mathscr{M}]^{(18)} = [\mathscr{M}]^{(16)}\mathbf{L}[\tilde{\mathscr{M}}]^{(2)} \qquad \text{and} \qquad [\mathscr{M}]^{(19)} = [\mathscr{M}]^{(18)}\mathbf{L}[\tilde{\mathscr{M}}]^{(1)}.$$

Another Example (Fig. 39.1b). Our first example dealt with a strongly connected graph, and we shall now apply the method to one that is not strongly connected; to save space, only the results of the iterations are given and the zeros are omitted from the squares.

$[\mathscr{M}]^{(1)} \ \mathbf{L} \ [\tilde{\mathscr{M}}]^{(1)}$

	A	B	C	D	E	F			A	B	C	D	E	F
A		AB				AF		A		B				F
B	BA		BC					B	A		C			
C		CB		CD				C		B		D		
=							**L**							
D			DC			DF		D			C			F
E	EA		EC	ED				E	A		C	D		
F	FA	FB	FC					F	A	B	C			

$$(39.1h)$$

	A	B	C	D	E	F
A		AFB	ABC AFC			
B				BCD		BAF
C	CBA					CDF
D	DFA	DCB DFB	DFC			
E		EAB ECB	EDC	ECD		EAF EDF
F	FBA	FAB FCB	FBC	FCD		

$$= \ = [\mathscr{M}]^{(2)}$$

$[\mathscr{M}]^{(2)} \ \mathbf{L} \ [\bar{\mathscr{M}}]^{(2)}$

$=$

	A	B	C	D	E	F
A		AFB	ABC AFC			
B				BCD		BAF
C	CBA					CDF
D	DFA	DCB DFB	DFC			
E		EAB ECB	EDC	ECD		EAF EDF
F	FBA	FAB FCB	FBC	FCD		

$\mathbf{L}$

	A	B	C	D	E	F
A		FB	BC FC			
B				CD		AF
C	BA					DF
D	FA	CB FB	FC			
E		AB CB	DC	CD		AF DF
F	BA	AB CB	BC	CD		

(39.1i)

<table>
<thead>
<tr><th></th><th>A</th><th>B</th><th>C</th><th>D</th><th>E</th><th>F</th></tr>
</thead>
<tbody>
<tr><td>A</td><td></td><td></td><td></td><td>AFBCD</td><td></td><td>ABCDF</td></tr>
<tr><td>B</td><td>BCDFA</td><td></td><td></td><td>BAFCD</td><td></td><td></td></tr>
<tr><td>C</td><td>CDFBA</td><td>CDFAB</td><td></td><td></td><td></td><td></td></tr>
<tr><td>D</td><td>DFCBA</td><td></td><td>DFABC</td><td></td><td></td><td>DCBAF</td></tr>
<tr><td>E</td><td>EDCBA ECDFA EDFBA</td><td>ECDFB EAFCB EDFAB EDFCB</td><td>EAFBC EDFBC</td><td>EABCD EAFCD</td><td></td><td>ECBAF</td></tr>
<tr><td>F</td><td></td><td></td><td></td><td>FABCD</td><td></td><td></td></tr>
</tbody>
</table>

$[\mathscr{M}]^{(4)}\ \mathbf{L}\ [\tilde{\mathscr{M}}]^{(1)}$

<table>
<thead>
<tr><th></th><th>A</th><th>B</th><th>C</th><th>D</th><th>E</th><th>F</th></tr>
</thead>
<tbody>
<tr><td>A</td><td></td><td></td><td></td><td>AFBCD</td><td></td><td>ABCDF</td></tr>
<tr><td>B</td><td>BCDFA</td><td></td><td></td><td>BAFCD</td><td></td><td></td></tr>
<tr><td>C</td><td>CDFBA</td><td>CDFAB</td><td></td><td></td><td></td><td></td></tr>
<tr><td>D</td><td>DFCBA</td><td></td><td>DFABC</td><td></td><td></td><td>DCBAF</td></tr>
<tr><td>E</td><td>EDCBA ECDFA EDFBA</td><td>ECDFB EAFCB EDFAB EDFCB</td><td>EAFBC EDFBC</td><td>EABCD EAFCD</td><td></td><td>ECBAF</td></tr>
<tr><td>F</td><td></td><td></td><td></td><td>FABCD</td><td></td><td></td></tr>
</tbody>
</table>

	A	B	C	D	E	F
A		B				F
B	A		C			
C		B		D		
D			C			F
E	A		C	D		
F	A	B	C			

L   (39.1j)

	A	B	C	D	E	F
A						
B						
C						
D						
E	ECDFBA EDFCBA	ECDFAB	EDFABC	EAFBCD		EDCBAF EABCDF
F						

$$= [\mathscr{M}]^{(5)},$$

which gives the seven Hamiltonian paths of the graph.

Latin multiplication can easily be programmed to a computer, and if suitably modified, it can be used to find paths which are subject to constraints such as binary relations supplementary to those which define the graph, or those which will enter if values are given to the arcs.

When the order of the graph is high, and it is not strongly connected, it is useful to decompose it into strongly connected subgraphs to find the Hamiltonian paths. For this purpose, Foulkes's algorithm (given in Part I, Section 4, and proved later) can be used, or another method given in Part IV, Section D.

C. DECOMPOSITION INTO STRONGLY CONNECTED SUBGRAPHS. FOULKES'S METHOD

We wish to consider the relation of equivalence[1] between the vertices X_i of the graph: there is a path from X_i towards X_j and back again, which means that a strongly connected subgraph will form an equivalence class.

If there is a path from X_i to X_j, and a reciprocal one, we have

$$X_j \in \hat{\Gamma} X_i \quad \text{and} \quad X_i \in \hat{\Gamma} X_j, \tag{39.2}$$

where

$$\hat{\Gamma}(X_i) = \{X_i\} \cup \Gamma(X_i) \cup \Gamma^2(X_i) \cup \cdots$$

is the transitive closure of X_i.

Let $[M]$ be the boolean matrix[2] of the given graph and $[1]$ the matrix unit of the same type as M.

The matrix

$$[M]^k = [M] \cdot [M] \cdot \cdots \cdot [M] \tag{39.3}$$

obtained by boolean multiplication[3] of $[M]$ by itself k times, is the boolean matrix of the graph $G_k = (\mathbf{X}, \Gamma^k)$, in other words, a matrix of

---

[1] Let us recall what is meant by the *equivalence class* of a set. In a subset **C**, in which the elements are considered equivalent in relation to a property, they must satisfy the following conditions where $\equiv$ represents equivalence

   (a)   Reflexiveness : $\forall x_i \in \mathbf{C} : x_i \equiv x_i$,
   (b)   Symmetry : $\forall x_i, x_j \in \mathbf{C} : [x_i \equiv x_j] \Rightarrow [x_j \equiv x_i]$,
   (c)   Transitiveness : $\forall x_i, x_j, x_k \in \mathbf{C} : [x_i \equiv x_j \text{ and } x_j \equiv x_k] \Rightarrow [x_i \equiv x_k]$.

[2] See Section 42 for the meaning of *boolean matrix* of a graph.
[3] For *sum* ($\dot{+}$) or *product* ($\cdot$) we use the properties:

$$1 \dot{+} 1 = 0 \dot{+} 1 = 1 \dot{+} 0 = 1, \quad 0 \dot{+} 0 = 0; \quad 1 \cdot 1 = 1, \quad 1 \cdot 0 = 0 \cdot 1 = 0 \cdot 0 = 0$$

if $[a]$ and $[b]$ are matrices with binary coefficients:

   $[a] \dot{+} [b] = [a \dot{+} b]$ (sum boolean matrix)
   $[a] \cdot [b] = [c]$ where $c_{ij} = \dot{\sum}_k a_{ik} b_{kj}$ (product boolean matrix where the sign $\dot{\sum}_k$ represents a boolean sum).

which the coefficient $c_{ij}^{(k)}$ is equal to 1 if there is a path of length K from X_i to X_j. Calculating:

$$([1] \dotplus [M])^2 = [1] \dotplus [M] \dotplus [M]^2 \qquad \text{(since this is a boolean process)} \qquad (39.4)$$

$$\vdots$$

$$([1] \dotplus [M])^R = [1] \dotplus [M] \dotplus [M]^2 \dotplus \cdots \dotplus [M]^R \qquad (39.5)$$

the coefficient $d_{ij}^{(R)}$ of this matrix equals 1 if

$$X_j \in \{X_i\} \cup \Gamma(X_i) \cup \Gamma^2(X_i) \cup \cdots \cup \Gamma^R(X_i), \qquad (39.6)$$

that is to say, if there is a route of length less than or equal to R from X_i to X_j, and 0 in the contrary case.

With N vertices, every elementary path has a length less than or equal to $N - 1$; hence, the matrix $([1] \dotplus [M])^{(N-1)}$ gives all the paths (elementary or otherwise) of this length. Since every route is the aggregate of elementary ones, matrix

$$([1] \dotplus [M])^{N+P-1}, \qquad P \geqslant 1,$$

must equal matrix $([1] \dotplus [M])^{(N-1)}$. Hence, we need only calculate the latter to find whether there is a path between every point X_i and another, X_j. To shorten the calculations, we successively take

$$([1] \dotplus [M])^2, \qquad ([1] \dotplus [M])^4, \ldots, ([1] \dotplus [M])^{2^k}$$

and stop when

$$([1] \dotplus [M])^{2^{k+1}} = ([1] \dotplus [M])^{2^k} \qquad \text{(or when} \quad 2^k \geqslant N - 1 \text{)}.$$

It is then a very simple matter to decompose the graph into classes of equivalence forming strongly connected subgraphs. Indeed,[1] if

$$\hat{\Gamma}(X_i) = \hat{\Gamma}(X_j), \qquad \text{then} \quad X_i \in \hat{\Gamma}(X_j) \qquad \text{and} \quad X_j \in \hat{\Gamma}(X_i), \qquad (39.8)$$

for $\hat{\Gamma}(X_i)$ contains X_i, $\hat{\Gamma}(X_j) = \hat{\Gamma}(X_i)$ involves $\hat{\Gamma}(X_j)$ which contains X_i. In the same way, $\hat{\Gamma}(X_j)$ contains X_j, then $\hat{\Gamma}(X_i) = \hat{\Gamma}(X_j)$ involves $\hat{\Gamma}(X_i)$ which includes X_j.

This shows that there are reciprocal paths between X_i and X_j. Hence, X_i and X_j belong to the same class of equivalence.

---

[1] The transitive closure of the graph is

$$\hat{\Gamma}(X_i) = \{X_i\} \cup \Gamma(X_i) \cup \Gamma^2(X_i) \cup \cdots \cup \Gamma^N(X_i),$$

where N is the number of vertices.

To find the Hamiltonian paths, if there are any, we proceed as follows. If two vertices X_i and X_j belong to different classes of equivalence, and if there is a path from X_i towards X_j (that is, if $X_j \in \hat{\Gamma}(X_i)$), then a path does not exist from X_j towards X_i ; otherwise, X_i and X_j would belong to the same class. In addition, if X_i belongs to class **C** and X_j to class **C′** there is a route from every vertex in **C** to every vertex in **C′** and none in the reverse direction.

By saying that **C** $<$ **C′** if **C** $\neq$ **C′**, and if there is a path from a vertex of **C** to a vertex of **C′**, we obtain a relation of a transitive and anti-symmetrical type. The existence of a "total order" between all the classes **C**₁ , **C**₂ ,..., **C**_r is a necessary but not sufficient condition for the existence of one or more Hamiltonian paths with origin in **C**₁ and termination in **C**_r . It will be explained shortly that a graph must contain at least one Hamiltonian path if it is *complete*, a condition which is sufficient but is clearly not necessary.

An easy way of finding the vertices belonging to the same class is to remember that if any two of them belong to the same class, in matrix $([1] \dotplus (M))^{2^k}$, the lines corresponding to each of them are the same. All we need do is to count the number of times 1 appears in each line and group the vertices with identical lines. The classes are then arranged according to the number of 1's, and Table 39.7 now assumes a *semi-triangular* appearance. If this matrix is strictly triangular in relation to the submatrices representing the classes (Fig. 39.2a), there is a total order and one or more Hamiltonian paths may exist (Fig. 39.2a); otherwise, (Fig. 39.2b, for example), there is no Hamiltonian path.

The enumeration of the Hamiltonian paths will be made by considering those in each class and joining them to the arcs which are part of the paths which cross all the classes that constitute a total order.

D. ANOTHER METHOD FOR DECOMPOSING A GRAPH WHICH IS NOT STRONGLY CONNECTED[1]

We have defined transitive closure:

$$\hat{\Gamma}(X_i) = \bigcup_{n=0}^{\infty} \Gamma^n(X_i), \tag{39.9}$$

with the conventional notation

$$\Gamma^0(X_i) = X_i . \tag{39.10}$$

[1] This method was made known to us by Y. Malgrange, Mathematical Engineer, *Compagnie des Machines Bull.*

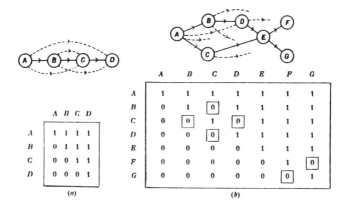

FIG. 39.2.

We have also defined

$$\Gamma^{-1}(X_i) = \{X_j \mid X_i \in \Gamma(X_j)\} \tag{39.11}$$

in the same way we now define

$$\Gamma^{-n} = \Gamma^{-1}(\Gamma^{-(n-1)}); \tag{39.12}$$

and, lastly, an inverse transitive closure $\hat{\Gamma}^-$:

$$\hat{\Gamma}^-(X_i) = \bigcup_{n=0}^{\infty} \Gamma^{-n}(X_i) \tag{39.13}$$

with the conventional notation $\Gamma^{-0}(X_i) = X_i$. If $X_i \in \hat{\Gamma}(X_j) \cap \hat{\Gamma}^-(X_j)$, there is a route from X_i to X_j and from X_j to X_i. That is to say, $X_i \equiv X_j$.

If X_j is an arbitrary point of $G = (\mathbf{X}, \Gamma)$, its equivalence class is the set:

$$\hat{\Gamma}(X_j) \cap \hat{\Gamma}^-(X_j) = \mathscr{C}(X_j). \tag{39.14}$$

The method to be used in practice is therefore the following:

(1)   Choose a vertex X_j and calculate $\hat{\Gamma}(X_j) \cap \hat{\Gamma}^-(X_j) = \mathscr{C}(X_j)$;

(2)   Suppress all the vertices $\mathscr{C}(X_j)$ of the graph;

(3)   In the remaining subgraph repeat (1) and (2) until all the vertices are eliminated.

Example.   There is a graph $G = (\mathbf{X}, \Gamma)$ represented by its boolean matrix (Table 39.1).

TABLE 39.1

	A	B	C	D	E	F	G	H	I		$\hat{\Gamma}(D) \cap \hat{\Gamma}^-(D)$
A			1			1					×
B	1		1								×
C		1			1			1			×
D					1				1	D	0
E				1				1		E	1
F							1		1	F	×
G								1	1	G	×
H						1				H	×
I						1		1		I	×

	A	B	C	D	E	F	G	H	I
$\hat{\Gamma}(D)$	×	×	×	0	1	2	3	2	1

To simplify the matrix, the zeros have been omitted.

Starting from an arbitrary point D, we add a supplementary line in which $\hat{\Gamma}(D)$ will be entered. In square D of this line we place a zero. Line D of the matrix gives us $\Gamma^1(D)$; we therefore enter 1 in squares E and I. Lines E and I give the vertices of $\Gamma^2(D)$ which have not yet been obtained, and they are marked with 2 in $\hat{\Gamma}(D)$, and so on.

Finally, there is no longer a new square to fill and $\hat{\Gamma}(D)$ is the set of marked squares.

On the right of the matrix we now add a separate column in which we shall construct $\hat{\Gamma}(D) \cap \hat{\Gamma}^-(D)$, and begin by crossing out squares A, B, and C, since these points do not belong to $\hat{\Gamma}(D)$. Besides, we need not consider lines A, B, C of the matrix, for if

$$X_i \notin \hat{\Gamma}(D), \quad \text{then} \quad \hat{\Gamma}^-(X_i) \cap \hat{\Gamma}(D) = \varnothing. \tag{39.15}$$

Indeed, if a point of $\hat{\Gamma}(D)$ were also in $\hat{\Gamma}^-(X_i)$ there would be a route from this vertex to X_i, and X_i would be in $\hat{\Gamma}(D)$, which contradicts the assumption. Hence, we place a zero in square D of the new column.

Where 1 appears in column D of the matrix, it is entered in the new column, in this case in E. If 2 is the column of the matrix corresponding to the vertex marked with 1 (here column E), we enter it in the supplementary column; but here there is none, and D and E form the class of D.

We now eliminate lines and columns D and E in the boolean matrix and make another arbitrary choice of a vertex, and so on, until every vertex has been used.

The advantage which this method has over that of Foulkes is that the boolean matrix need not be carried to successive powers (boolean multiplication and addition), a process which can require lengthy calculations in graphs of a high type. Nevertheless, it demands a certain knowledge of the theory of sets, and for this reason we were not free to use it for the introductory problems in Section D.

E. König's Theorem

A graph which is complete contains at least one Hamiltonian path (see [H2, p. 109] for the proof).

F. Dirac's Theorem

If a graph is symmetrical, connected and without a loop, and if

$$| \Gamma X_i | \geqslant \tfrac{1}{2} | \mathbf{X} |, \qquad \forall X_i \in \mathbf{X}, \tag{39.16}$$

there is at least one Hamiltonian circuit.

G. Finding the Hamiltonian Circuits

We first find the Hamiltonian paths in the given graph and then close them, if that is possible, by an arc belonging to the graph. In Fig. 39.3 we have shown that a Hamiltonian path does not always produce a

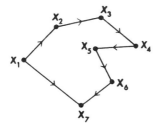

Fig. 39.3.

Hamiltonian circuit. Thus X_1 to X_7 is a Hamiltonian path, but the cycle in which this path returns to X_1 is not a Hamiltonian circuit.

The method of latin multiplication is particularly useful for finding hamiltonian circuits. It is sufficient to find $[\mathscr{M}]^{(n-1)}$, where n is the type of the graph, and then $[\tilde{\mathscr{M}}]^{(n-1)}$; we then calculate[1]

$$[\mathscr{M}]^*_{(n)} = [\tilde{\mathscr{M}}]^{(n-1)} \mathbf{L} [\tilde{\mathscr{M}}]^{(1)}. \tag{39.17}$$

In each latin sequence of $[\mathscr{M}]^*_{(n)}$ we need only add as the initial letter the one which is the index of the line, in order to obtain Hamiltonian paths; the process can also be simplified by *a priori* eliminations.

Note. The space devoted in this book to the enumeration of Hamiltonian paths and circuits should not cause surprise, for it is probably the aspect of modern graphs which is of the greatest practical importance today.

H. PROBLEM OF THE TRAVELING SALESMAN

We will use the theory of graphs to deal with this problem which was given in Volume I (p. 73).[2]

Given a strongly connected graph $G = (\mathbf{X}, \mathbf{U})$ in which a real number $c(u_i)$ is assigned to each $u_i \in \mathbf{U}$, let us assume that at least one Hamiltonian circuit $h_j = [u_{j_1}, u_{j_2}, ..., u_{j_n}]$ exists in G, and that $\mathbf{H}$ is the set of Hamiltonian circuits. Let $c(h_j) = c(u_{j_1}) + c(u_{j_2}) + \cdots + c(u_{j_n})$. Which is the circuit $h_j \in \mathbf{H}$ for which $c(h_j)$ is minimal? (There may be several solutions, and it would also be possible to seek the maximal.)

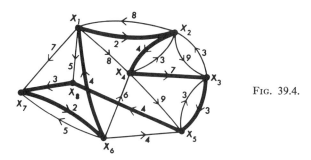

FIG. 39.4.

---

[1] Any latin matrix $[\mathscr{M}]^{(k)}$ in the sequences of which the initial letters are suppressed will be called $[\tilde{\mathscr{M}}]^{(k)}$.

[2] Volume I (English language ed.): A. Kaufmann, *Methods and Models of Operations Research.* Prentice-Hall, Englewood Cliffs, New Jersey, 1963.

Figure 39.4 provides an example. The circuit with heavy lines is the minimal Hamiltonian circuit, for which $c(h) = 29$.

This problem can be stated in a different form: find a cyclic permutation $P(1, i_2, i_3, ..., i_n)$ of the integers from 1 to n, which minimizes the quantity

$$Z = c_{1i_2} + c_{i_2 i_3} + \cdots + c_{i_n 1}, \tag{39.17}$$

where the

$$c_{\alpha\beta}, \quad \alpha = 1, i_2, i_3, ..., i_n, \quad \beta = i_2, i_3, ..., i_n, 1$$

are real numbers (positive, zero, or negative).

We have only recently discovered that this problem can be solved by finding the optimal solution of a linear program with integral variables, but the calculations become very difficult as soon as the number of vertices exceeds 10, even when a number of computers are used. An empirical method consists in considering the problem of attribution corresponding to the graph, seeking the optimal solutions formed by matrices of permutation, and finally looking for an optimal cyclic matrix among them. This process of regressive enumeration can be very long, and may be inapplicable when there is a large number of vertices. A variant of this problem, the search for an optimal Hamiltonian path, presents the same difficulties as the previous one.

I. HAMILTONIAN CHAIN AND CYCLE

A chain is Hamiltonian when it meets each vertex once and once only. If we treat a nondirected graph as a symmetrical graph, we must look for a Hamiltonian path in the latter.

A cycle is Hamiltonian if it meets each vertex once and once only. The search for such a cycle is also carried out in the symmetrical graph which has the equivalence of the nonorientated graph.

40. Center and Radius of a Graph

A. PROPERTIES

The concept of separation was defined in Section 36, and the following properties are obvious:

$$d(X_i, X_j) + d(X_j, X_k) \geqslant d(X_i, X_k), \quad X_i, X_j, X_k \in \mathbf{X} \tag{40.1}$$

$$d(X_i, X_j) = d(X_j, X_i) \tag{40.2}$$

if the graph is symmetrical.

Using the concept of separation as our basis, we can define other concepts.

B. DISTANCE BETWEEN TWO VERTICES

If the graph is symmetrical, the separation

$$d(X_i , X_j) = d(X_j , X_i)$$

between two vertices is called the *distance*, and is expressed as $\lambda(X_i , X_j)$.

C. SEPARATION OF A VERTEX

The *separation of* X_i is represented by the number

$$s(X_i) = \underset{X_j \in \mathbf{X}}{\text{MAX}}\, d(X_i , X_j). \tag{40.3}$$

This separation is also called *the number associated with a vertex*.

D. CENTERS

If $\text{MIN}_{X_i \in \mathbf{X}}\, s(X_i)$ is a finite number, the point of minimal separation is a *center* of the graph. Hence, a center of the graph is vertex X_0 such that its separation $s(X_0)$ has the value

$$s(X_0) = \underset{X_i \in \mathbf{X}}{\text{MIN}} \left[\underset{X_j \in \mathbf{X}}{\text{MAX}}\, d(X_i , X_j) \right]. \tag{40.4}$$

There may be several points which satisfy this equation, but an infinite graph may have no center.

E. POINTS ON THE PERIPHERY

A point on the periphery is a point of maximal separation, and is therefore a point such that $s(X_p)$ has the value

$$s(X_p) = \underset{X_i \in \mathbf{X}}{\text{MAX}} \left[\underset{X_j \in \mathbf{X}}{\text{MAX}}\, d(X_i , X_j) \right]. \tag{40.5}$$

This maximum may not exist and the point on the periphery then corresponds to $s(X_p) = \infty$.

F. RADIUS

If the graph has a center X_0, the quantity

$$\rho = s(X_0), \tag{40.6}$$

is called the *radius*.

If the graph has no center, we state that $\rho = \infty$.

G. TOTAL FINITE GRAPH

A finite graph $G = (\mathbf{X}, \Gamma)$ is total if $\forall X_i$, $X_j \in \mathbf{X}$, there is a route from X_i towards X_j and/or from X_j towards X_i.

Every total finite graph has a center.

Such a graph must not be confused with a complete graph or a strongly connected one. A finite strongly connected graph is total.

The difference between a finite strongly connected graph and a finite total graph must be clearly understood: in the former there is a path from X_i to X_j, and in the latter a path from X_i to X_j and/or from X_j to X_i.

41. Network (Strongly Connected Graph without a Loop)

A. DEFINITIONS AND PROPERTIES

A strongly connected graph without a loop and with more than one vertex is called a *network*.

In a network, for each point X_i, there are at least two incident arcs connected to it; one terminally and the other initially. In other words, every vertex has at least one path arriving at it and at least one leaving it.

B. NODE, ANTINODE

In a network, a vertex which has more than two oriented arcs is called a *node*.

Example (Fig. 41.1).   Vertices A, B, D, E, and F are nodes.

A vertex of the network which has two arcs connected to it, one terminally and the other initially, is called an *antinode*.

Example.   Vertices C and G.

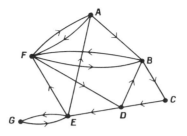

FIG. 41.1.

C. BRANCH

A *branch* is a path in which the first and last vertices alone are nodes.

Example (Fig. 41.1). (E, F), (B, F), (F, B), (B, C, D), and (E, G, E) are branches.

D. ROSE

A network with a single node is called a *rose*.

Example (Fig. 41.2).

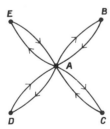

FIG. 41.2.

E. TRACK

In a network a path of minimal length between X_i and X_j is called the *track from X_i to X_j*. A network always possesses at least one track.

Various properties of nodes, antinodes, branches, roses, and tracks will be found in [H2].

F. DIAMETER

In a finite network the *diameter* is the distance δ of the longest track of the graph;

$$\delta = \underset{X_i \in \mathbf{X}}{\text{MAX}} \left[\underset{X_j \in \mathbf{X}}{\text{MAX}} \, d(X_i, X_j) \right] \tag{41.1}$$

is therefore the separation of a peripheral point.

Example (Fig. 41.1). The diameter is 3, as we can verify by making a table of the distance between the vertices.

G. THEOREM OF NETWORKS

If n is the number of vertices, m the number of arcs, and δ the value of the diameter, then

$$n \leqslant m \leqslant n(n-1),$$

$$\delta \leqslant n - 1.$$

In fact, $n \leqslant m$, since for each vertex X_i there is at least one corresponding arc connected to it initially; $m \leqslant n(n-1)$, for m is at least equal to the number of arrangements of the n vertices taken in pairs; $\delta \leqslant n - 1$, for the longest track contains $\delta + 1$ separate vertices; hence $n \geqslant \delta + 1$.

42. Matrix Associated with a Graph

A. DEFINITION

Given a graph $G = (\mathbf{X}, \mathbf{U})$ with vertices $X_1, X_2, ..., X_n$, let us assume

$$m_{ij} = 0 \qquad (X_i, X_j) \notin \mathbf{X}$$

$$m_{ij} = 1 \qquad (X_i, X_j) \in \mathbf{X}.$$

The square matrix formed by the m_{ij} terms is called the *associated matrix* of graph G, and is represented by the sign $[M]$.

Example. In many problems we shall consider operations undertaken with boolean matrices, the chief ones being "product" and "sum" in accordance with the meaning given to them by Boole (Figure 42.1 and Table 42.1).

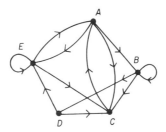

FIG. 42.1.

TABLE 42.1

	A	B	C	D	E
A	0	1	1	0	1
B	0	1	1	1	0
C	1	0	0	0	0
D	0	0	1	0	1
E	1	0	1	0	1

Transposed of [M]. This is the matrix [M]' obtained by permutating each line with each corresponding column, which is the same as reversing the direction of the arrows in G (Fig. 42.2).

Complement of [M]. The complementary matrix of [M] associated with graph G is the matrix [$\bar{M}$] obtained in the following manner:

$$\bar{m}_{ij} = 0 \quad \text{if} \quad m_{ij} = 1$$
$$\bar{m}_{ij} = 1 \quad \text{if} \quad m_{ij} = 0.$$

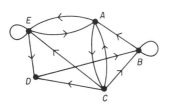

FIG. 42.2.   Graph G.

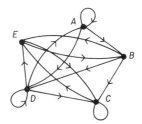

FIG. 42.3.   Graph $\bar{G}$.

Hence we obtain $[\bar{M}]$ from $[M]$ simply by replacing 0 by 1, and reciprocally. The graph $\bar{G}$ corresponding to $[\bar{M}]$ is called "the complementary graph of G." It is easily obtained by drawing arcs where they do not exist in G and by suppressing those which exist in G (Fig. 42.3).

By taking $\bar{\Gamma}$ for the multivocal mapping corresponding to $\bar{G}$, that is, $\bar{G} = (\mathbf{X}, \bar{\Gamma})$, we have

$$\forall X_i \in \mathbf{X}: \qquad \bar{\Gamma}X_i = \mathbf{C}_{\mathbf{X}}\,(\Gamma X_i). \tag{42.1}$$

B. PRINCIPAL PROPERTIES OF MATRICES ASSOCIATED WITH A GRAPH

$[M]$ is symmetrical ($\forall_i$, $\forall_j$, $m_{ij} = m_{ji}$) if and only if G is symmetrical.

$[M]$ is antisymmetrical ($\forall_i$, $\forall_j$, $m_{ij} + m_{ji} \leqslant 1$) if and only if G is antisymmetrical.

$[M]$ is complete ($\forall_i$, $\forall_j$, $1 \leqslant m_{ij} + m_{ji} \leqslant 2$) if and only if G is complete.

If we find the matricial product $[M] \times [M]$, that is $[M]^2$, this matrix with elements $m_{ij}^{(2)}$ gives the number of routes of length 2 between X_i and X_j. Stated in a more general form, matrix $[M]^\lambda$ ($\lambda = 1, 2, 3,...$) gives by its elements $m_{ij}^{(\lambda)}$, the number of routes of length λ between X_i and X_j.

C. EXAMPLE (Fig. 42.4)

We have calculated matrices $[M]$, $[M]^2$, $[M]^3$, and $[M]^4$.

$$
[M] =
\begin{array}{c}
\\ A \\ B \\ C \\ D \\ E \\ F \\ G
\end{array}
\begin{array}{c}
A\ B\ C\ D\ E\ F\ G \\
\left[
\begin{array}{ccccccc}
0 & 1 & 0 & 0 & 1 & 0 & 1 \\
0 & 1 & 1 & 0 & 0 & 0 & 0 \\
0 & 0 & 0 & 1 & 1 & 0 & 0 \\
0 & 1 & 0 & 0 & 0 & 0 & 1 \\
0 & 1 & 1 & 0 & 0 & 0 & 0 \\
0 & 0 & 0 & 1 & 0 & 1 & 1 \\
1 & 0 & 0 & 0 & 0 & 0 & 0
\end{array}
\right],
\end{array}
\tag{42.2}
$$

$$
[M]^2 =
\begin{array}{c}
\\ A \\ B \\ C \\ D \\ E \\ F \\ G
\end{array}
\begin{array}{c}
A\ B\ C\ D\ E\ F\ G \\
\left[
\begin{array}{ccccccc}
1 & 2 & 2 & 0 & 0 & 0 & 0 \\
0 & 1 & 1 & 1 & 1 & 0 & 0 \\
0 & 2 & 1 & 0 & 0 & 0 & 1 \\
1 & 1 & 1 & 0 & 0 & 0 & 0 \\
0 & 1 & 1 & 1 & 1 & 0 & 0 \\
1 & 1 & 0 & 1 & 0 & 1 & 2 \\
0 & 1 & 0 & 0 & 1 & 0 & 1
\end{array}
\right],
\end{array}
\tag{42.3}
$$

$$[M]^3 = \begin{array}{c} \\ A \\ B \\ C \\ D \\ E \\ F \\ G \end{array} \begin{array}{ccccccc} A & B & C & D & E & F & G \\ \left[\begin{array}{ccccccc} 0 & 3 & 2 & 2 & 3 & 0 & 1 \\ 0 & 3 & 2 & 1 & 1 & 0 & 1 \\ 1 & 2 & 2 & 1 & 1 & 0 & 0 \\ 0 & 2 & 1 & 1 & 2 & 0 & 1 \\ 0 & 3 & 2 & 1 & 1 & 0 & 1 \\ 2 & 3 & 1 & 1 & 1 & 1 & 3 \\ 1 & 2 & 2 & 0 & 0 & 0 & 0 \end{array}\right] \end{array} \qquad (42.4)$$

$$[M]^4 = \begin{array}{c} \\ A \\ B \\ C \\ D \\ E \\ F \\ G \end{array} \begin{array}{ccccccc} A & B & C & D & E & F & G \\ \left[\begin{array}{ccccccc} 1 & 8 & 6 & 2 & 2 & 0 & 2 \\ 1 & 5 & 4 & 2 & 2 & 0 & 1 \\ 0 & 5 & 3 & 2 & 3 & 0 & 2 \\ 1 & 5 & 4 & 1 & 1 & 0 & 1 \\ 1 & 5 & 4 & 2 & 2 & 0 & 1 \\ 3 & 7 & 4 & 2 & 3 & 1 & 4 \\ 0 & 3 & 2 & 2 & 3 & 0 & 1 \end{array}\right] \end{array} \qquad (42.5)$$

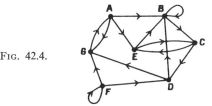

Fig. 42.4.

Hence, in $[M]^4$ we find five routes of length 4 from C to B; these are

$$\{C, D, G, A, B\}, \quad \{C, D, B, B, B\}, \quad \{C, E, B, B, B\},$$

$$\{C, E, A, B, B\}, \quad \{C, E, C, D, B\}.$$

In a graph G there is route of length λ if and only if $[M]^\lambda \neq [0]$; there are no circuits if and only if $[M]^\lambda = [0]$, after a certain stage.

BOOLEAN ADDITION AND MULTIPLICATION OF BOOLEAN MATRICES. We have $[a]$ and $[b]$, two boolean matrices of m lines and n columns with elements

$$a_{ij}, \quad b_{ij}, \quad i = 1, 2, \dots, m, \quad j = 1, 2, \dots, n.$$

The *boolean sum* of [a] and [b] is a boolean matrix $m \times n$ called [s] with elements

$$s_{ij}, \qquad i = 1, 2,..., m, \qquad j = 1, 2,..., n$$

such that

$$
\begin{aligned}
s_{ij} &= 0 \quad \text{if} \quad a_{ij} = 0, \qquad b_{ij} = 0 \\
s_{ij} &= 1 \quad \text{if} \quad a_{ij} = 0, \qquad b_{ij} = 1 \\
s_{ij} &= 1 \quad \text{if} \quad a_{ij} = 1, \qquad b_{ij} = 0 \\
s_{ij} &= 1 \quad \text{if} \quad a_{ij} = 1, \qquad b_{ij} = 1.
\end{aligned}
\tag{42.6}
$$

We shall state

$$s_{ij} = a_{ij} \dotplus b_{ij}, \tag{42.7}$$

and

$$[s] = [a] \dotplus [b]. \tag{42.8}$$

Given two boolean matrices [a] and [b], the first $m \times r$, the second $r \times n$:

$$a_{ij}, \qquad i = 1, 2,..., m, \qquad j = 1, 2,..., r;$$
$$b_{jk}, \qquad j = 1, 2,..., r, \qquad k = 1, 2,..., n;$$

we call the *boolean product* of [a] and [b] a boolean matrix $m \times n$ called [p] with elements

$$p_{ij}, \qquad i = 1, 2,..., m, \qquad j = 1, 2,..., n$$

such that

$$c_{ij} = a_{i1} \cdot b_{ij} \dotplus a_{i2} \cdot b_{2j} \dotplus \cdots \dotplus a_{ik} \cdot b_{kj} = \overset{\dotplus}{\sigma}_k a_{ik} b_{kj}, \tag{42.9}$$

where

$$
\begin{aligned}
\forall k: \quad a_{ik} \cdot b_{kj} &= 0 \quad \text{if} \quad a_{ik} = 0, \qquad b_{kj} = 0, \\
a_{ik} \cdot b_{kj} &= 0 \quad \text{if} \quad a_{ik} = 0, \qquad b_{kj} = 1, \\
a_{ik} \cdot b_{kj} &= 0 \quad \text{if} \quad a_{ik} = 1, \qquad b_{kj} = 0, \\
a_{ik} \cdot b_{kj} &= 1 \quad \text{if} \quad a_{ik} = 1, \qquad b_{kj} = 1.
\end{aligned}
\tag{42.10}
$$

Given a square boolean matrix [a] of type $n \times n$ with elements

$$a_{ij}, \qquad i, j = 1, 2,..., n,$$

we term the *boolean square* of $[a]$ a boolean matrix $[p]$ with $n \times n$ elements

$$p_{ij}, \qquad i, j = 1, 2, ..., n,$$

which is also referred to as $[a]^{(2)}$, such that

$$p_{ij} = \overset{+}{\underset{k}{\sigma}} a_{ik} \cdot a_{kj}. \qquad (42.11)$$

By stating

$$p_{ij} = a_{ij}^{(2)}, \qquad i, j = 1, 2, ..., n, \qquad (42.12)$$

We define a *boolean cube* of $[a]$, and thence a *boolean rth power* of $[a]$ $(r = 4, 5, 6,...)$:

$$a_{ij}^{(3)} = \overset{+}{\underset{k}{\sigma}} a_{ik}^{(2)} \cdot a_{kj} \qquad (42.13)$$

$$\vdots$$

$$a_{ij}^{(r)} = \overset{+}{\underset{k}{\sigma}} a_{ik}^{(r-1)} \cdot a_{kj}. \qquad (42.14)$$

$[a]^{(r)}$ will represent the boolean rth power of $[a]$.

We shall not give further examples of these boolean processes, since they were explained earlier.[1]

If $[M]$ is the boolean matrix of the graph $G = (\mathbf{X}, \Gamma)$, the boolean power $[M]^{(\alpha)}$ is associated with a graph which possesses an arc (X_i, X_j) if and only if there is a route of length α from X_i to X_j.

Given two graphs $G_1 = (\mathbf{X}, \Gamma_1)$ and $G_2 = (\mathbf{X}, \Gamma_2)$ with respective boolean matrices $[M_1]$ and $[M_2]$, then[2]

$[M_1] \overset{+}{} [M_2]$ corresponds to the graph $G^{(1)} = (\mathbf{X}, \Gamma_1 \cup \Gamma_2)$ (42.15)

$[M_1] \cdot [M_2]$ corresponds to the graph $G^{(2)} = (\mathbf{X}, \Gamma_2 \cdot \Gamma_1)$,   (42.16)

where

$$(\Gamma_1 \cup \Gamma_2) X_i = (\Gamma_1 X_i) \cup (\Gamma_2 X_i) \qquad (42.17)$$

and

$$(\Gamma_2 \cdot \Gamma_1) X_i = \Gamma_2 (\Gamma_1 X_i). \qquad (42.18)$$

---

[1] For properties of boolean algebra, see reference [H4].

[2] We are dealing here with boolean sums $(+)$ and products $(\cdot)$. See note (3), p. 280.

D. EXAMPLES (Figs. 42.5–8)

$$[M_1] = \begin{array}{c} \\ A \\ B \\ C \\ D \end{array} \begin{array}{cccc} A & B & C & D \\ \begin{bmatrix} 0 & 1 & 0 & 1 \\ 1 & 0 & 0 & 1 \\ 0 & 1 & 0 & 1 \\ 0 & 0 & 1 & 1 \end{bmatrix} \end{array}, \tag{42.19}$$

$$[M_2] = \begin{array}{c} \\ A \\ B \\ C \\ D \end{array} \begin{array}{cccc} A & B & C & D \\ \begin{bmatrix} 1 & 0 & 1 & 0 \\ 1 & 0 & 0 & 0 \\ 0 & 0 & 0 & 1 \\ 1 & 0 & 1 & 0 \end{bmatrix} \end{array}. \tag{42.20}$$

$$[M_1] \dotplus [M_2] = \begin{array}{c} \\ A \\ B \\ C \\ D \end{array} \begin{array}{cccc} A & B & C & D \\ \begin{bmatrix} 1 & 1 & 1 & 1 \\ 1 & 0 & 0 & 1 \\ 0 & 1 & 0 & 1 \\ 1 & 0 & 1 & 1 \end{bmatrix} \end{array} \tag{42.21}$$

$$[M_1] \cdot [M_2] = \begin{array}{c} \\ A \\ B \\ C \\ D \end{array} \begin{array}{cccc} A & B & C & D \\ \begin{bmatrix} 1 & 0 & 1 & 0 \\ 1 & 0 & 1 & 0 \\ 1 & 0 & 1 & 0 \\ 1 & 0 & 1 & 1 \end{bmatrix} \end{array}. \tag{42.22}$$

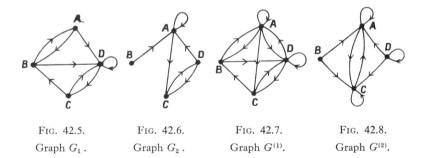

FIG. 42.5.　　　FIG. 42.6.　　　FIG. 42.7.　　　FIG. 42.8.

Graph G_1.　　Graph G_2.　　Graph $G^{(1)}$.　　Graph $G^{(2)}$.

As an exercise, the reader can find the $(\Gamma_2 \cdot \Gamma_1)X_i$ in (42.22). For example,

$$\Gamma(\Gamma_1 D) = \Gamma_2\{C, D\} - \{A, C, D\}. \tag{42.23}$$

43. Incidence Matrix

A. INCIDENCE MATRIX OF THE ARCS

If u_1, u_2 ,..., u_n are the arcs of a graph $G = (\mathbf{X}, \Gamma)$ without a loop and with vertices X_1, X_2 ,..., X_m , let us assume

$$u_{ij} = \quad 1 \qquad \text{if } X_i \text{ is the initial extremity of } u_j \text{,}$$

$$u_{ij} = -1 \qquad \text{if } X_i \text{ is the terminal extremity of } u_j \text{.} \qquad (43.1)$$

$$u_{ij} = \quad 0 \qquad \text{if } X_i \text{ is not an extremity of } u_j \text{.}$$

The matrix $[u]$ of the elements u_{ij} , $i = 1, 2,..., m$; $j = 1, 2,..., n$, is called the *incidence matrix* of the graph G.

Example (Fig. 43.1).

$$[u] = \begin{array}{c} \\ X_1 \\ X_2 \\ X_3 \\ X_4 \\ X_5 \end{array} \begin{array}{c} \begin{array}{cccccccc} u_1 & u_2 & u_3 & u_4 & u_5 & u_6 & u_7 & u_8 \end{array} \\ \left[\begin{array}{cccccccc} 1 & -1 & 0 & 0 & 1 & 0 & -1 & 1 \\ -1 & 1 & 1 & 0 & 0 & 0 & 0 & 0 \\ 0 & 0 & -1 & 1 & -1 & 1 & 0 & 0 \\ 0 & 0 & 0 & -1 & 0 & 0 & 0 & 0 \\ 0 & 0 & 0 & 0 & 0 & -1 & 1 & -1 \end{array} \right] \end{array} \qquad (43.2)$$

B. INCIDENCE MATRIX OF THE LINKS

If $\bar{u}_1$, $\bar{u}_2$,..., $\bar{u}_p$ are the links and X_1 , X_2 ,..., X_m the vertices of a graph $G = (\mathbf{X}, \Gamma)$ without a loop, and if we assume:

$$u_{ij} = 1 \qquad \text{if } X_i \text{ is an extremity of } u_j \text{,}$$

$$u_{ij} = 0 \qquad \text{if this not the case,}$$

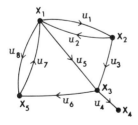

FIG. 43.1.

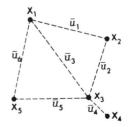

FIG. 43.2.

the matrix $[\bar{u}]$ with elements

$$\bar{u}_{ij}, \qquad i = 1, 2, ..., m; \qquad j = 1, 2, ..., p,$$

is called the *incidence matrix of the links* of graph G.

Example (Fig. 43.2).   The links correspond to the arcs of Fig. 43.1.

$$[\bar{u}] = \begin{array}{c} \\ X_1 \\ X_2 \\ X_3 \\ X_4 \\ X_5 \end{array} \begin{array}{cccccc} \bar{u}_1 & \bar{u}_2 & \bar{u}_3 & \bar{u}_4 & \bar{u}_5 & \bar{u}_6 \\ \left[\begin{array}{cccccc} 1 & 0 & 1 & 0 & 0 & 1 \\ 1 & 1 & 0 & 0 & 0 & 0 \\ 0 & 1 & 1 & 1 & 1 & 0 \\ 0 & 0 & 0 & 1 & 0 & 0 \\ 0 & 0 & 0 & 0 & 1 & 1 \end{array}\right] \end{array} \qquad (43.3)$$

In Reference [H2] the reader will find a number of theorems relating to the incidence matrices $[u]$ or $[\bar{u}]$.

44. Tree. Arborescence

A. TREE. DEFINITIONS

This is a concept considered in the nondirected study of finite 1-graphs. The name *tree* is given to a finite graph without a cycle and with at least two vertices. The generalization of this concept for cases, whether they are connected or not, is called a *forest*, which is therefore a set of nonconnected trees.[1]

Figure 44.1 represents a tree.

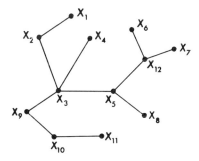

FIG. 44.1.

[1] It is possible to consider the concept of a tree as based on the arcs; it is then a concept in the directed study of graphs. In this case a graph does not always contain an *oriented tree* forming a partial graph of the original graph.

We can define a tree in various ways. Given a graph H with at least two vertices, one of the following properties is sufficient to define a tree:

(1) H is connected and without a cycle.

(2) H is without a cycle and has $n - 1$ links, n being the number of vertices.

(3) H is connected and has $n - 1$ links if there are n vertices.

(4) H has no cycle, but by adding a link between two nonadjacent vertices, we create one, and only one, cycle.

(5) H is connected, unless any link is suppressed.

(6)   Every pair of vertices is connected by one, and only one, chain.

B. DETERMINING A PARTIAL TREE IN A GRAPH

A graph $G = (\mathbf{X}, \mathbf{\bar{U}})$ contains a partial graph which is a tree if, and only if, it is connected. Such a tree is called a *partial tree*.

To construct a partial tree from a connected graph, we seek a link, the suppression of which will not *disconnect* the graph. If such a link does not exist, the graph is a tree; if it exists, we suppress it and look for a new link to suppress, and so on.

Example (Fig. 44.2).   After the successive suppression of links 1, 2,..., 7, the remaining links form a tree.

C. TOTAL VALUE OF A PARTIAL TREE

If a real value $1(\bar{u})$ is given to each link $\bar{u} \in \mathbf{\bar{U}}$ of a connected graph $G = (\mathbf{X}, \mathbf{\bar{U}})$, we give the term *total value of the partial tree* $H = (\mathbf{X}, \mathbf{\bar{V}})$ to the sum

$$S = \sum_{\bar{u} \in \mathbf{\bar{V}}} l(\bar{u}). \tag{44.1}$$

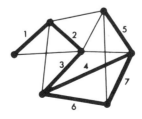

FIG. 44.2.

D. Finding a Partial Tree of Minimal Total Value

KRUSKAL'S ALGORITHM. This remarkably simple algorithm was given in Section 5. To prove it, we proceed by stages, each time choosing the link with the lowest value which does not form a cycle with the links already chosen. By this process, we eventually manage to choose a set of links

$$\bar{\mathbf{U}}_{n-1} = \{\bar{u}_1, \bar{u}_2, ..., \bar{u}_{n-1}\}$$

of $n-1$ links and the graph $(\mathbf{X}, \bar{\mathbf{U}}_{n-1})$ is a tree of minimal value.

In fact, we can arrange for all the links to have different values, and if, for example,

$$l(\bar{u}_1) = l(\bar{u}_2) = l(\bar{u}_3),$$

we make the change:

$$l(\bar{u}_1) \rightarrow l'(\bar{u}_1) = l(\bar{u}_1) + \epsilon,$$
$$l(\bar{u}_2) \rightarrow l'(\bar{u}_2) = l(\bar{u}_2) + 2\epsilon, \qquad (44.2)$$
$$l(\bar{u}_3) \rightarrow l'(\bar{u}_3) = l(\bar{u}_3) + 3\epsilon,$$

giving ϵ a fairly low value so as to avoid any inversion of the order of values.

In the same way, we can obtain a complete graph merely by adding links $\bar{u}_k'$ of sufficient values that

$$l(\bar{u}_k') > \sum_{\bar{u}_i \in \bar{\mathbf{U}}} l(\bar{u}_i). \qquad (44.3)$$

In the new graph, there is one, and only one, tree $H = (\mathbf{X}, \bar{\mathbf{U}}_{n-1})$ of minimal total value; and in accordance with the property which enables us to construct a partial tree, $\bar{\mathbf{U}}_{n-1}$ does not contain any links $\bar{u}_k' \notin \bar{\mathbf{U}}$ if we take (44.3) into account.

If all the links have different values, there can only be one tree H. The above process can also be carried out inversely by the successive elimination of links with the highest value, which do not disconnect the graph when they are suppressed.

SOLLIN'S ALGORITHM [H3].[1] In a connected graph $G = (\mathbf{X}, \bar{\mathbf{U}})$, we assign a number $l(\bar{u}) \geqslant 0$ to each link $\bar{u}$, and then find a partial tree $H = (\mathbf{X}, \bar{\mathbf{V}})$ of G, with minimal value

$$\sum_{\bar{u} \in \bar{\mathbf{V}}} l(\bar{u}).$$

---

[1] See reference in Section 5, p. 31.

LEMMA 1.   If the values $l(\bar{u})$ of the links are all different, the graphs $A(\mathbf{X}, \bar{\mathbf{V}})$, obtained by taking for each link $\bar{v}_i \in G(\mathbf{X}, \bar{\mathbf{U}})$, the one which connects vertex X_i to its nearest neighbor, (viz. the link with the lowest value) are trees.

By construction, graph $A(\mathbf{X}, \bar{\mathbf{V}})$ is connected, and if it contains n vertices, it also contains n arcs[1] if we agree to orientate each link $\bar{v}$ (considered by construction) in the direction $X_1 \to X_i$ when the choice of arc v_1 with the minimal value $l(v_1)$ is made, beginning at X_1. This graph contains n arcs, since the construction ends when any vertex is reached for a second time.

If these n arcs form n separate links, the graph contains one, and only one, cycle. Indeed, since $A(\mathbf{X}, \bar{\mathbf{V}})$ is connected and has n links, it becomes a tree in accordance with property (3), (see p. 300), if one link is taken from it, and satisfies property (4), (p. 300), before this elimination.

First Case.   The cycle is formed by arcs with the same orientation.

By construction, if we traverse this cycle in the direction of the arcs, beginning with the arc of minimal value, we should have

$$l(v_1) < l(v_2) < l(v_3) < \cdots < l(v_{n-1}) < l(v_n). \tag{44.3.1}$$

Since we are concerned with a cycle, and since arc v_n connects X_n to X_1, by construction, we should have

$$l(v_n) < l(v_1) \quad \text{and} \quad l(v_n) > l(v_{n-1}) > l(v_1), \tag{44.3.2}$$

whence the contradiction with (44.31).

Second Case.   The cycle includes arcs with opposite orientation.

By construction, one arc, and only one, should leave each vertex; hence two arcs with opposite directions are necessarily merged, since the value of an arc does not depend on its orientation. In this case, the number of links is equal to the number of arcs diminished by unity, and this is sufficient to prove that $A(\mathbf{X}, \bar{\mathbf{V}})$ is a tree.

This first lemma has certain properties such that:

(1)   A tree $A(\mathbf{X}, \bar{\mathbf{U}})$ is the minimal connected one, and includes only one node;

---

[1] We must not be surprised at finding the concept of arcs in the following proofs. The algorithm provides for the connection of a vertex to the one nearest to it, and it is advisable when making the drawing to mark the vertex from which the "link" was drawn so that no vertex of departure is omitted.

(2)   The number of arcs is equal to the number of vertices, and such a tree $A(\mathbf{X}, \bar{\mathbf{V}})$ can only have two arcs with opposite orientation which are merged;

(3)   A link $\bar{u}_0$ of $G(\mathbf{X}, \bar{\mathbf{V}})$ which does not belong to $A(\mathbf{X}, \bar{\mathbf{V}})$, but which forms a cycle in this tree, has a length $l(\bar{u}_0)$ greater than that of any of the links of $A(\mathbf{X}, \bar{\mathbf{V}})$, which together with u_0 form the cycle being considered.

Indeed, we can choose in the relation (44.3.1): $\bar{v}_n = \bar{u}_0$, which shows that $l(\bar{u}_0) > l(v_1)$. It should also be noted that as the first case of a cycle cannot be contained in $A(\mathbf{X}, \bar{\mathbf{V}})$, the suppression of link $\bar{v}_n$ leaves two arcs of opposite orientation attached to link $\bar{v}_{n-1}$.

LEMMA 2.   Since no vertex of $G(\mathbf{X}, \mathbf{U})$ can be joined to another vertex except by the link with the lowest value, the trees $A(\mathbf{X}, \bar{\mathbf{V}})$ are the only ones formed; the case of equal values being necessarily excluded by the change in (44.2) used in the proof of Kruskal's algorithm.

Proof of the method.   We notice that the algorithm is first applied to the vertices to obtain trees $A(\mathbf{X}, \bar{\mathbf{V}})$ which are nonconnected between each other and that each tree $A(\mathbf{X}, \bar{\mathbf{V}})$ is next "narrowed" to one vertex[1]; the set of these vertices forms a graph $G_1(\mathbf{X}, \bar{\mathbf{U}})$ to which the same algorithm is applied. This gives a new set of nonconnected trees $B(\mathbf{X}, \bar{\mathbf{V}})$ to which the same process is applied, and so on. After a certain number of repetitions, we obtain a single tree $H(\mathbf{X}, \bar{\mathbf{V}})$ connecting all the vertices $X_i \in \mathbf{X}$. We say that this tree is of minimal length.

Let $F = (\mathbf{X}, \bar{\mathbf{V}})$ be a tree of minimal length and let us show that $F = H$.

If $F \neq H$, let $\bar{v}_k$ be the first link of H which is not in F. Graph $F \cup \{\bar{v}_k\}$ contains one and only one cycle, [property (4), p. 300]; in this cycle, there is a link $\bar{u}_0$ which belongs to F and not to H. The link $\bar{u}_0$ is such that $l(\bar{u}_0) > l(\bar{v}_k)$ by the third paragraph of Lemma 1. If we assume

$$W = (F \cup \{\bar{v}_k\}) - \{\bar{u}_0\},$$

graph $W = (\mathbf{X}, \overline{\mathbf{W}})$ has no cycle and contains $n - 1$ links; it is therefore a tree accordancing to property (2), p. 300. Its value differs from that of F only by the change of $l(\bar{u}_0)$ into $l(\bar{v}_k)$. Since $l(\bar{u}_0) > l(\bar{v}_k)$, W would have a smaller value than F, so that F is the minimal tree, whence the contradiction.

---

[1] The narrowed subgraph becomes a vertex with the result that the new graph to be considered is then a multigraph.

E. FINDING A FUNDAMENTAL SYSTEM OF INDEPENDENT CYCLES
   IN A CONNECTED GRAPH

To find such a system, we construct a partial tree of the graph and then find one cycle, and one only, which will pass along each one of the links that does not belong to the tree. If the graph is not connected, we carry out the same process for each of its connected components and obtain the independent cycles starting with the forest.

Example (Fig. 44.3).   The cyclomatic number of this graph is

$$\nu(G) = m - n + 1 = 11 - 6 + 1 = 6. \tag{44.4}$$

There are six independent cycles. In Fig. 44.3 we have made an arbitrary choice of a tree, and constructed the system of independent cycles which corresponds to it.

This concept is of particular value in electrical networks, for it enables us to evaluate basic mesh currents.

F. ARBORESCENCE

This is a concept in the oriented study of finite graphs.
A finite graph $G = (\mathbf{X}, \mathbf{U})$ is an *arborescence* of root $X_1 \in \mathbf{X}$ if

(1)   Each $X_i \neq X_1$ is the terminal extremity of one arc only;

(2) X_1 is not the terminal extremity of any arc;

(3)   The graph does not contain a circuit.

It should be noted that: Any arborescence is a tree; $\forall X_i \in \mathbf{X}$, there is a route from X_1 to X_i .

Example (Figs. 44.4 and 44.5).

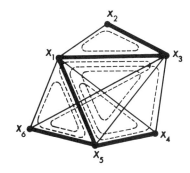

FIG. 44.3.

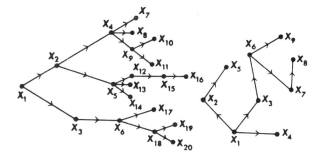

FIG. 44.4.   FIG. 44.5.

G. THE BOTT–MAYBERRY THEOREM[1]

Among the important properties of partial graphs which constitute arborescences,[2] mention must be made of this theorem because of its particular value in calculating the very large determinants enountered in intersecting (input–output) matrices in management.

Let us consider a square matrix $[a]$ of order $n - 1$ with elements a_{ij}, $i, j = 2, 3,..., n$, and assume $[a]$ is such that

$$a_{ij} \leqslant 0, \quad \forall i \neq j \tag{44.5}$$

and

$$a_{jj} \geqslant -\sum_{\substack{i=2 \\ i \neq j}}^{n} a_{ij}, \quad \forall j. \tag{44.5'}$$

Let us form a new matrix such that

$$[b] = \begin{array}{c} \\ (1) \\ (2) \\ (3) \\ \\ (n) \end{array} \begin{array}{cccc} (1) \quad (2) \quad (3) \qquad (n) \\ \begin{bmatrix} 0 & \lambda_2 & \lambda_3 & \cdots & \lambda_n \\ 0 & 0 & -a_{23} & \cdots & -a_{2n} \\ 0 & -a_{32} & 0 & \cdots & -a_{3n} \\ & & & \vdots & \\ 0 & -a_{n2} & -a_{n3} & \cdots & 0 \end{bmatrix} \end{array}, \tag{44.6}$$

where

$$\lambda_j = a_{jj} + \sum_{\substack{i=2 \\ i \neq j}}^{n} a_{ij} \geqslant 0, \quad j = 2, 3,..., n. \tag{44.7}$$

[1] See R. Bott and J. P. Mayberry, Matrices and trees, in *Economic Activity Analysis*, p. 391, Wiley, 1954.
[2] See Reference [H2], p. 150.

We can consider matrix (44.6) as representing the values attached to the arcs of a graph $G = (\mathbf{X}, \Gamma)$ which forms a network such that $\Gamma^{-1}X_1 = \varnothing$, where the vertices are numbered like the lines and columns of the matrix.

Let $\mathbf{A}_1$ be the set of arborescences $\mathbf{H}_k = (\mathbf{X}, \varDelta_k)$ formed in G and of root X_1, and let us call:

$$C(\mathbf{H}_k) = \prod_{\substack{X_i \in \mathbf{H} \\ X_j \in \varDelta_k X_i}} C(X_i, X_j) \tag{44.8}$$

where

$$C(X_i, X_j) = b_{ij}, \tag{44.9}$$

then:

$$\det a_{ij} = \sum_k{}' C(\mathbf{H}_k). \tag{44.10}$$

Expressed differently, it is sufficient to enumerate all the separate arborescences of G, to calculate for each one the product of all the values it contains, and then to find the sum of these products; the result gives us the value of *the determinant of matrix* [a]. The characteristic form of [a] is exactly the same as the encountered in Léontief's matrices (economic intersecting matrices), and the above theorem was intended to facilitate the checking and calculation of the determinants of such matrices.

Example.   Let

$$[a] = \begin{bmatrix} 3 & -3 & -1 \\ -1 & 7 & 0 \\ -2 & -2 & 2 \end{bmatrix}. \tag{44.11}$$

We have

$$[b] = \begin{bmatrix} 0 & 0 & 2 & 1 \\ 0 & 0 & 3 & 1 \\ 0 & 1 & 0 & 0 \\ 0 & 2 & 2 & 0 \end{bmatrix}. \tag{44.12}$$

The network G associated with [b] is given in Fig. 44.6. If we look for the set $\mathbf{A}_1$ of the arborescences of root X_1, we find that it contains the six arborescences shown in Fig. 44.7, whence

$$\det[a] = 2 + 2 + 6 + 4 + 2 + 4 = 20,$$

which can easily be checked.

Lack of space prevents our giving the very elegant proof, which will be found in [H2].

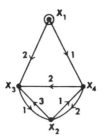

FIG. 44.6.

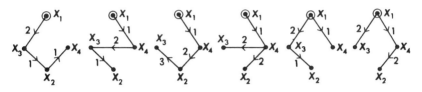

$2 \times 1 \times 1 = 2$, $1 \times 2 \times 1 = 2$, $1 \times 2 \times 3 = 6$, $1 \times 2 \times 2 = 4$, $1 \times 2 \times 1 = 2$, $2 \times 1 \times 2 = 4$.

FIG. 44.7.

45. Eulerian Chain and Cycle

A. DEFINITION

A chain or cycle of a p-graph is eulerian if it uses each link once and only once. In other words, we can draw the graph without lifting our pen from the paper or going over the same line twice.

Example (Fig. 45.1).   This 1-graph has an eulerian cycle:

$$[X_1, X_2, X_8, X_7, X_6, X_3, X_4, X_5, X_1, X_8, X_6, X_5, X_7, X_1];$$

the path is shown by numbered arrows.

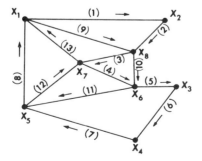

FIG. 45.1.

B. THE CELEBRATED PROBLEM OF THE BRIDGES OF KÖNIGSBERG

The river Pregel flows through Königsberg and surrounds the island of Kneiphof. Is it possible for a person taking a walk to cross each of the seven bridges (Fig. 45.3) once, and once only, and return to his starting point? (See Figs. 45.2 and 45.3.)

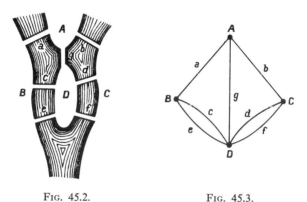

FIG. 45.2.   FIG. 45.3.

As Euler has proved, this celebrated problem cannot be solved. To understand the reason we should study the 2-graph of Fig. 45.3, in which each region A, B, C, D is a vertex, and each of the bridges a, b,..., g is a link. We shall then recognize the impossibility of finding an eulerian chain.

C. THEOREM

A p-graph contains an eulerian chain if and only if it is connected, and if the number of vertices of uneven degree is 0 or 2.

The following proof is given in Reference [H2]:"If there is an eulerian chain μ, the p-graph G is obviously connected; also, the two terminal vertices of μ (if they are separate) alone are of uneven degree, so that the condition is necessary."

We shall now prove the following proposition: "if there are two vertices A and B of uneven degree, there is an eulerian chain from A to B; if there are not two such vertices, there is an eulerian cycle."

Let us assume that this proposition is true for graphs with less than m links, and prove that it is also true when there are m links. To do so, we shall assume that G has two vertices of uneven degree, A and B.

Chain μ can be defined by a traveller who sets off from A in any direction, and does not traverse the same link twice. If the traveller reaches a vertex $X_i \neq B$, he will have used an uneven number of links touching X_i, and can therefore leave by an unused link; when he can no longer leave, he must be at B. But, in this arbitrary journey μ from A to B, all the links may not have been used, and in that case a partial graph will remain in which all the vertices are of even degree.

If C_1, C_2,..., C_k are the components of G' with at least one link, they contain by assumption eulerian cycles μ_1, μ_2,...; and as G is connected, chain μ successively encounters all the components C_i at vertices $X_1 \in C_1$, $X_2 \in C_2$,..., $X_k \in C_k$, in this order. We now consider the chain

$$\mu[A, X_1] + \mu_1 + \mu[X_1, X_2] + \mu_2 + \cdots + \mu[X_k, B],$$

and find that it is an eulerian chain from A to B.

Hence, we have proved that the condition is sufficient.

Example (Fig. 45.3).   The 2-graph which represents the bridges of Königsberg does not have an eulerian chain because it contains 4 vertices of uneven degree.

D. DRAWING OF AN EULERIAN CYCLE. FLEURY'S ALGORITHM

This algorithm can be used for any connected p-graph $G = (\mathbf{X}, \mathbf{\bar{U}})$ in which *all the vertices are of even degree*.

(a)   Start from any vertex X_i and then number and eliminate every link which is traversed;

(b)   Never use a link $\bar{u}$ which is an *isthmus*, in other words a link, the suppression of which would create two connected components with at least one link.

Example (Fig. 45.4).   Let us begin with a link of (A, B); then a link of (B, C); then (C, E), a link of (E, F); then (F, A), (A, E), the other

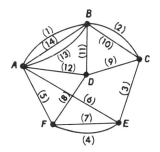

FIG. 45.4.

link of (F, E); then (F, D), (D, C), the other link of (C, B); then (B, D), (A, D), a link of (A, B); and, finally, the last link of (B, A).

E. EULERIAN CIRCUIT

In a p-applied graph an *eulerian circuit* is a circuit which uses every arc once, and once only.

F. PSEUDOSYMMETRICAL GRAPH

A graph is *pseudosymmetrical* if as many arcs arrive at each vertex as there are arcs leaving it; in other words, if

$$\forall X_i \in \mathbf{X}: |\mathbf{U}_{X_i}^+| = |\mathbf{U}_{X_i}^-|. \tag{45.1}$$

Example (Fig. 45.5).

$$|\mathbf{U}_{X_1}^+| = 2, \quad |\mathbf{U}_{X_1}^-| = 2; \quad |\mathbf{U}_{X_2}^+| = 3, \quad |\mathbf{U}_{X_2}^-| = 3;$$

$$|\mathbf{U}_{X_3}^+| = 2, \quad |\mathbf{U}_{X_3}^-| = 2; \quad |\mathbf{U}_{X_4}^+| = 2, \quad |\mathbf{U}_{X_4}^-| = 2.$$

This graph is pseudosymmetrical.

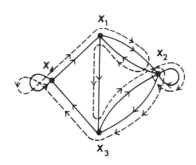

FIG. 45.5.

G. FUNDAMENTAL THEOREM OF EULERIAN CIRCUITS

A graph contains an eulerian circuit if and only if it is connected and pseudosymmetrical. The proof is the same as for eulerian cycles.

Example (Fig. 45.5).   This graph, which is connected and pseudosymmetrical, contains an eulerian circuit shown by a dotted line.

46. Point and Set of Articulation. Connectivity Number

A. POINT OF ARTICULATION. NONARTICULATED GRAPH

In a graph $G = (\mathbf{X}, \bar{\mathbf{U}})$, which is connected, we say that a vertex X_i is a *point of articulation* if the subgraph obtained by suppression of X_i is not connected.

Example. In Fig. 46.1 the vertices X_1 and X_5, and in Fig. 46.2, X_4, X_3, X_5, and X_9 are all points of articulation.

If the point of articulation is connected by a single link to each component of this subgraph, it is called *simple* as, for example, X_4 in Fig. 46.2.

`Nonarticulated Graph or Star.`   This is a connected graph which does not have a point of articulation.

Example (Fig. 46.3).

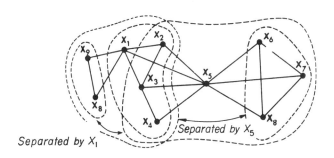

FIG. 46.1.

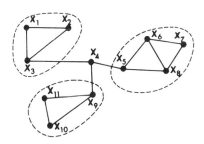

FIG. 46.2.

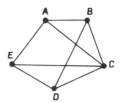

FIG. 46.3.

B. SET OF ARTICULATION

In a connected graph $G = (\mathbf{X}, \bar{\mathbf{U}})$ where $|\mathbf{X}| = n$, a subset $\mathbf{A} \subset \mathbf{X}(\mathbf{A} \neq \varnothing)$ is by definition a *set of articulation* if the subgraph produced by $\mathbf{C} \mathbf{A_X}$ is not connected.[1]

If $\mathbf{A}$ is reduced to a single vertex the *set of articulation* is a point of articulation.

Example (Fig. 46.4). $\{X_5, X_9\}$ is a set of articulation.

C. FINDING THE SETS OF MINIMAL ARTICULATION[2]

Using the graph $G = (\mathbf{X}, \Gamma)$ shown in Fig. 46.4, we shall try to find the set of minimal articulation which will enable us to separate it into two disjoint subgraphs, one containing X_6, and the other X_2 and X_3.

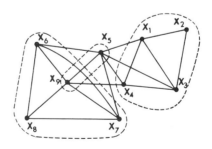

FIG. 46.4.

---

[1] In the case of a nonconnected graph, $\varnothing$ is also considered as an articulated set.
[2] We owe this method to Malgrange, who has been cited earlier.

This means that $\mathbf{X}$ must be divided into three subsets: $\mathbf{X}_1$, $\mathbf{X}_2$, and $\mathbf{A}$, such that

$$X_6 \in \mathbf{X}_1 ; \qquad X_2, X_3 \in \mathbf{X}_2 ; \qquad \Gamma(\mathbf{X}_1) \cap \mathbf{X}_2 = \varnothing ;$$
$$\Gamma(\mathbf{X}_2) \cap \mathbf{X}_1 = \varnothing ; \qquad |\mathbf{A}| \text{ minimal.} \tag{46.1}$$

Set $\mathbf{A}$ is an articulated set of the graph, for G is connected and any chain joining a point of $\mathbf{X}_1$ to a point of $\mathbf{X}_2$ must pass through a vertex of $\mathbf{A}$.

Let $[M]$ be the boolean matrix of G and $[\bar{M}] = [1] - [M]$ the complementary matrix. The submatrix of $[\bar{M}]$, defined by the lines corresponding to $\mathbf{X}_1$ and the columns corresponding to $\mathbf{X}_2$, has all its elements equal to 1 in accordance with the relations (46.1.3) and (46.1.4) given above. This *complete* submatrix is not the submatrix of any other complete submatrix, for if it were,

$$\mathbf{A} = [\mathbf{X} - (\mathbf{X}_1 \cup \mathbf{X}_2)]$$

would not be minimal; indeed, it would be possible to add at least one point such as $X_i \in \mathbf{X}_1$ to $\mathbf{X}_1 \cup \mathbf{X}_2$, so that the relations (46.1.3) and (46.1.4) would still be true, and we would then have

$$\mathbf{A}' = [\mathbf{X} - (\mathbf{X}_1 \cup \mathbf{X}_2 \cup X_i)] \subset\subset \mathbf{A}.$$

	1	2	3	4	5	6	7	8	9
1		1	1	1	1				
2	1		1						
3	1	1		1	1				
4	1		1		1				1
$[M] = $ 5	1		1	1		1	1		1
6					1		1	1	1
7				1	1		1		1
8						1	1		1
9				1	1	1	1	1	

$$\tag{46.2}$$

	1	2	3	4	5	6	7	8	9
1	1					1	1	1	1
2		1		1	1	1	1	1	1
3			1			1	1	1	1
4		1		1		1	1	1	
$[\bar{M}] = $ 5		1			1		1	1	1
6	1	1	1	1		1			
7	1	1	1	1	1		1		
8	1	1	1	1	1			1	
9	1	1	1		1				1

$$(46.3)$$

This complete submatrix is therefore maximal or *primary* (see Section 48). Hence, all we need find are all the primary submatrices of $[\bar{M}]$ which define two subsets $\mathbf{X}_1$ and $\mathbf{X}_2$, such that $X_6 \in \mathbf{X}_1$ and $\{X_2, X_3\} \subset \mathbf{X}_2$. These can be calculated by the algorithm given in Section 48.

We obtain the matrices:

$$[M'], \quad \mathbf{X}_1' = \{X_1, X_2, X_3, X_4\}; \quad \mathbf{X}_2' = \{X_6, X_7, X_8\}, \quad (46.4)$$

whence

$$\mathbf{A}' = \{X_5, X_9\} \quad \text{and} \quad |\mathbf{A}'| = 2;$$

$$[M''], \quad \mathbf{X}_1'' = \{X_1, X_2, X_3\}; \quad \mathbf{X}_2'' = \{X_6, X_7, X_8, X_9\}, \quad (46.5)$$

whence

$$\mathbf{A}'' = \{X_4, X_5\} \quad \text{and} \quad |\mathbf{A}''| = 2.$$

Hence there are two minimal articulated subsets which allow us to separate X_6, on the one hand, and X_2, X_3, on the other. They are

$$\mathbf{A}' = \{X_5, X_9\} \quad \text{and} \quad \mathbf{A}'' = \{X_4, X_5\}. \quad (46.6)$$

We have shown $\mathbf{A}'$ in Fig. 46.4.

D. Set of Isolation

In a graph $G = (\mathbf{X}, \mathbf{\bar{U}})$ where $|\mathbf{X}| = n$, every set of $n - 1$ elements is a set of isolation.

E. Connectivity Number of a Graph. h-Connected Graph

The *connectivity number* of a graph $G = (\mathbf{X}, \mathbf{\bar{U}})$ is the minimal number of elements of a set which is articulated or isolated for G, and is shown as $\omega(G)$. If h is a nonnegative integer, a graph G is said to be h-connected if its connectivity $\omega(G) \geqslant h$.

Example. The graph of Fig. 46.4 is 2-connected. It contains isolated sets, the smallest of which are

$$\{X_5, X_9\}, \quad \{X_1, X_3\}, \quad \{X_4, X_5\}.$$

When a graph G has n vertices and is h-connected, this means:

(1)   That it is connected;
(2)   That $h < n$;
(3)   That G has no articulated set $\mathbf{A}$ with $|\mathbf{A}| < h$.

A graph G is 2-connected if and only if it is nonarticulated.

F. Properties of h-Connected Graphs

(1)   In an h-connected graph the degree of every vertex is greater than or equal to h;

(2)   If G is an h-connected graph, the graph G', obtained by suppressing a vertex, is $(h - 1)$-connected;

(3)   If G is an h-connected graph, graph G', obtained by suppressing k links ($k < h$) is $(h - k)$-connected;

(4)   A necessary and sufficient condition for a graph to be h-connected is that two arbitrary vertices X_i and X_j can be joined by h-separate elementary chains which have nothing in common except the vertices X_i and X_j (Whitney's theorem).

47. Planar Graphs

A. Definition

A p-graph $G = (\mathbf{X}, \mathbf{\bar{U}})$ is *planar* if it can be represented in such a way that all the vertices are separate points, the links are simple curves, and that no two links encounter each other except at their extremities. Such a representation of G is called a *planar topological graph*.

Examples (Fig. 47.1 and 47.2).

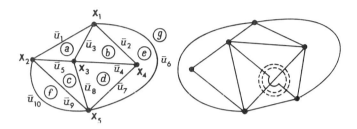

FIG. 47.1.   Planar graph.   FIG. 47.2.   Nonplanar graph.

B. FACES OF A PLANAR CONNECTED p-GRAPH

The links of a planar p-graph define planar surfaces and are called *faces*; the exterior surface without bounds is called the *infinite face*.

Example (Fig. 47.1): a, b, c, d, e, and f are faces; g is the infinite face.

C. THEOREM CONCERNING PLANAR p-GRAPHS

(1)   In a planar p-graph G, the contours of the different finite faces form a fundamental foundation of independent cycles.[1] This property makes it very easy to construct a system of independent linear currents in an electrical network.

(2)   *Euler's Formula.*   If a planar p-graph has n vertices, m links and f faces, we have

$$n - m + f = 2. \qquad (47.1)$$

Indeed, if ν is the cyclomatic number, $f = \nu + 1$, since we must count the infinite face. In accordance with (35.2), $\nu = m - n + 1$; that is, $f - 1 = m - n + 1$, whence $n - m + f = 2$ (see Fig. 47.3).

(3)   In any planar 1-graph there is a vertex X_i which is less than or equal to 5 in degree.

(4)   In a geographical map (in other words, a set **X** of countries where we assume $(X_i, X_j) \in$ **U**, if the countries X_i and X_j have a

---

[1] See Sections 35 and 44.

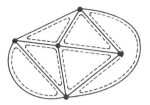

FIG. 47.3.   Independent cycles based on the faces.

common frontier, this graph being planar and symmetrical), there is at least one face with a countour which contains a number of links less than or equal to 5.

(5)   A planar 1-graph is 5-chromatic (see proof in [H2]). It has not so far been possible to prove that every 1-graph is 4-chromatic, but there is reason to think it is.

(6)   *Kuratowski's Theorem.*   A p-graph is planar if it does not include any partial subgraph of type I or II (Fig. 47.4) in which the vertices can be added in arbitrary positions and numbers (see [H2]).

D. THE DUAL OF A PLANAR CONNECTED p-GRAPH

If we have such a graph G and form a corresponding planar connected q-graph[1] G^*, which is without any isolated vertices and is the *dual* of G, G^* is defined as follows:

"Inside each face r of G we place a vertex X_r^* of G^*; for each link $\bar{u}_k$ of G we draw a corresponding link $\bar{v}_m$ of G^* to connect vertices X_r^* and X_s^* which correspond to the faces r and s on either side of $\bar{u}_k$."

Example (Fig. 47.5).   The dual is shown by a broken line. We should remember to place a vertex of G^* on the infinite face.

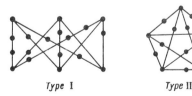

Type I   Type II

FIG. 47.4.

---

[1] According to the particular case, q can be equal to or different from p. For example, in Fig. 47.5, G is a 3-graph and G^* is a 2-graph.

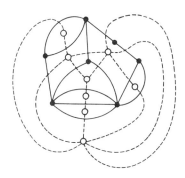

FIG. 47.5.

E. PROPERTIES OF THE DUAL

Given a planar connected p-graph without any isolated vertex and its dual G^*:

(a) G and G^* have equal cyclomatic numbers;

(b)   If G has a point of articulation, G^* also has one.

F. DUALITY OF THE PROBLEMS OF THE MINIMAL FLOW AND OF THE OPTIMAL FLOW WHEN THE GRAPH IS PLANAR AND NONDIRECTED

Given a planar nondirected graph G and its dual G^*, the two following problems can be treated as one:

Problem of the Maximal Flow between Two Vertices A and B.   Find a cut $\bar{\mathbf{V}}$ in G of minimal capacity; that is to say, minimize

$$c(\bar{\mathbf{V}}) = \sum_{\bar{v} \in \bar{\mathbf{V}}} c(\bar{v}); \qquad (47.2)$$

or (which amounts to the same thing), pass a maximal flow through the network.

Problem of the Chain of Minimal Value between Two Points X_0 and X_n.   Find in G^* the minimal chain $\bar{\mathbf{V}}^*$ from X_0 to X_n. In other words, minimize

$$l(\bar{\mathbf{V}}^*) = \sum_{\bar{v}^* \in \bar{\mathbf{V}}^*} l(\bar{v}^*). \qquad (47.3)$$

Figure 47.6 gives an example of this property. The unbroken lines represent a planar nondirected transport network, in which the entry is

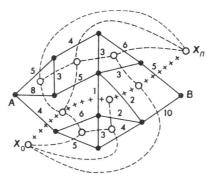

FIG. 47.6.

at A and the exit at B, and the broken lines shown its dual in which X_0 and X_n have been placed on different sides of the network which connects A with B.

The links with unbroken and broken lines have the same value, as shown at their intersection in the drawing.

The chain of minimal value between X_0 and X_n, denoted by crosses, is equal to 12, and the maximal flow from A to B (or from B to A) is also equal to 12.

48. Mapping a Graph into a Graph

A. DEFINITION. IMBEDDING

Taking graph $G_2 = (\mathbf{Y}, \Delta)$ of Fig. 48.2 and graph $G_1 = (\mathbf{X}, \Gamma)$ of Fig. 48.1, we call the *mapping of G_2 into G_1* a set of oriented pairs such as, for example, $\{(A, \beta), (B, \gamma), (D, \alpha), (E, \delta)\}$ (Fig. 48.3), and such that

$$\Delta\beta \subset \Gamma A, \quad \Delta\gamma \subset \Gamma B, \quad \Delta\alpha \subset \Gamma D, \quad \Delta\delta \subset \Gamma E. \qquad (48.1)$$

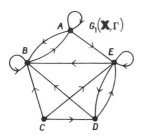

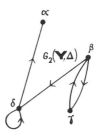

FIG. 48.1.   FIG. 48.2.

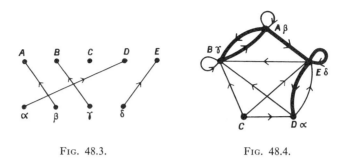

FIG. 48.3.   FIG. 48.4.

More generally, given two graphs $G_2 = (\mathbf{Y}, \Delta)$ and $G_1 = (\mathbf{X}, \Gamma)$ such that $|\mathbf{Y}| \leqslant |\mathbf{X}|$ (Fig. 48.4), we say that it is possible to map G_2 into G_1 if there is a set of oriented pairs[1]:

$$\mathbf{E} = \{(X_{i_1}, Y_{j_1}), (X_{i_2}, Y_{j_2}),..., (X_{i_m}, Y_{j_m})\} \qquad (48.2)$$

where $m = |\mathbf{Y}|$ and such that

$$\Delta Y_{j_r} \subset \Gamma X_{i_r}, \qquad \forall Y_{j_r} \in \mathbf{Y}. \qquad (48.3)$$

A more synthetic definition lies in saying that a graph G_2 can be mapped into a graph G_1 if there is at least one partial subgraph $g_1'(G_1)$ which is isomorphic to G_2.

Set $\mathbf{E}$ forms a mapping or *imbedding* of G_2 into G_1. With two graphs G_1 and G_2, there may be zero, one, or several imbeddings. In our example (Figs. 48.1 and 48.2), it will be found that there are several solutions.

B. FINDING IMBEDDINGS

In the last resort, the problem which confronts us consists of finding the partial subgraphs[2] of G_1 which are isomorphic (in this case identical) to G_2 by a permutation of the elements of $\mathbf{Y}$ in those of $\mathbf{X}$. This is shown in Fig. 48.4.

The concept we have defined can be applied to a variety of practical problems such as the rational distribution of machines in a factory, the

---

[1] When $|\mathbf{Y}| = |\mathbf{X}|$ we customarily say that it is possible to find a *mapping of* G_2 *into* G_1. An example will show the meaning of the following notation:

$$\Delta\beta = \{\gamma, \delta\}, \qquad \Gamma A = \{A, B, E\},$$
$$\{(B, \gamma), (E, \delta)\} \subset \{(A, \beta), (B, \gamma), (E, \delta)\}.$$

[2] If $|\mathbf{Y}| = |\mathbf{X}|$ it will be a case of a partial graph.

coordination of the different departments in a business, the wiring of machines, the search for analogies, etc.

Finding the mappings of a graph into another one usually proves a difficult problem unless there are only a few vertices in the graph onto which an imbedding is to be made; in which case the imbeddings can be found by combinatorial enumeration, if there are any. A generalized analytic method for finding them has not yet been discovered, but in some cases elimination will provide a fairly rapid solution.

In the graphs of Figs. 48.1 and 48.2, for instance, elimination is easy if we begin by finding the pairs which correspond to symmetrical arcs (if the nonsymmetrical arcs were in a minority we should begin with them). Let us note the principle that in any method of elimination we begin more or less arbitrarily with certain classes, and then pass from sets to increasingly smaller subsets by intersections or inclusions.

In our very limited example we consider the symmetrical arcs (β, γ), (γ, β), and find there are four possible imbeddings:

$$[(D, \beta), (E, \gamma)], \qquad [(D, \gamma), (E, \beta)], \qquad [(A, \beta), (B, \gamma)], \qquad [(A, \gamma), (B, \beta)]$$

for β and γ in D and E, or A and B. But when the graphs have a larger number of vertices, the elimination must be an automatic process carried out by means of an algorithm such as the one we shall now introduce. Based on different premises to the above elimination, it enables us to perform successive automatic suppressions and thereby find the imbeddings in the subsets of partial subgraphs, subsets which contain far fewer parts than the graph on which the imbedding is being made.

C. Malgrange's Algorithm. Obtaining Primary Submatrices

The explanation of this algorithm requires certain preliminary definitions.

In a matrix with binary coefficients (0 or 1) (that is to say, a boolean matrix), a submatrix in which all the elements are 1, is called a *complete submatrix*.[1] Thus in Fig. 48.5 there are 7 primary matrices in matrix [M]. The *cover* of a boolean matrix is a set of complete submatrices covering all the coefficients which have a value of 1 in this matrix.

If **I** is the set of lines and **J** the set of columns of a boolean matrix, each complete submatrix is defined by the oriented pair of subsets

---

[1] A *primary submatrix* is a complete submatrix which is not contained in any other.

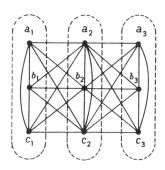

FIG. 48.5.   Graph G_1.

$(\mathbf{I}_p , \mathbf{J}_q)$, with $\mathbf{I}_p \subset \mathbf{I}$ and $\mathbf{J}_q \subset \mathbf{J}$. We now show that the operations $\cup$ and $\cap$ which cause any two complete submatrices of a boolean matrix $[M]$,

   (1) $[M_1]$ defined by the oriented pair $(\mathbf{I}_1 , \mathbf{J}_1)$, and

   (2) $[M_2]$ defined by the oriented pair $(\mathbf{I}_2 , \mathbf{J}_2)$,

to correspond to the two submatrices

$$[M_1] \cup [M_2] = [M'] \quad \text{defined by the oriented pair} \quad (\mathbf{I}_1 \cup \mathbf{I}_2 , \mathbf{J}_1 \cap \mathbf{J}_2), \qquad (48.4)$$

$$[M_1] \cap [M_2] = [M''] \quad \text{defined by the oriented pair} \quad (\mathbf{I}_1 \cap \mathbf{I}_2 , \mathbf{J}_1 \cup \mathbf{J}_2) \qquad (48.5)$$

are internal[1] operations in set $\mathbf{M}^*$ of the complete submatrices of $[M]$.

The alternate and exhaustive application of the following two rules to the complete submatrices of a cover

$$C = \{[M_1], [M_2],..., [M_p]\}$$

enables us to obtain all the primary submatrices of matrix $[M]$ in a finite number of iterations.

First Rule. Suppress every submatrix $[M_k]$ contained in another submatrix $[M_l]$ of the cover C.

Second Rule.   Place beside C the submatrices obtained by operations $\cup$ and $\cap$ applied to all the pairs of matrices $[M_k]$, $[M_l]$ remaining in the

---

[1] "A law of internal composition" for a set $\mathbf{E}$ is a mapping of $\mathbf{E} \times \mathbf{E}$ into $\mathbf{E}$. The operation "sum" for integers n_1 and $n_2 : n_1 + n_2$ gives an integer and is an internal operation on the set of integers. In (48.4) and (48.5) the operations $\cup$ and $\cap$ on the elements of $\mathbf{M}^*$ provide elements of this set.

TABLE 48.1

$$[M] =$$

	a	b	c	d	e	f
A	0	1	0	1	1	1
B	1	0	1	0	0	0
C	0	1	0	1	1	0
D	0	1	1	0	0	1

	b	d	e	f
A	1	1	1	1

	b	d	e
A	1	1	1
C	1	1	1

	b	f
A	1	1
D	1	1

	a	c
B	1	1

	b	c	f
D	1	1	1

	b
A	1
C	1
D	1

	c
B	1
D	1

cover (except, to avoid an unending process, any complete submatrix contained in a submatrix which has already appeared in C).

Example. Let us set out to calculate the primary submatrices of the boolean matrix of Table 48.1.

Stage 1. Let us choose as cover:

$$[M_1] = A$$

	b	d	e	f
	1	1	1	1

, $[M_2] = B$

	a	c
	1	1

,

$$[M_3] = C$$

	b	d	e
	1	1	1

, $[M_4] = D$

	b	c	f
	1	1	1

Stage 2 (Second Rule). Let us calculate the unions and inter-sections

$$\mathbf{I}_1 \cup \mathbf{I}_2 = \{A, B\}, \qquad \mathbf{J}_1 \cap \mathbf{J}_2 = \varnothing,$$
$$\mathbf{I}_1 \cup \mathbf{I}_3 = \{A, C\}, \qquad \mathbf{J}_1 \cap \mathbf{J}_3 = \{b, d, e\},$$

whence another submatrix

$$[M_5] = \begin{array}{c|c|c|c} & b & d & e \\ \hline A & 1 & 1 & 1 \\ \hline C & 1 & 1 & 1 \end{array},$$

$$\mathbf{I}_1 \cup \mathbf{I}_4 = \{A, D\}, \qquad \mathbf{J}_1 \cap \mathbf{J}_4 = \{b, f\}, \tag{48.6}$$

whence a further submatrix

$$[M_6] = \begin{array}{c|c|c} & b & f \\ \hline A & 1 & 1 \\ \hline D & 1 & 1 \end{array},$$

$$\mathbf{I}_2 \cup \mathbf{I}_3 = \{B, C\}, \qquad \mathbf{J}_2 \cap \mathbf{J}_3 = \varnothing,$$
$$\mathbf{I}_2 \cup \mathbf{I}_4 = \{B, D\}, \qquad \mathbf{J}_2 \cap \mathbf{J}_4 = \{c\},$$

whence a further submatrix

$$[M_7] = \begin{array}{c|c} & c \\ \hline B & 1 \\ \hline D & 1 \end{array},$$

$$\mathbf{I}_3 \cup \mathbf{I}_4 = \{C, D\}, \qquad \mathbf{J}_3 \cap \mathbf{J}_4 = \{b\},$$

whence a final submatrix

$$[M_8] = \begin{array}{c|c} & b \\ \hline C & 1 \\ \hline D & 1 \end{array}.$$

As all the intersections $\mathbf{I}_i \cap \mathbf{I}_j$ are empty $\forall i, j$, it is useless to calculate $\mathbf{J}_i \cup \mathbf{J}_j$.

Stage 3 (First Rule).   The new cover is

$$\{[M_1], [M_2], [M_4], [M_5], [M_6], [M_7], [M_8]\};$$

$[M_3]$, contained in $[M_5]$, has been eliminated.

Stage 4 (Second Rule).   (We shall show all the details of the evaluations although it is clear that some submatrices have already been obtained or are empty.)

$$\mathbf{I}_1 \cup \mathbf{I}_5 = \{A, C\}, \qquad \mathbf{J}_1 \cap \mathbf{J}_5 = \{b, d, e\} \quad \text{gives } [M_5];$$
$$\mathbf{I}_1 \cup \mathbf{I}_6 = \{A, D\}, \qquad \mathbf{J}_1 \cap \mathbf{J}_6 = \{b, f\} \quad \text{gives } [M_6];$$
$$\mathbf{I}_1 \cup \mathbf{I}_7 = \{A, B, D\}, \qquad \mathbf{J}_1 \cap \mathbf{J}_7 = \varnothing \, ;$$
$$\mathbf{I}_1 \cup \mathbf{I}_8 = \{A, C, D\}, \qquad \mathbf{J}_1 \cap \mathbf{J}_8 = \{b\},$$

with a new submatrix

	b
A	1
$[M_9] = C$	1
D	1

$$\mathbf{I}_2 \cup \mathbf{I}_5 = \{A, B, C\}, \qquad \mathbf{J}_2 \cap \mathbf{J}_5 = \varnothing \, ;$$
$$\mathbf{I}_2 \cup \mathbf{I}_6 = \{A, B, D\}, \qquad \mathbf{J}_2 \cap \mathbf{J}_6 = \varnothing \, ;$$
$$\mathbf{I}_2 \cup \mathbf{I}_7 = \{B, D\}, \qquad \mathbf{J}_2 \cap \mathbf{J}_7 = \{c\} \qquad \text{gives } [M_7];$$
$$\mathbf{I}_2 \cup \mathbf{I}_8 = \{B, C, D\}, \qquad \mathbf{J}_2 \cap \mathbf{J}_8 = \varnothing \, ;$$
$$\mathbf{I}_4 \cup \mathbf{I}_5 = \{A, C, D\}, \qquad \mathbf{J}_4 \cap \mathbf{J}_5 = \{b\} \qquad \text{gives } [M_9];$$
$$\mathbf{I}_4 \cup \mathbf{I}_6 = \{A, D\}, \qquad \mathbf{J}_4 \cap \mathbf{J}_6 = \{b, f\} \qquad \text{gives } [M_6];$$
$$\mathbf{I}_4 \cup \mathbf{I}_7 = \{B, D\}, \qquad \mathbf{J}_4 \cap \mathbf{J}_7 = \{c\} \qquad \text{gives } [M_7];$$
$$\mathbf{I}_4 \cup \mathbf{I}_8 = \{C, D\}, \qquad \mathbf{J}_4 \cap \mathbf{J}_8 = \{b\} \qquad \text{gives } [M_8] \text{ contained in } [M_9];$$
$$\mathbf{I}_5 \cup \mathbf{I}_6 = \{A, C, D\}, \qquad \mathbf{J}_5 \cap \mathbf{J}_6 = \{b\} \qquad \text{gives } [M_9];$$
$$\mathbf{I}_5 \cup \mathbf{I}_7 = \{A, B, C, D\}, \qquad \mathbf{J}_5 \cap \mathbf{J}_7 = \varnothing \, ;$$
$$\mathbf{I}_5 \cup \mathbf{I}_8 = \{A, C, D\}, \qquad \mathbf{J}_5 \cap \mathbf{J}_8 = \{b\} \qquad \text{gives } [M_9];$$
$$\mathbf{I}_6 \cup \mathbf{I}_7 = \{A, B, D\}, \qquad \mathbf{J}_6 \cap \mathbf{J}_7 = \varnothing \, ;$$
$$\mathbf{I}_6 \cup \mathbf{I}_8 = \{A, C, D\}, \qquad \mathbf{J}_6 \cap \mathbf{J}_8 = \{b\} \qquad \text{gives } [M_9];$$
$$\mathbf{I}_7 \cup \mathbf{I}_8 = \{B, C, D\}, \qquad \mathbf{J}_7 \cap \mathbf{J}_8 = \varnothing \, ;$$

$\mathbf{I}_1 \cap \mathbf{I}_5 = \{A\}$, $\mathbf{J}_1 \cup \mathbf{J}_5 = \{b, d, e, f\}$ gives $[M_1]$;

$\mathbf{I}_1 \cap \mathbf{I}_6 = \{A\}$, $\mathbf{J}_1 \cup \mathbf{J}_6 = \{b, d, e, f\}$ gives $[M_1]$;

$\mathbf{I}_1 \cap \mathbf{I}_7 = \varnothing$, $\mathbf{I}_1 \cap \mathbf{I}_8 = \varnothing$;

$\mathbf{I}_2 \cap \mathbf{I}_5 = \varnothing$, $\mathbf{I}_2 \cap \mathbf{I}_6 = \varnothing$;

$\mathbf{I}_2 \cap \mathbf{I}_7 = \{B\}$, $\mathbf{J}_2 \cup \mathbf{J}_7 = \{a, c\}$ gives $[M_2]$;

$\mathbf{I}_2 \cap \mathbf{I}_8 = \varnothing$; $\mathbf{I}_4 \cap \mathbf{I}_5 = \varnothing$;

$$(48.7)$$

$\mathbf{I}_4 \cap \mathbf{I}_6 = \{D\}$ $\mathbf{J}_4 \cup \mathbf{J}_6 = \{b, c, f\}$ gives $[M_4]$;

$\mathbf{I}_4 \cap \mathbf{I}_7 = \{D\}$, $\mathbf{J}_4 \cup \mathbf{J}_7 = \{b, c, f\}$ gives $[M_4]$;

$\mathbf{I}_4 \cap \mathbf{I}_8 = \{D\}$, $\mathbf{J}_4 \cup \mathbf{J}_8 = \{b, c, f\}$ gives $[M_4]$;

$\mathbf{I}_5 \cap \mathbf{I}_6 = \{A\}$, $\mathbf{J}_5 \cup \mathbf{J}_6 = \{b, d, e, f\}$ gives $[M_1]$;

$\mathbf{I}_5 \cap \mathbf{I}_7 = \varnothing$;

$\mathbf{I}_5 \cap \mathbf{I}_8 = \{C\}$, $\mathbf{J}_5 \cup \mathbf{J}_8 = \{b, d, e\}$ gives $[M_3]$ contained in $[M_5]$;

$\mathbf{I}_6 \cap \mathbf{I}_7 = \{D\}$, $\mathbf{J}_6 \cup \mathbf{J}_7 = \{b, c, f\}$ gives $[M_4]$;

	b	f
	1	1

$\mathbf{I}_6 \cap \mathbf{I}_8 = \{D\}$, $\mathbf{J}_6 \cup \mathbf{J}_8 = \{b, f\}$ gives D ⎡1 1⎤ contained in $[M_4]$;

	b	c
	1	1

$\mathbf{I}_7 \cap \mathbf{I}_8 = \{D\}$, $\mathbf{J}_7 \cup \mathbf{J}_8 = \{b, c\}$ gives D ⎡1 1⎤ contained in $[M_4]$.

Stage 5 (First Rule).   The new cover is

$$\{[M_1], [M_2], [M_4], [M_5], [M_6], [M_7], [M_9]\};\qquad (48.8)$$

$[M_8]$ has been eliminated because it was contained in $[M_9]$.

Stage 6 (Second Rule).   The calculation of the unions and inter-sections shows that it is no longer possible to find any complete sub-matrices which are not equal to elements or contained in the elements of the previous cover; hence it is a cover which gives the set of the primary matrices.

D. An Example of Finding the Mapping of a Graph into
    a Graph. Problem of the Chorus Girls[1]

The somewhat frivolous character of this example should not make
us overlook the many important practical problems which can be
treated in a similar manner.

The presentation of an act featuring a chorus of girls involves certain
difficulties for the producer. In particular, if the girls are a talkative
group he may have to find an arrangement which will minimize the risk
of loud voices being audible to the audience. Knowing the girls who are
the most prone to chatter, he will try to place them together so that they
will not need to raise their voices to communicate with one another.

The nine girls will appear on the stage in three rows, and their posi-
tioning will be shown as a symmetrical graph (their affinities being
assumed as bilateral). The symmetrical arcs are represented by dark
lines which are nonoriented. Each vertex stands for a girl, and the
connections between the vertices show acceptable sounds of conversation (Fig. 48.5). This graph will be called G_1.

Another graph G_2 (Fig. 48.6) shows the affinities between the different
girls, in other words, the likely exchanges of conversation. The problem
which confronts us is therefore to attempt the mapping of G_2 on G_1.[2]

In G_1 the index points i are all connected together as well as to the points
$i + 1$ and $i - 1$, but no point 1 is joined to a point 3. The set of points i
forms "zone" i. We can now realize that a necessary condition (and a sufficient one, since there are only three zones) of mapping is that we should
be able to find two subsets **A** and **B** of three points in G_2, each such that

$$(a \in \mathbf{A}) \quad \text{and} \quad (b \in \mathbf{B}) \Rightarrow d(a, b) \geqslant 2$$

(we say in this case that $d(\mathbf{A}, \mathbf{B}) \geqslant 2$).

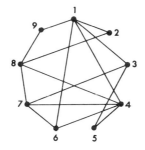

FIG. 48.6.   Graph G_2.

[1] See Y. Malgrange, *Utilisation des calculateurs électroniques en recherche opérationelle*, Publications Bull, Paris, 1961.

[2] Here G_2 has the same number of vertices as G_1; hence it is a mapping "on."

Let us now consider the matrix $[M]$ of distances greater than or equal to 2 associated with G_2, a matrix whose coefficients c_{ij} have the following values:

$$c_{ij} = 1 \quad \text{if} \quad d(X_i, X_j) \geqslant 2$$

$$c_{ij} = 0 \quad \text{if} \quad d(X_i, X_j) < 2.$$

(48.9)

If we find two sets **A** and **B** such that $d(\mathbf{A}, \mathbf{B}) \geqslant 2$, the submatrix of $[M]$ defined by the pair $(\mathbf{A}, \mathbf{B})$ is complete, and we need only find all the primary submatrices of $[M]$ and retain those which answer to the condition

$$|\mathbf{A}| \geqslant 3 \quad \text{and} \quad |\mathbf{B}| \geqslant 3.$$

(48.10)

In our example, the matrix of distances $\geqslant 2$ is given by Table 48.2. The only primary matrix satisfying the above conditions is the one defined by the sets

$$\mathbf{A} = \{2, 8, 9\} \quad \text{and} \quad \mathbf{B} = \{3, 5, 6\}.$$

(48.11)

TABLE 48.2

	1	2	3	4	5	6	7	8	9
1					1		1	1	
2			1	1	1	1	1		1
3		1		1		1		1	1
4		1	1						1
5	1	1				1	1	1	1
6		1	1		1			1	1
7	1	1			1				1
8	1		1		1	1			
9		1	1	1	1	1	1		

Figure 48.7 shows the only possible mapping.

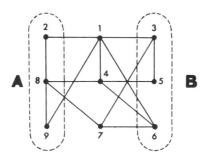

FIG. 48.7.

CHAPTER V

MATHEMATICAL PROPERTIES OF DYNAMIC PROGRAMMING

49. Introduction

The various examples of dynamic programming which were given in Chapter II should have convinced the reader of the great possibilities of this method, but we have deliberately avoided mentioning the difficulties which arise when it is treated analytically. These are of two kinds:

(1) The passage from one stage to another, when the variables are considered continuously, introduces delicate questions of topology in carrying forward the optimal values. One method of overcoming this difficulty is to change the continuous sets into discrete denumerable ones, and in many practical problems the latter prove satisfactory; at the same time, it deprives us of the power of discussion and analysis which is a part of differential calculus.

(2) The valuable concept of an *economic horizon* enters into many problems and often leads to a study of convergence on an infinite horizon. In a case of uncertainty we should decide whether the ergodic assumption is satisfied, and discover the reason if it is not. This question, however, is only of theoretical interest, for in practice the phenomena are not stationary because of the continuous evolution of the surroundings, as well as for many other reasons. It is the introduction of the realistic concept of a rate of interest which renders the process convergent in actual economic problems.

50. Formulas for Multistage Optimization

A. FUNCTION WHICH CAN BE SEPARATED INTO STAGES

Let us consider the function of $N + 1$ variables:

$$F(x_0, x_1, x_2, ..., x_{N-1}, x_N) = v_1(x_0, x_1) + v_2(x_1, x_2) + \cdots + v_N(x_{N-1}, x_N),$$

$$(50.1)$$

which, by assumption, can be decomposed into a sum of N elementary functions

$$v_n(x_{n-1}, x_n), \qquad n = 1, 2, ..., N.$$

We intend to find the maximum (or minimum, depending on the case) of function F, with the knowledge that each magnitude x_n can vary in a domain which depends only on x_0 and x_{n+1}, whatever value between 1 and N is possessed by n.

A reader with some knowledge of advanced mathematics will have no difficulty in seeing that it is possible to generalize this definition, as well as the optimization formula which follows, in cases where the x_n terms are vectors

$$x_n = \{x_{1n}, x_{2n}, ..., x_{k_n n}\}$$

with k_n components.

The particular form (50.1) imposed on F, and the nature of the domains of the variables, make it possible for us to consider a system with N stages for which $v_n(x_{n-1}, x_n)$, $n = 1, 2, ..., N$, would be the function of value associated with each stage, and F the function associated with the set of stages.

B. POLICIES AND SUBPOLICIES

A set of particular values such as

$$x_j, \quad x_{j+1}, \quad x_{j+2}, ..., x_k, \qquad 0 \leqslant j < k \leqslant N, \tag{50.2}$$

is called a *subpolicy*, whereas a *policy* is a set of $N + 1$ values

$$x_0, \quad x_1, ..., x_{N-1}, \quad x_N. \tag{50.3}$$

In problems where the decomposition into stages can be carried out in a variety of ways (see Section 12), the system is called *nonordered*, and where the nature of the system is sequential it is called *ordered*.

C. MULTISTAGE OPTIMIZATION

To find the maximal or minimal value of F, we shall use the theorem of optimality which was proved in Section 10, and which is valid in cases of variables which are discrete or continuous in an interval.

Let us consider stages 1 and 2 together, and let us use $f_{0,2}(x_0, x_2)$ for the optimal value of the sum

$$v_1(x_0, x_1) + v_2(x_1, x_2),$$

when x_1 varies in its domain which depends only on x_0 and x_2 . Hence we have for our optimization, in which the formulas obtained can be used equally for finding the maximum or the minimum:

$$f_{0,2}(x_0 , x_2) = \underset{x_1 \in \mathbf{X}_1(x_0,x_2)}{\text{MAX}} [v_1(x_0 , x_1) + v_2(x_1 , x_2)] \qquad (50.4)$$

where $x_1 \in \mathbf{X}_1(x_0 , x_2)$ shows that x_1 belongs to a set of values $\mathbf{X}_1$ which depends only on x_0 and x_2 . The value or values of x_1 which optimize

$$v_1(x_0 , x_1) + v_2(x_1 , x_2)$$

will define the optimal subpolicies for stages 1 and 2 and for the oriented pair x_0 and x_2 which are being considered.

Now let us take stages 1, 2, and 3 together, with $f_{0,3}(x_0 , x_3)$ as the optimal value of the sum

$$v_1(x_0 , x_1) + v_2(x_1 , x_2) + v_3(x_2 , x_3)$$

when x_1 and x_2 are made to vary in their respective domains. In accordance with the theorem of optimality, we have

$$f_{0,3}(x_0 , x_3) = \underset{x_2 \in \mathbf{X}_2(x_0,x_3)}{\text{MAX}} [f_{0,2}(x_0 , x_2) + v_3(x_2 , x_3)], \qquad (50.5)$$

where $x_2 \in \mathbf{X}_2(x_0 , x_3)$ have the meaning given above. The value or values of x_1 obtained earlier, and the value of x_2 which optimizes

$$f_{0,2}(x_0 , x_2) + v_3(x_2 , x_3)$$

will define the optimal subpolicies for x_0 and x_3 .

More generally, and with the same notation, we can state

$$f_{0,n}(x_0 , x_n) = \underset{x_{n-1} \in \mathbf{X}_{n-1}(x_0,x_n)}{\text{MAX}} [f_{0,n-1}(x_0 , x_{n-1}) + v_n(x_{n-1} , x_n)], \quad (50.6)$$

with

$$f_{0,1}(x_0 , x_1) = v_1(x_0 , x_1).$$

This enables us to calculate the successive optimal subpolicies for stages 1 and 2 together, and then for 1, 2, and 3 combined, up to stages 1, 2,..., $N-1$, N combined. In other words, the optimal policies or policy

$$F^*(x_0 , x_N) = \underset{x_{N-1} \in \mathbf{X}_{N-1}(x_0,x_N)}{\text{MAX}} [f_{0,N-1}(x_0 , x_{N-1}) + v_N(x_{N-1} , x_N)]. \quad (50.7)$$

Optimization has been carried out in the direction $n = 0$ to $n = N$, but it could be performed in the reverse direction.

In cases where x_N is not given, but only its domain of variation, we should then find

$$F^*(x_0) = \underset{x_N \in \mathbf{X}_N}{\text{MAX}} F^*(x_0, x_N), \tag{50.8}$$

which is the maximum corresponding to every possible value of x_N. In the same way, if x_0 is not known,

$$F^*(x_N) = \underset{x_0 \in \mathbf{X}_0}{\text{MAX}} F^*(x_0, x_N). \tag{50.9}$$

Finally, if both x_0 and x_N are unknown, then

$$F^* = \underset{\substack{x_0 \in \mathbf{X}_0 \\ x_N \in \mathbf{X}_N}}{\text{MAX}} F^*(x_0, x_N). \tag{50.10}$$

The generalization which we have just given for a sum can be extended to other types of function besides (50.1), for example:

$$F(x_0, x_1, ..., x_{N-1}, x_N) = v_1(x_0, x_1) \cdot v_2(x_1, x_2) \cdot \cdots \cdot v_N(x_{N-1}, x_N) \tag{50.11}$$

and also

$$F(x_0, x_1, ..., x_{N-1}, x_N) = v_1(x_0, x_1) * v_2(x_1, x_2) * \cdots * v_N(x_{N-1}, x_N) \tag{50.12}$$

where the symbol $(\cdot)$ represents a multiplication, and $(*)$ a product in the sense given to them by Borel.[1]

Generalization is also possible when the domain of each x_n, $n = 1$, 2,..., $N - 1$ does not depend only on x_0 and x_{n+1} but on other variables as well. However, all the advantages resulting from the simplicity of dynamic programming may be lost if the structure of the functional relations becomes over complicated.

51. Convergence of a Dynamic Program in a Limited Future

A. ECONOMIC HORIZON

Generally, where sequential optimization has to be performed, the stages represent intervals of time or *periods*, and the set of all the periods is called the *economic horizon*. This horizon may be closed on the left and on the right, which means that the number of periods is finite. It may also be open to the left and closed on the right, or open on both sides, neither of which cases is of value to us, for the most distant past does not count. Finally, it may be closed on the left and open to the right, which means that we start from the present moment towards a future

---

[1] See A. Kaufmann, *Méthodes et Modèles de la Recherche Opérationnelle,* Vol. I, p. 249. Dunod, Paris, 1962.

which is not limited. It is this case that we shall consider, and shall endeavor to show how we can overcome the difficulties which are encountered when the future is not limited and two policies are judged by a certain criterion.

Let $F(x_0, x_1, ..., x_n, ...)$ be the function of value given to a system containing an infinite number of stages.

B. FIRST CASE: REAL CONVERGENCE

If the given function converges towards a finite limit L_x for every infinite policy $x = \{x_0, x_1, x_2, ..., x_n, ...\}$ which satisfies the constraints, we choose the policy x^* which optimizes F. A convergence of this kind is rarely met with in practice, since it means that the function of value for each period approaches zero as the number of future periods becomes increasingly large, and a theoretical example is given by Bellman [I1] where this actually takes place. Nevertheless, as we shall see later on, the introduction of a rate of interest can restore the situation to that of the first case given above.

C. SECOND CASE: CONVERGENCE OF THE AVERAGE

The given function does not converge, but in order to optimize the function we shall adopt the criterion:

$$f(x_0, x_1, x_2, ..., x_n, ...) = \lim_{N \to \infty} \frac{F(x_0, x_1, x_2, ..., x_N)}{N} \qquad (51.1)$$

when it converges towards a value l_x for every policy

$$x = \{x_0, x_1, x_2, ..., x_n, ...\}.$$

It is now possible to compare the different policies and to choose the best one. In fact, we are led to compare concepts which are not necessarily coherent among themselves. Nevertheless, by agreeing that the nature of the problem allows us to cut the horizon into arbitrary slices or intervals of time, within which the same phenomenon makes its appearance, a comparison of the subpolicies which are obtained by this method can have a meaning and can be carried out. For example, assuming that we can compare the two cyclic subpolicies which follow:

Stage	0	1	2	3	4	5	6	7	8	9	10	⋯
Cyclic subpolicy for 3 stages	x_0	x_1	x_2	x_3	x_1	x_2	x_3	x_1	x_2	x_3	x_1	⋯
Cyclic subpolicy for 4 stages	x_0	x_1	x_2	x_3	x_4	x_1	x_2	x_3	x_4	x_1	x_2	⋯

The subpolicies such as

$$\{x_1{}', x_2{}', x_3{}' \mid x_0\} \quad \text{and} \quad \{x_1{}'', x_2{}'', x_3{}'', x_4{}'' \mid x_0\}$$

(or all the other cyclic subpolicies compatible with the problem) give
values:

$$f_{0,3}(x_0, x_1{}', x_2{}', x_3{}') \quad \text{and} \quad f_{0,4}(x_0, x_1{}'', x_2{}'', x_3{}'', x_4{}''),$$

the comparison of which has a meaning.

This is typical of various problems concerned with the operation of
machinery where we have to find the optimal cycles for its replacement
or maintenance.

D. THIRD CASE: CONVERGENCE OF THE DISCOUNTED VALUE

The given function is not convergent, but the introduction of a rate
of interest allows us to define another function which can be convergent.
For example, assuming that F is of the form:

$$F(x_0, x_1, x_2, ..., x_n, ...) = v_1(x_0, x_1) + v_2(x_1, x_2) + \cdots + v_n(x_{n-1}, x_n) + \cdots \tag{51.2}$$

and that we adopt a rate of interest α, we can assume

$$F_{\mathrm{act}}(x_0, x_1, x_2, ..., x_n, ...)$$
$$= v_1(x_0, x_1) + \frac{v_2(x_1, x_2)}{(1 + \alpha)} + \cdots + \frac{v_n(x_{n-1}, x_n)}{(1 + \alpha)^{n-1}} + \cdots. \tag{51.3}$$

If this new function is convergent for all suitable subpolicies we can
find its optimum by dynamic programming, considering first N stages
and then letting $N \to \infty$.

The economic coherence of the criterion chosen in cases where the
horizon is infinite is a question with which economists are greatly concerned.[1]

E. EXTENDED HORIZON

It is important to draw the reader's attention to a final precaution:
when a function of value N is optimized for a finite number of periods
we must find out what would happen in "an extended horizon," in
other words, a few periods after time N. Indeed, unless the value of x_N
is imposed, the value obtained by resolving the program for N periods

---

[1] The reader is referred to: G. Lesourne, *Technique économique et gestion industrielle*.
Dunod, Paris, 1959; and P. Massé, *Le Choix des investissements*. Dunod, Paris, 1959.

can lead to disastrous results for periods $N + 1, N + 2, \ldots$. Even a summary and intuitive study of the extended horizon usually enables us to decide a final[1] value for x_N, or an interval of values within which we should remain.

F. CASE OF A STATIONARY PHENOMENON

Let us consider the special case of a function such as (51.2) where we should have

$$v_n(x_{n-1}, x_n) = v(x_{n-1}, x_n), \tag{51.4}$$

let

$$F(x_0, x_1, x_2, \ldots, x_n, \ldots) = v(x_0, x_1) + v(x_1, x_2) + \cdots + v(x_{n-1}, x_n) + \cdots. \tag{51.5}$$

A large number of valuable simplifications can then be introduced; if the function $v(x_i, x_{i+1})$ is a suitable one, a repetitive algorithm can be established so that the calculations become automatic.

The stationary case with discounting:

$$F_{\text{act}}(x_0, x_1, x_2, \ldots, x_n, \ldots) = v(x_0, x_1) + \frac{v(x_1, x_2)}{(1 + \alpha)} + \cdots + \frac{v(x_{n-1}, x_n)}{(1 + \alpha)^{(n-1)}} + \cdots \tag{51.6}$$

leads to useful results, especially in a Markovian case of uncertainty (see Section 63).

52. Difficulty of the Final Calculations

It is important for the reader to realize that dynamic programming does not necessarily lead to simple calculations. To show the difficulties that can arise, we will return to the problem treated in Section 13, this time with $S = 20$ but with the same notation.

Period 6. In the case where $S = 9$, we have

$$x_6 = d_6 = 4, \tag{52.1}$$

---

[1] In a macroeconomic plan, x_N is nearly always a vector with a very large number of components which are the objectives to be attained (or surpassed, or not surpassed, or placed within bounds, etc.) in N periods. In a case of this kind, the study of the extended horizon can prove very difficult.

and

$$v_6(x_5, x_6) = p_6 a_6 = 110 - x_5 ; \tag{52.2}$$

but the bounds are no longer the same:

$$\max[d_5, x_4 - d_4] \leqslant x_5 \leqslant \min[S, x_6 + d_5],$$

or

$$\max[7, x_4 - 2] \leqslant x_5 \leqslant \min[20, 11],$$

or again

$$\max[7, x_4 - 2] \leqslant x_5 \leqslant 11. \tag{52.3}$$

Periods 6 and 5 Combined.

$$f_{6,5}(x_4) = \underset{\max[7,x_4-2]\leqslant x_5\leqslant 11}{\text{MIN}} [v_5(x_4, x_5) + v_6(x_5, x_6)]. \tag{52.4}$$

The quantity $p_5 a_5$ was calculated in Section 13, and we deduce

$$f_{6,5}(x_4) = \underset{\max[7,x_4-2]\leqslant x_5\leqslant 11}{\text{MIN}} [150 + 10x_5 - 20x_4]. \tag{52.5}$$

Let us take

$$x_5 = \max[7, x_4 - 2], \tag{52.6}$$

with the condition $x_5 \leqslant 11$, whence $x_4 \leqslant 13$, and

$$f_{6,5}(x_4) = 150 + 10 \max[7 - 2x_4, -x_4 - 2]. \tag{52.7}$$

Let us calculate the bounds of x_4 :

$$\max[2, x_3 - 3] \leqslant x_4 \leqslant \min[20, \max(7, x_4 - 2) + 2]. \tag{52.8}$$

The second inequality is expressed:

$$x_4 \leqslant \min[20, \max(9, x_4)]; \tag{52.9}$$

let

$$x_4 \leqslant 20.$$

But, we have, further:

$$x_5 = \max(7, x_4 - 2) \leqslant 11, \tag{52.10}$$

whence lastly

$$\max[2, x_3 - 3] \leqslant x_4 \leqslant 13. \tag{52.11}$$

Periods 6, 5, and 4 Combined.

$$f_{6,5,4}(x_3) = \min_{\max[2,x_3-3]\leqslant x_4\leqslant 13} [v_4(x_3, x_4) + f_{6,5}(x_4)]$$

$$= \text{MIN}[150 + 10\max(7 - 2x_4, -x_4 - 2)$$
$$+ 17(x_4 - x_3 + 3)]$$

$$= \text{MIN}[201 - 17x_3 + \max(70 - 3x_4, 7x_4 - 20)]$$

$$= 201 - 17x_3 + \min_{\max[2,x_3-3]\leqslant x_4\leqslant 13} [\max(70 - 3x_4, 7x_4 - 20)]. \quad (52.12)$$

We must take

$$x_4 = \max[9, \max(2, x_3 - 3)] = \max[9, x_3 - 3], \quad (52.13)$$

with the condition

$$x_4 \leqslant 13, \quad \text{whence} \quad x_3 \leqslant 16;$$

we then have

$$f_{6,5,4}(x_3) = 201 - 17x_3 + \max[63 - 20, 7\max(2, x_3 - 3) - 20]$$

$$= 201 - 17x_3 + \max[43, \max(-6, 7x_3 - 41)]$$

$$= 201 - 17x_3 + \max[43, -6, 7x_3 - 41]$$

$$= \max[244 - 17x_3, 160 - 10x_3]. \quad (52.14)$$

Let us calculate the bounds of x_3 :

$$\max[3, x_2 - 5] \leqslant x_3 \leqslant \min[20, x_4 + 3], \quad (52.15)$$

that is,

$$\max[3, x_2 - 5] \leqslant x_3 \leqslant \min[20, 3 + \max(9, x_3 - 3)], \quad (52.16)$$

or

$$\max[3, x_2 - 5] \leqslant x_3 \leqslant \min[20, \max(12, x_3)], \quad (52.17)$$

which gives

$$x_3 \leqslant 20.$$

But we should have $x_3 \leqslant 16$, and we have finally

$$\max[3, x_2 - 5] \leqslant x_3 \leqslant 16. \quad (52.18)$$

Periods 6, 5, 4, and *3 Combined.*

$$f_{6,5,4,3}(x_2) = \underset{\max[3,x_2-5]\leqslant x_3\leqslant16}{\text{MIN}} [v_3(x_2, x_3) + f_{6,5,4}(x_3)]$$

$$= \text{MIN}[\max(244 - 17x_3, 160 - 10x_3) + 13(x_3 - x_2 + 5)]$$

$$= 65 - 13x_2 + \underset{\max[3,x_2-5]\leqslant x_3\leqslant16}{\text{MIN}} [\max(244 - 4x_3, 160 + 3x_3)].$$

$$(52.19)$$

We must take

$$x_3 = \max[12, \max(3, x_2 - 5)]$$

$$= \max[12, x_2 - 5], \qquad (52.20)$$

with $x_3 \leqslant 16$, whence $x_2 \leqslant 21$.

$$f_{6,5,4,3}(x_2) = 65 - 13x_2 + \max[160 + 36, 160 + 3\max(3, x_2 - 5)]$$

$$= \max[261 - 13x_2, 210 - 10x_2]. \qquad (52.21)$$

Let us calculate the bounds of x_2 :

$$\max[5, x_1 - 8] \leqslant x_2 \leqslant \min[20, x_3 + 5] \qquad (52.22)$$

or

$$\max[5, x_1 - 8] \leqslant x_2 \leqslant 20; \qquad (52.23)$$

the condition $x_3 \leqslant 21$ does not enter.

Periods 6, 5, 4, 3, and 2 Combined.

$$f_{6,5,4,3,2}(x_1) = \underset{\max[5,x_1-8]\leqslant x_2\leqslant20}{\text{MIN}} [v_2(x_1, x_2) + f_{6,5,4,3}(x_2)]$$

$$= \text{MIN}[\max(261 - 13x_2, 210 - 10x_2) + 18(x_2 - x_1 + 8)] \quad (52.24)$$

$$= 144 - 18x_1 + \underset{\max[5,x_1-8]\leqslant x_2\leqslant20}{\text{MIN}} [\max(261 + 5x_2, 210 + 8x_2)].$$

We take $x_2 = \max[5, x_1 - 8]$:

$$f_{6,5,4,3,2}(x_1) = 144 - 18x_1 + \max[261 + 5\max(5, x_1 - 8),$$
$$210 + 8\max(5, x_1 - 8)]$$

$$= 144 - 18x_1 + \max[286, 5x_1 + 221]$$

$$= \max[430 - 18x_1, 365 - 13x_1] \qquad (52.25)$$

(we have observed the condition $x_1 \leqslant 20$).

Let us calculate the bounds of x_1 :

$$\max[8, 2] \leqslant x_1 \leqslant \min[20, x_2 + 8] \qquad (52.26)$$

$$8 \leqslant x_1 \leqslant \min[20, \max(13, x_1)] \qquad (52.27)$$

$$8 \leqslant x_1 \leqslant 20. \qquad (52.28)$$

Set of the 6 Periods.

$$F^* = \underset{8 \leqslant x_1 \leqslant 20}{\text{MIN}} [v_1(x_0, x_1) + f_{6,5,4,3,2}(x_1)]$$

$$= \text{MIN}[\max(430 - 18x_1, 365 - 13x_1) + 11(x_1 - 2)]$$

$$= \underset{8 \leqslant x_1 \leqslant 20}{\text{MIN}} [\max(408 - 7x_1, 343 - 2x_1)] \qquad (52.29)$$

We take $x_1 = 20$.

$$F^* = \max[408 - 140, 343 - 40]$$
$$= \max[268, 303]$$
$$= 303. \qquad (52.30)$$

We can now calculate the x_i, a_i, and s_i terms:

$$
\begin{array}{lll}
x_1 = 20, & a_1 = 18, & s_1 = 12 \\
x_2 = \max[5, 12] = 12, & a_2 = 0, & s_2 = 7 \\
x_3 = \max[12, 7] = 12, & a_3 = 5, & s_3 = 9 \\
x_4 = \max[9, \ 9] = 9, & a_4 = 0, & s_4 = 7 \\
x_5 = \max[7, \ 7] = 7, & a_5 = 0, & s_5 = 0 \\
x_6 = 4, & a_6 = 4, & s_6 = 0.
\end{array}
\qquad (52.31)
$$

As was to be expected, the optimal expenditure for the set of 6 periods is inferior to that case $S = 9$.

$$F^* = 303 \qquad \text{instead of} \quad F^* = 357. \qquad (52.32)$$

In this particular case, the result shown in Fig. 52.1 could have been obtained directly by intuition. We notice that p_6 is the smallest of the p_i terms, and as $x_6 = 4$, we take $a_6 = 4$. Of the other p_i terms, p_1 is the minimum; hence, we shall put the maximum (18) in period 1, and the rest in period 3, for which $a_3 = 5$. This type of reasoning would, however, become much less easy in a more complicated case.

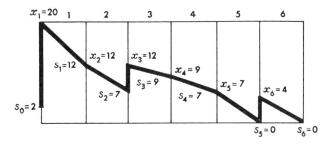

FIG. 52.1.

It should be noted that the difficulties which we experienced were in finding literal expressions for f. If we had used a computer, the programming would have been such as to give us the numerical values of f for all the values of the variable at each stage. For example, if we consider the equation (52.4) it will be seen that for every value of x_4, the lower bound of variation of x_5 can be calculated by comparing two numbers, an operation which the computer can perform very quickly. With a computer, in fact, the problem is not any more difficult than it was with $S = 9$, but is merely more lengthy since the number of integral values of the variables to be considered is greater.

53. Uncertain Development Using a Markovian Chain

A. Résumé of Finite Markovian Chains[1]

In a system with M possible states E_1, E_2,..., E_M, where M is finite, we shall use $p_i(n)$ as the probability that the state will be E_i at time n. The $p_i(n)$ terms, $i = 1, 2,..., M$, form the set of *state vectors* which describe the system for all the future times to be considered.

Let us now assume the *transition* from one state to another *depends solely on those two states*. In a more general form, let us assume that with every oriented pair (E_i, E_j), it is possible to associate a probability p_{ij} that the system will be in state E_j at time $n + 1$ if it was in E_i at time n. A graph, called *the graph of the transitions*, can be drawn for these probabilities (Fig. 53.1).

[1] The reader will find a more advanced treatment in A. Kaufmann and R. Cruon, *La Programmation Dynamique et ses Applications*, Dunod, Paris, 1965.

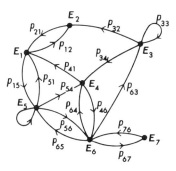

FIG. 53.1.

Assuming that the probabilities $p_i(0)$, $i = 1, 2,..., M$ are known (first-state vector), we then have a *Markovian chain* governed by the equations:

$$p_j(n + 1) = \sum_{i=0}^{M} p_i(n) \cdot p_{ij}, \qquad j = 1, 2,..., M; \qquad (53.1)$$

or, using matricial symbols:

$$[p(n + 1)] = [p(n)][\mathscr{M}]. \qquad (53.2)$$

$[\mathscr{M}]$ is the square matrix formed by the elements p_{ij} such that:

$$p_{ij} \geqslant 0, \qquad i, j = 1, 2,..., M, \qquad (53.3)$$

$$\sum_{j=1}^{M} p_{ij} = 1, \qquad i = 1, 2,..., M. \qquad (53.4)$$

The state vectors $p_j(n)$, $j = 1, 2,..., M$, are such that

$$p_j(n) \geqslant 0, \qquad j = 1, 2,..., M, \qquad (53.5)$$

$$\sum_{j=1}^{M} p_j(n) = 1. \qquad (53.6)$$

Any matrix with the properties (53.3) and (53.4) is called a *stochastic matrix*. The probabilities p_{ij} are the *probabilities of transition*, and the matrix is often referred to as the *transition matrix*.

A Markovian chain in its entirety is defined by the stochastic matrix $[\mathscr{M}]$ and the vector $[p(0)]$.

The matrix of transition may depend on time n, which means that the probabilities may also depend on it, and we then have

$$p_j(n+1) = \sum_{i=1}^{M} p_i(n) \cdot p_{ij}(n), \qquad j = 1, 2,..., M; \qquad (53.7)$$

and, since the $p_i(0)$ and $p_{ij}(n)$ terms are given for each i, j, and n, we say that the chain is *nonstationary*. Unless the contrary is stated, the only chains which we shall consider here will be stationary.

B. Dynamic Matrix and Equation

Let

$$[\mathcal{M}] = \begin{bmatrix} p_{11} & p_{12} & p_{13} & \cdots & p_{1M} \\ p_{21} & p_{22} & p_{23} & \cdots & p_{2M} \\ p_{31} & p_{32} & p_{33} & \cdots & p_{3M} \\ & & & \vdots & \\ p_{M1} & p_{M2} & p_{M3} & \cdots & p_{MM} \end{bmatrix}. \qquad (53.8)$$

The matrix

$$[\mathcal{D}] = [\mathcal{M}] - [1], \qquad (53.9)$$

where $[1]$ is the unit matrix of order M (of the same order as $[\mathcal{M}]$), and is called the *dynamic matrix*. The elements of $[\mathcal{D}]$ will be called d_{ij} :

$$[\mathcal{D}] = \begin{bmatrix} d_{11} & d_{12} & d_{13} & \cdots & d_{1M} \\ d_{21} & d_{22} & d_{23} & \cdots & d_{2M} \\ d_{31} & d_{32} & d_{33} & \cdots & d_{3M} \\ & & & \vdots & \\ d_{M1} & d_{M2} & d_{M3} & \cdots & d_{MM} \end{bmatrix} \qquad (53.10)$$

$$= \begin{bmatrix} p_{11} - 1 & p_{12} & p_{13} & \cdots & p_{1M} \\ p_{21} & p_{22} - 1 & p_{23} & \cdots & p_{2M} \\ p_{31} & p_{32} & p_{33} - 1 & \cdots & p_{3M} \\ & & & \vdots & \\ p_{M1} & p_{M2} & p_{M3} & \cdots & p_{MM} - 1 \end{bmatrix};$$

hence it is such that

$$0 \leqslant d_{ij} \leqslant 1, \qquad i \neq j; \qquad (53.11)$$

$$-1 \leqslant d_{ij} \leqslant 0, \qquad i = j; \qquad (53.12)$$

$$\sum_j d_{ij} = 0, \qquad i = 1, 2,..., M. \qquad (53.13)$$

Equation (53.2) can then be expressed as

$$[p(n + 1)] - [p(n)] = [p(n)][\mathscr{D}];\qquad\qquad(53.14)$$

it is in the form of a matricial equation with finite differences.

C. PROPERTIES AND STRUCTURES OF STOCHASTIC MATRICES

(1)   If $[\mathscr{M}]$ is a stochastic matrix, $[\mathscr{M}]^r$, $r = 0, 1, 2, 3$, is also one.

(2)   If all the lines of a stochastic matrix are identical, then

$$[\mathscr{M}]^r = [\mathscr{M}],\qquad r = 1, 2, 3,....\qquad\qquad(53.15)$$

(3)   If $[\mathscr{M}]$ is of the form[1] $\left[\begin{smallmatrix} A & O \\ O & D \end{smallmatrix}\right]$, where A and D are square sub-matrices, the system which begins in a state contained in A can never be in one included in D, and conversely. In such a case, we say that $[\mathscr{M}]$ is *reducible* or *decomposable*; the two subsets are *isolated* and form disjoint subgraphs. Naturally, the generalization can be extended to more than two subsets of states.

(4)   If $[\mathscr{M}]$ is of the form[1] $\left[\begin{smallmatrix} A & O \\ C & D \end{smallmatrix}\right]$, where A and D are square sub-matrices, the probability that the system will be in one of the states of D decreases monotonically as the number of transitions r increases. The passage from a state of D to a state of A is possible, but the contrary transition is not. The states of D are called *transitory*, those of A, *recurrent*. This can also be expressed as a generalization for more than two subsets of states.

The form of the matrix is retained whatever the power $r = 1, 2, 3,...$ to which it is raised.

(5)   If $[\mathscr{M}]$ is of the form[1] $\left[\begin{smallmatrix} O & B \\ C & O \end{smallmatrix}\right]$ in which O are square submatrices, all the even powers of it will give matrices of the form $\left[\begin{smallmatrix} A & O \\ O & D \end{smallmatrix}\right]$ and all the uneven powers matrices of the type $\left[\begin{smallmatrix} O & B \\ C & O \end{smallmatrix}\right]$. The system will oscillate between the two subsets of states and one of this type is called *periodic*. This concept can be generalized for more than two subsets of states which form cycles.

D. ERGODIC PROPERTY IN A MARKOVIAN CHAIN

If

$$\lim_{r \to \infty} [\mathscr{M}]^r = [\tilde{\mathscr{M}}],\qquad\qquad(53.16)$$

---

[1] Eventually, after a new and suitable numbering of the states.

where $[\mathcal{M}]$ is a stochastic matrix with no zero element we call the system *ergodic* and say it is *stable in probability*, or possesses a *permanent regime*, and that a matrix which has this property is an *ergodic matrix*. It can be proved that a matrix which is neither periodic nor decomposable is ergodic.

If $[\tilde{\mathcal{M}}]$ is such that all its lines are identical, the system is *completely ergodic*,[1] and in this case, where n is sufficiently large, the state of the system no longer depends on the initial state. Indeed,

$$[p(n)] = [p(0)][\mathcal{M}]^n \tag{53.17}$$

$$\lim_{n\to\infty}[p(n)] = \lim_{n\to\infty}[p(0)][\mathcal{M}]^n,$$

$$= [p(0)] \cdot \lim_{n\to\infty}[\mathcal{M}]^n,$$

$$= [p(0)][\tilde{\mathcal{M}}],$$

$$= [p]. \tag{53.18}$$

where $[p]$ is one of the identical lines of $[\tilde{\mathcal{M}}]$.

It can be proved that a necessary and sufficient condition[2] for a matrix to be completely ergodic is that it must only contain one subset of recurrent states,[3] and that the matrix of the latter should not be periodic. Figures 53.2 and 53.3 show examples of this. The precise definition of a subset of recurrent states can be given as follows:

Given the set **E** of the states of the finite Markovian chain and $\mathbf{E}_r \subset \mathbf{E}$, this subset $\mathbf{E}_r$ is recurrent:

(1)   If the subgraph $G(\mathbf{E}_r, \Gamma_r)$ is strongly connected;

(2)   If the subgraph, formed by adding a subset of **E** which is disjoint from $\mathbf{E}_r$, is no longer strongly connected;

(3)   If in the graph $G(\mathbf{E}, \Gamma)$ of the Markovian chain there is no arc which is initially connected to $\mathbf{E}_r$.

[1] In *Les phénomènes d'attente* (Dunod, Paris, 1961), written in collaboration with R. Cruon, we have used a different terminology, using "reducible" for a matrix of the form $[{}^{AO}_{OB}]$ or $[{}^{AO}_{CB}]$, (see footnote 2), and "ergodic" for a matrix which is neither "periodic" nor "reducible" in Feller's sense. Since the publication of Howard's work [16], the terminology which we have used here seems to have been accepted in recent works, so that we have introduced the terms "ergodic" and "completely ergodic" as Howard defines them. Nevertheless, this terminology does not seem entirely satisfactory, and in a new work, *La programmation dynamique* (Dunod), R. Cruon and I have shown that a more precise definition of the properties of Markovian chains can be given by recourse to certain properties of graphs.

[2] W. Feller, *An Introduction to Probability Theory and its Applications.* Wiley, New York, 1950. See also [16].

[3] We also speak of the "subset of persistent states."

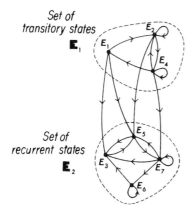

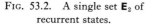

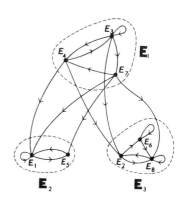

FIG. 53.2.   A single set $\mathbf{E}_2$ of   FIG. 53.3.   Two sets of recurrent states
recurrent states. $\mathbf{E}_2$ and $\mathbf{E}_3$.

E. SEEKING THE PROBABILITY LIMITS

In Section 55 we shall explain a generalized method introduced by Howard [I6] for finding matrix $[\mathcal{M}]$ (when it exists), as well as the vector limit $[p]$. For the moment, we shall content ourselves with giving a simple algebraic method which is restricted to cases where we know in advance that the system is completely ergodic. In such cases, the equations for p_i, $i = 1, 2,..., M$, with

$$[p][\mathcal{D}] = [0]; \tag{53.19}$$

with account being taken of the property:

$$p_1 + p_2 + \cdots + p_M = 1, \tag{53.20}$$

constitute a system of equations which always possesses a single solution. Indeed, if $\lim_{n \to \infty}[p(n)]$ exists and is equal to $[p]$, Eq. (53.14) at once gives us (53.19).

Example.   Let

$$[\mathcal{M}] = \begin{bmatrix} 0.3 & 0.4 & 0.3 \\ 0 & 0.2 & 0.8 \\ 0.5 & 0.5 & 0 \end{bmatrix}, \tag{53.21}$$

whence

$$[\mathcal{D}] = \begin{bmatrix} -0.7 & 0.4 & 0.3 \\ 0 & -0.8 & 0.8 \\ 0.5 & 0.5 & -1 \end{bmatrix}. \tag{53.22}$$

$[p][\mathscr{D}] = [0]$ gives, if we retain (for example) only the first two lines and incorporate $p_1 + p_2 + p_3 = 1$:

$$[p_1 \quad p_2 \quad p_3] \begin{bmatrix} -0.7 & 0.4 & 0.3 \\ 0 & -0.8 & 0.8 \\ 1 & 1 & 1 \end{bmatrix} = [0 \quad 0 \quad 1]. \tag{53.23}$$

Whence

$$p_1 = 40/151, \qquad p_2 = 55/151, \qquad p_3 = 56/151, \tag{53.24}$$

and

$$[p] = [40/151 \quad 55/151 \quad 56/151], \tag{53.25}$$

$$[\tilde{\mathscr{M}}] = \begin{bmatrix} 40/151 & 55/151 & 56/151 \\ 40/151 & 55/151 & 56/151 \\ 40/151 & 55/151 & 56/151 \end{bmatrix}. \tag{53.26}$$

54. The z-Transform

A. DEFINITION

To prove the principal properties of stationary Markovian decision chains and to study the permanent regimes in various problems, a generalization of the concept of the generating function can be most useful. In the form used here, we shall consider a biunivocal functional transformation known as the z-*transform*.

Let us consider a function $f(n)$ defined and univocal for integral nonnegative values of n, and assume that whatever the value of n, this function satisfies the condition:

$$\exists a \geqslant 0, \qquad |f(n)| \leqslant a^n. \tag{54.1}$$

The series[1]

$$f^*(z) = \sum_{n=0}^{\infty} f(n) \cdot z^n, \tag{54.2}$$

---

[1] The transformation

$$f^*(z) = (1 - z) \sum_{n=0}^{\infty} f(n) \cdot z^n,$$

has the advantage over (54.2) of giving 1 as the z-transform of (54.3), and will prove more convenient in certain cases.

called the "z-transform" of $f(n)$, is then uniformly convergent at least for

$$|z| < 1/a,$$

so that it is *holomorphic*[1] in this domain.

If the above conditions are satisfied, a biunivocal correspondence exists between $f^*(z)$ and $f(n)$, that is, for every function $f(n)$ there is one, and only one, corresponding function $f^*(z)$. Reciprocally, every function $f^*(z)$ which is holomorphic for $|z| < 1/a$, can be expanded into an integral series in this domain, and the coefficients of this expansion define a single function $f(n)$.

FIG. 54.1.

B. EXAMPLES

Example 1 (Fig. 54.2).   Let

$$f(n) = 0, \qquad n \notin N$$
$$\quad = 1, \qquad n \in N \tag{54.3}$$

The conditions laid down are satisfied for every $a \geqslant 1$, and we have

$$f^*(z) = \sum_{n=0}^{\infty} f(n) \cdot z^n = \sum_{n=0}^{\infty} z^n$$

$$= 1 + z + z^2 + z^3 + \cdots + z^n + \cdots \tag{54.4}$$

$$= \frac{1}{1-z}.$$

Hence the z-tranform of $f(n)$ is $f^*(z) = 1/(1-z)$.

---

[1] We say that a function $w(z)$, where z is the complex variable $z = x + jy$, is holomorphic in a domain if it is uniform and analytic in this domain. See A. Kaufmann and R. Douriaux, *Les fonctions de la variable complexe*, Eyrolles, Paris, 1962.

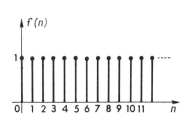

FIG. 54.2.

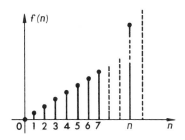

FIG. 54.3.

Example 2 (Fig. 54.3).   Let

$$f(n) = 0, \qquad n \notin N$$
$$= n, \qquad n \in N \tag{54.5}$$

Then

$$f^*(z) = \sum_{n=0}^{\infty} nz^n = z \sum_{n=0}^{\infty} nz^{n-1} = z \frac{d}{dz} \sum_{n=0}^{\infty} z^n, \tag{54.6}$$

whence

$$f^*(z) = \frac{z}{(1-z)^2}. \tag{54.7}$$

C. PRINCIPAL PROPERTIES OF THE z-TRANSFORM

(1) if $f(n) = f_1(n) + f_2(n)$,   then $f^*(z) = f_1^*(z) + f_2^*(z)$;   (54.8)

(2) if $F(n) = kf(n)$,   then $F^*(z) = kf^*(z)$;   (54.9)

(3) if $F(n) \begin{cases} = 0, \\ = f(n-1), \end{cases}$ $\begin{matrix} n < 1, \\ n \geqslant 1, \end{matrix}$ then $F^*(z) = zf^*(z)$;   (54.10)

(4) if $F(n) = f(n+1)$,   then $F^*(z) = \dfrac{f^*(z) - f(0)}{z}$;   (54.11)

(5) if $F(n) = \alpha^n f(n)$,   then $F^*(z) = f^*(\alpha z)$;   (54.12)

(6) if $F(n) = \displaystyle\sum_{i=0}^{n} f(i)$,   then $F^*(z) = \dfrac{f^*(z)}{1-z}$;   (54.13)

(7) $\displaystyle\lim_{N \to \infty} \frac{1}{N} \sum_{n=0}^{N-1} f(n) = \lim_{z \to 1}(1-z)f^*(z)$,   (54.14)

in the sense that, if one of these limits exists, the other also exists and is equal to it. The first limit represents the *Cesaro limit* of $f(n)$; we know that if $f(n)$ has a limit in the strict sense, the limit defined by Cesaro also exists and is equal to it.[1] Hence we also have

(8)   if $f(n)$ has a limit when $n \to \infty$:

$$\lim_{n\to\infty} f(n) = \lim_{z\to 1}(1 - z)f^*(z) \qquad (54.14)'$$

D. TABLE OF THE PRINCIPAL z-TRANSFORMS (Table 54.1)

TABLE 54.1

$f(n)$	$f^*(z)$	
$f(n) = 0, \quad n \notin N$ $= 1, \quad n \in N$	$\dfrac{1}{1 - z}$	(54.15)
$f(n) = 0, \quad n \notin N$ $= \alpha^n \quad n \in N$	$\|\alpha\| \leqslant 1 \quad \dfrac{1}{1 - \alpha z}$	(54.16)
$f(n) = 0, \quad n \notin N$ $= n\alpha^n, \quad n \in N$	$\|\alpha\| \leqslant 1 \quad \dfrac{\alpha a}{(1 - \alpha z)^2}$	(54.17)
$f(n) = 0, \quad n \notin N$ $= \alpha n, \quad n \in N$	$\dfrac{\alpha z}{(1 - z)^2}$	(54.18)
$f(n) = 0, \qquad n \notin N$ $= C_{n+p-1}^{n} \alpha^{n}, \quad n \in N$	$\|\alpha\| \leqslant 1 \quad \dfrac{1}{(1 - \alpha z)^p}$ $p = 1, 2, 3, \dots .$	(54.19)
$f(n) = 0, \qquad n \notin N$ $= C_{n+p-2}^{n-1} \alpha^{n}, \quad n \in N$	$\|\alpha\| \leqslant 1 \quad \dfrac{\alpha z}{(1 - \alpha z)^p}$ $p = 1, 2, 3, \dots .$	(54.20)
$f(n) = 0, \quad n \neq 0$ $= 1, \quad n = 0$	1	(54.21)
$f(n) = 0, \quad n \neq \alpha$ $1, \quad n = \alpha$	α Integral positive z^α	(54.22)

[1] A sequence u_n, $n = 0, 1, 2,\dots$ has a limit in the sense given to the term by Cesaro if

$$\lim_{N\to\infty} \frac{1}{N} \sum_{n=0}^{N-1} u_n \text{ exists.}$$

E. NOTE

In point of fact the z-transform constitutes a veritable "symbolic calculus" or "operations calculus" similar to the Carson-Laplace transform[1] for functions which are continuous in a interval, though the z-transform deals with series.

A number of authors have shown the value of the z-transform or "impulse calculus," especially Cuenod [I33], in several important works. As we have explained, the z-transform is merely the use of the generating function, which is due to Laplace as much as is the one which bears his name. In a few years time usage will be fixed, and we should like to see the z-transform known as *the Laplace-transform for integers* and the one which now bears his name as *the Laplace-transform for continuous variables*.

The theorems and properties of the present Laplace transform can easily be transposed to fit the z-transform, with $1 - z$ being substituted for p.

By combining the two transforms, we obtain a double transform *continuous over t* and *discrete over n* which provides an excellent method for treating the differential and integral equations so often encountered in stochastic and other processes. This is what we mean by the double transform:

Given a function $f(n, t)$, $n = 0, 1, 2, 3,...$, continuous in an interval for $0 < t < \infty$, null for $t < 0$, and satisfying Laplace's conditions for integral convergence, we state

$$f^*(z, t) = \sum_{n=0}^{\infty} f(n, t) z^n, \qquad (54.23)$$

$$F^*(z, p) = \sum_{n=0}^{\infty} \mathscr{L} f(n, t) z^n, \qquad (54.24)$$

where

$$\mathscr{L} f(n, t) = \int_{0}^{\infty} f(n, t) e^{-pt} \, dt, \qquad n = 0, 1, 2, 3,... . \qquad (54.25)$$

Then $F^*(z, p)$ is the double transform described.

---

[1] See M. Denis-Papin and A. Kaufmann, *Cours de calcul opérationnel appliqué*, 3rd ed., Albin-Michel, Paris, 1960.

Using the Carson-Laplace transform we should have

$$f^*(z, t) = (1 - z) \sum_{n=0}^{\infty} f(n, t)z^n, \qquad (54.26)$$

$$F^*(z, p) = (1 - z) \sum_{n=0}^{\infty} \mathscr{L}f(n, t)z^n, \qquad (54.27)$$

where

$$\mathscr{L}f(n, t) = p \int_{0}^{\infty} f(n, t)e^{-pt}\, dt, \qquad n = 0, 1, 2, 3, 4,... \qquad (54.28)$$

In this way we can contruct an operations or symbolic calculus, in which important theorems can be proved, and which can be of great value.

But the z-transform itself can be very useful in matricial calculus, as we shall show in the next section.

Other discrete transforms are used in the study of series, of which the following are the most important:

Type I : $f_I^*(z) = \sum_{n=0}^{\infty} f(n)z^n$ (the one we have just considered);   (54.29)

Type II : $f_{II}^*(z) = \sum_{n=0}^{\infty} f(n)z^{-n}$ (z-transform with negative powers);   (54.30)

Type III: $f_{III}^*(z) = \sum_{n=0}^{\infty} f(n)\dfrac{z^n}{n!}$ (exponential transform).   (54.31)

As examples, we give a few elementary results in Table 54.2.

TABLE 54.2

$f(n)$	$f_I^*(z)$	$f_{II}^*(z)$	$f_{III}^*(z)$
$f(n) = 0, \quad n < 0$ $= 1, n = 0, 1, 2, 3, ...$	$f_I^*(z) = \dfrac{1}{1 - z}$	$f_{II}^*(z) = \dfrac{1}{z - 1}$	$f_{III}^*(z) = e^z$
$f(n) = 0, n < 0$ $= n, n = 0, 1, 2, 3, ...$	$f_I^*(z) = \dfrac{z}{(1 - z)^2}$	$f_{II}^*(z) = \dfrac{z}{(z - 1)^2}$	$f_{III}^*(z) = ze^z$

The Type III transform enables us to study series where the convergence cannot be found by the first two, since a factorial series converges more quickly than a geometric one. Other kinds of discrete transforms can be created according to the requirements of the problem. In the following pages we shall only use Type I transforms, though we might equally have made use of Type II.

55. Using the *z*-Transform to Study Markovian Chains[1]

A. FINDING PROBABILITY LIMITS BY THE *z*-TRANSFORM

Let us refer to Eq. (53.2):

$$[p(n+1)] = [p(n)][\mathcal{M}], \qquad n = 0, 1, 2, 3, \ldots. \tag{55.1}$$

Let us assume that $[p(0)]$ is given; then, by (54.11) and with the assumption that $[\mathcal{M}]$ is formed of elements which are independent of n (stationary process),

$$\frac{[p^*(z)] - [p(0)]}{z} = [p^*(z)][\mathcal{M}]; \tag{55.2}$$

or again,

$$[p^*(z)]([1] - z[\mathcal{M}]) = [p(0)]. \tag{55.3}$$

Let us assume

$$[\mathcal{C}(z)] = [1] - z[\mathcal{M}] = \begin{bmatrix} 1 - zp_{11} & -zp_{12} & \cdots & -zp_{1M} \\ -zp_{21} & 1 - zp_{22} & \cdots & -zp_{2M} \\ & & \vdots & \\ -zp_{M1} & -zp_{M2} & \cdots & 1 - zp_{MM} \end{bmatrix}. \tag{55.4}$$

Thus

$$[p^*(z)] = [p(0)][\mathcal{C}(z)]^{-1}. \tag{55.5}$$

If we examine (55.4) we see that the equation $|\mathcal{C}(z)| = 0$ has roots in z which are the inverse of the values of $[\mathcal{M}]$.

In [17] a proof will be found that any stochastic matrix possesses real or complex conjugate values which have a modulus less than or equal to 1. There is at least one root equal to 1, since the determinant of (55.4)

[1] See Reference [16].

is erased if we make $z = 1$. In the case of a completely ergodic matrix there is only one root $z = 1$, but the other roots may be simple or multiple.

B. FIRST EXAMPLE

Let

$$[\mathcal{M}] = \begin{bmatrix} 0.7 & 0.2 & 0.1 \\ 0.7 & 0 & 0.3 \\ 0.4 & 0.3 & 0.3 \end{bmatrix} \tag{55.6}$$

$$[\mathcal{C}(z)] = \begin{bmatrix} 1 - 0.7z & -0.2z & -0.1z \\ -0.7z & 1 & -0.3z \\ -0.4z & -0.3z & 1 - 0.3z \end{bmatrix} \tag{55.7}$$

$$|\mathcal{C}(z)| = 0.06z^3 - 0.06z^2 - z + 1 \tag{55.8}$$
$$= (1 - z)(1 - \sqrt{0.06}z)(1 + \sqrt{0.06}z).$$

$$\mathcal{C}(z)]^{-1} = \frac{1}{|\mathcal{C}(z)|} \begin{bmatrix} 1 - 0.3z - 0.09z^2 & 0.2z - 0.03z^2 & 0.1z + 0.06z^2 \\ 0.7z - 0.09z^2 & 1 - z + 0.17z^2 & 0.3z - 0.14z^2 \\ 0.4z + 0.21z^2 & 0.3z - 0.13z^2 & 1 - 0.7z - 0.14z^2 \end{bmatrix}. \tag{55.9}$$

Let us expand $[\mathcal{C}(z)]^{-1}$ with simple rational fractions:

$$[\mathcal{C}(z)]^{-1} = \frac{1}{1 - z}[\mathcal{A}] + \frac{1}{1 - \sqrt{0.06}z}[\mathcal{B}] + \frac{1}{1 + \sqrt{0.06}z}[\mathcal{C}]. \tag{55.10}$$

We shall obtain the coefficients of $[\mathcal{A}]$ by multiplying the two members of (55.10), respectively, by $1 - z$, and making $z = 1$ in the two members. It follows:

$$[\mathcal{A}] = \begin{bmatrix} 61/94 & 17/94 & 16/94 \\ 61/94 & 17/94 & 16/94 \\ 61/94 & 17/94 & 16/94 \end{bmatrix}. \tag{55.11}$$

In the same way, we obtain the coefficient of $[\mathcal{B}]$ and of $[\mathcal{C}]$ by multiplying these two members, respectively, by $1 - \sqrt{0.06}z$ and by substituting $z = 1\sqrt{0.06}$ in them, then by $1 + \sqrt{0.06}z$, substituting $z = -1/\sqrt{0.06}$.

Let us return to (55.5):

$$[p^*(z)] = [p(0)] \left(\frac{1}{1 - z}[\mathcal{A}] + \frac{1}{1 - \sqrt{0.06}z}[\mathcal{B}] + \frac{1}{1 + \sqrt{0.06}z}[\mathcal{C}] \right). \tag{55.12}$$

Using (54.15) and (54.16), it follows that

$$[p(n)] = [p(0)]\{[\mathscr{A}] + (\sqrt{0.06})^n[\mathscr{B}] + (-\sqrt{0.06})^n[\mathscr{C}]\}. \qquad (55.13)$$

When $n \to \infty$: $(\sqrt{0.06})^n \to 0$, $(-\sqrt{0.06})^n \to 0$, and lastly:

$$\lim_{n \to \infty}[p(n)] = [p(0)][\mathscr{A}] = [61/94 \quad 17/94 \quad 16/94]. \qquad (55.14)$$

As a generalization, if the matrix is ergodic,

$$[\tilde{\mathscr{M}}] = \lim_{z \to 1}[(1 - z)[\mathscr{C}(z)]^{-1} = [\mathscr{A}], \qquad (55.15)$$

which will produce an expansion in simple rational fractions.

C. SECOND EXAMPLE

Let

$$[\mathscr{M}] = \begin{bmatrix} 1 & 0.8 & 0.2 \\ 1 & 0 & 0 \\ 1 & 0 & 0 \end{bmatrix} \qquad (55.16)$$

$$[\mathscr{C}(z)] = \begin{bmatrix} 1 & -0.8z & -0.2z \\ -z & 1 & 0 \\ -z & 0 & 1 \end{bmatrix} \qquad (55.17)$$

$$[\mathscr{C}(z)]^{-1} = \frac{1}{(1-z)(1+z)}\begin{bmatrix} 1 & 0.8z & 0.2z \\ z & 1 - 0.2z^2 & 0.2z^2 \\ z & 0.8z^2 & 1 - 0.8z^2 \end{bmatrix}$$

$$= \begin{bmatrix} 0 & 0 & 0 \\ 0 & 0.2 & -0.2 \\ 0 & -0.8 & 0.8 \end{bmatrix} + \frac{1}{1-z}\begin{bmatrix} 0.5 & 0.4 & 0.1 \\ 0.5 & 0.4 & 0.1 \\ 0.5 & 0.4 & 0.1 \end{bmatrix}$$

$$+ \frac{1}{1+z}\begin{bmatrix} 0.5 & -0.4 & -0.1 \\ -0.5 & 0.4 & 0.1 \\ -0.5 & 0.4 & 0.1 \end{bmatrix} \qquad (55.18)$$

Turning back to (54.15, 16, and 21), we find as the inverse transform for $n > 0$:

$$[\mathscr{M}]^n = \begin{bmatrix} 0.5 & 0.4 & 0.1 \\ 0.5 & 0.4 & 0.1 \\ 0.5 & 0.4 & 0.1 \end{bmatrix} + (-1)^n\begin{bmatrix} 0.5 & -0.4 & -0.1 \\ -0.5 & 0.4 & 0.1 \\ -0.5 & 0.4 & 0.1 \end{bmatrix}; \qquad (55.19)$$

that is to say,

$$[\mathscr{M}]^{2r} = \begin{bmatrix} 1 & 0 & 0 \\ 0 & 0.8 & 0.2 \\ 0 & 0.8 & 0.2 \end{bmatrix} \tag{55.20}$$

and

$$[\mathscr{M}]^{2r-1} = \begin{bmatrix} 0 & 0.8 & 0.2 \\ 1 & 0 & 0 \\ 1 & 0 & 0 \end{bmatrix}, \qquad r = 1, 2, 3, \dots. \tag{55.21}$$

For $n = 0$ we find:

$$[\mathscr{M}]^0 = \begin{bmatrix} 0 & 0 & 0 \\ 0 & 0.2 & -0.2 \\ 0 & -0.8 & 0.8 \end{bmatrix} + \begin{bmatrix} 1 & 0 & 0 \\ 0 & 0.8 & 0.2 \\ 0 & 0.8 & 0.2 \end{bmatrix} = \begin{bmatrix} 1 & 0 & 0 \\ 0 & 1 & 0 \\ 0 & 0 & 1 \end{bmatrix}. \tag{55.22}$$

$[\mathscr{M}]^n$ does not approach a limit when $n \to \infty$ [(55.20) and (55.21)].

56. Markovian Chain with Transition Values

A. VALUES OF TRANSITIONS

Let us assume that a value or revenue R_{ij} is assigned to each transition $E_i \to E_j$ and that the Markovian decision chain is stationary, both as to the probabilities p_{ij} and the values R_{ij} . Hence, taking $\bar{q}_i$ as the expected revenue of a stage when we are in state E_i at time n, we have

$$\bar{q}_i = \sum_{j=1}^{M} p_{ij} R_{ij} = p_{i1} R_{i1} + p_{i2} R_{i2} + \dots + p_{iM} R_{iM} . \tag{56.1}$$

The vector with the $\bar{q}_i$ terms for components will be called $\{\bar{q}\}$, and $\bar{v}_i(N - n, N)$ will represent the expected revenue for n periods from time $N - n$ to N. We then have

$$\bar{v}_i(N - n - 1, N) = \bar{q}_i + \sum_{j=1}^{M} p_{ij} \bar{v}_j(N - n, N)$$

$$i = 1, 2, \dots, M; \qquad n = 0, 1, 2, \dots, N - 1, \tag{56.2}$$

where

$$\bar{v}_j(N, N) = \bar{v}_{0j} , \qquad j = 1, 2, \dots, M \tag{56.3}$$

represents the value given to the system at time N and in state j.

Using matricial symbols and taking $\{\bar{v}(N - n - 1, N)\}$ for the vector with components $\bar{v}_i(N - n - 1, N)$, (56.2) will be expressed:

$$\{\bar{v}(N - n - 1, N)\} = \{\bar{q}\} + [\mathscr{M}]\{\bar{v}(N - n, N)\}, \qquad n = 0, 1, 2, \dots, N - 1. \tag{56.4}$$

To simplify the equation, let

$$\bar{v}_i(n) = \bar{v}_i(N - n, N), \qquad \begin{cases} i = 1, 2,..., M, \\ n = 0, 1, 2,..., N. \end{cases} \qquad (56.5)$$

where $\bar{v}_i(0) = \bar{v}_i(N, N) = \bar{v}_{0j}$ represents the value of the system in (56.2). With this easier notation (56.4) becomes

$$\{\bar{v}(n + 1)\} = \{\bar{q}\} + [\mathcal{M}]\{\bar{v}(n)\}, \qquad n = 0, 1, 2,..., N - 1. \qquad (56.6)$$

B. ANALYSIS OF A MARKOVIAN DECISION CHAIN WITH TRANSITION VALUES

Let us take the z-transform of this equation, using (54.10) and (54.15)

$$\frac{1}{z}\{\bar{v}^*(z) - v(0)\} = \frac{1}{1 - z}\{\bar{q}\} + [\mathcal{M}]\{\bar{v}^*(z)\} \qquad (56.7)$$

whence

$$([1] - z[\mathcal{M}])\{\bar{v}^*(z)\} = \frac{z}{1 - z}\{\bar{q}\} + \{v(0)\}. \qquad (56.8)$$

Let us again assume

$$[\mathscr{C}(z)] = [1] - z[\mathcal{M}], \qquad (56.9)$$

then

$$\{\bar{v}^*(z)\} = \frac{z}{1 - z}[\mathscr{C}(z)]^{-1}\{\bar{q}\} + [\mathscr{C}(z)]^{-1}\{v(0)\}. \qquad (56.10)$$

By expanding $[\mathscr{C}(z)]^{-1}$ with simple rational fractions, we obtain the behavior of the process when $n \to \infty$, in other words, as the number of periods considered becomes increasingly large.

C. CASE OF A COMPLETELY ERGODIC MATRIX

Let

$$[\mathcal{M}] = \begin{bmatrix} 0.5 & 0.5 \\ 0.7 & 0.3 \end{bmatrix}, \qquad (56.11)$$

$$[R] = \begin{bmatrix} 500 & 150 \\ 200 & -400 \end{bmatrix}, \qquad (56.12)$$

$$\{v(0)\} = \begin{Bmatrix} 0 \\ 0 \end{Bmatrix}; \qquad (56.12)'$$

whence

$$\{\bar{q}\} = \begin{Bmatrix} p_{11}r_{11} + p_{12}r_{12} \\ p_{21}r_{21} + p_{22}r_{22} \end{Bmatrix} = \begin{Bmatrix} (0.5)(500) + (0.5)(150) \\ (0.7)(200) + (0.3)(-400) \end{Bmatrix} = \begin{Bmatrix} 325 \\ 20 \end{Bmatrix}. \quad (56.13)$$

$$[\mathscr{C}(z)] = [1] - z[\mathscr{M}] = \begin{Bmatrix} 1 - 0.5z & -0.5z \\ -0.7z & 1 - 0.3z \end{Bmatrix} \quad (56.14)$$

$$[\mathscr{C}(z)]^{-1} = \frac{1}{(1-z)(1+0.2z)} \begin{bmatrix} 1 - 0.3z & 0.5z \\ 0.7z & 1 - 0.5z \end{bmatrix}$$

$$= \frac{1}{1-z} \begin{bmatrix} \frac{7}{12} & \frac{5}{12} \\ \frac{7}{12} & \frac{5}{12} \end{bmatrix} + \frac{1}{1+0.2z} \begin{bmatrix} \frac{5}{12} & -\frac{5}{12} \\ -\frac{7}{12} & \frac{7}{12} \end{bmatrix} \quad (56.15)$$

$$\{\bar{v}^{*}(z)\} = \left(\frac{z}{(1-z)^2} \begin{bmatrix} \frac{7}{12} & \frac{5}{12} \\ \frac{7}{12} & \frac{5}{12} \end{bmatrix} \right.$$

$$+ \frac{z}{(1-z)(1+0.2z)} \begin{bmatrix} \frac{5}{12} & -\frac{5}{12} \\ -\frac{7}{12} & \frac{7}{12} \end{bmatrix} \right) \begin{Bmatrix} 325 \\ 20 \end{Bmatrix}$$

$$= \left(\frac{z}{(1-z)^2} \begin{bmatrix} \frac{7}{12} & \frac{5}{12} \\ \frac{7}{12} & \frac{5}{12} \end{bmatrix} + \frac{10}{12} \frac{1}{1-z} \begin{bmatrix} \frac{5}{12} & -\frac{5}{12} \\ -\frac{7}{12} & \frac{7}{12} \end{bmatrix} \right.$$

$$\left. - \frac{10}{12} \frac{1}{1+0.2z} \begin{bmatrix} \frac{5}{12} & -\frac{5}{12} \\ -\frac{7}{12} & \frac{7}{12} \end{bmatrix} \right) \begin{Bmatrix} 325 \\ 20 \end{Bmatrix}. \quad (56.16)$$

Using (54.15) and (54.18):

$$\{\bar{v}(n)\} = \left(n \begin{bmatrix} \frac{7}{12} & \frac{5}{12} \\ \frac{7}{12} & \frac{5}{12} \end{bmatrix} + \frac{10}{12} \begin{bmatrix} \frac{5}{12} & -\frac{5}{12} \\ -\frac{7}{12} & \frac{7}{12} \end{bmatrix} \right.$$

$$\left. - \frac{10}{12}(-0.20)^n \begin{bmatrix} \frac{5}{12} & -\frac{5}{12} \\ -\frac{7}{12} & \frac{7}{12} \end{bmatrix} \right) \begin{Bmatrix} 325 \\ 20 \end{Bmatrix}. \quad (56.17)$$

If n is sufficiently large:

$$\{\bar{v}(n)\} = \left(n \begin{bmatrix} \frac{7}{12} & \frac{5}{12} \\ \frac{7}{12} & \frac{5}{12} \end{bmatrix} + \frac{10}{12} \begin{bmatrix} \frac{5}{12} & -\frac{5}{12} \\ -\frac{7}{12} & \frac{7}{12} \end{bmatrix} \right) \begin{Bmatrix} 325 \\ 20 \end{Bmatrix}$$

$$= \begin{Bmatrix} 197.92n + 105.90 \\ 197.92n - 148.26 \end{Bmatrix}. \quad (56.18)$$

Hence the average revenue for a period, provided n is sufficiently large, is independent of the initial state and is equal to 197.92, for

$$\lim_{n\to\infty} \frac{1}{n}\{\bar{v}(n)\} = \begin{cases} 197.92 \\ 197.92 \end{cases}. \tag{56.19}$$

In general, if we consider a stochastic matrix $[\mathcal{M}]$ which is fully ergodic and of type $[M]$, with which is associated a matrix of the revenues $[R]$, we shall, by using (56.1), by proceeding as above, and by taking into account the property that any matrix such as $[\mathcal{C}(z)]^{-1}$ has at least one value equal to 1 (and one only if it is fully ergodic), obtain

$$[\mathcal{C}(z)]^{-1} = \frac{1}{1-z}[\tilde{\mathcal{M}}] + [S(z)], \tag{56.20}$$

where $[S(z)]$ is a matrix with the following components: Let $\alpha_r^{(1)}$, $\alpha_s^{(2)},..., \alpha_v^{(p)}$ be the reciprocals of the simple roots of $|\mathcal{C}(z)| = 0$, of the double roots, ..., of the roots of order p. Then

$$[S(z)] = \sum_{r=1}^{k} \frac{1}{1-\alpha_r^{(1)}z}[S_r^{(1)}] \quad \text{(if there are } k \text{ roots of order 1)}$$

$$+ \sum_{s=1}^{l} \frac{1}{1-\alpha_s^{(2)}z}[S_s^{(2)}] + \sum_{s=1}^{l} \frac{1}{(1-\alpha_s^{(2)}z)^2}[T_s^{(2)}]$$

$$\text{(if there are } l \text{ roots of order 2)}$$

$$+ \sum_{v=1}^{m} \frac{1}{1-\alpha_v^{(p)}z}[S_v^{(p)}] + \sum_{v=1}^{m} \frac{1}{(1-\alpha_v^{(p)}z)^2}[T_v^{(p)}] + \cdots$$

$$+ \sum_{v=1}^{m} \frac{1}{(1-\alpha_v^{(p)}z)^p}[W_v^{(p)}] \quad \text{(if there are } m \text{ roots of order } p),$$

$$\tag{56.21}$$

where $\alpha_r^{(1)}$, $\alpha_s^{(2)}$, $\alpha_v^{(p)}$ all have a modulus between 0 and 1, and if a complex root exists, its conjugate also exists among the set of roots and is of the same order.

Hence, by substituting (56.20) in (56.10) it follows that

$$\{\bar{v}^*(z)\} = \left(\frac{z}{(1-z)^2}[\tilde{\mathcal{M}}] + \frac{z}{1-z}[S(z)]\right)\{\bar{q}\} + \left(\frac{1}{1-z}[\tilde{\mathcal{M}}] + [S(z)]\right)\{v(0)\}.$$

$$\tag{56.22}$$

Let us turn to the following equation:

$$\frac{z}{(1-z)(1-\alpha z)^p} = \frac{1}{(1-\alpha)^p} \cdot \frac{1}{1-z} + \frac{N(z)}{(1-\alpha z)^p} \cdots, \quad (56.23)$$

where $N(z)$ is a polynomial of z lower in degree than p, which does not have $1/\alpha$ as a root.

If we take the reciprocal transform of (56.22), and if we consider a sufficiently large number of periods n, all the reciprocals of the terms such as

$$\frac{1}{(1-\alpha z)^p}$$

will disappear in accordance with (54.19) since $|\alpha| < 1$, and we have

$$\{\bar{v}(n)\} = (n[\tilde{\mathcal{M}}] + [\mathcal{N}])\{\bar{q}\} + [\tilde{\mathcal{M}}]\{v(0)\}, \quad (56.24)$$

where

$$[\mathcal{N}] = [S(z)]_{z=1}. \quad (56.25)$$

The average revenue for the period will therefore be

$$\{\gamma\} = \lim_{n \to \infty} \frac{1}{n}\{\bar{v}(n)\} = [\tilde{\mathcal{M}}]\{\bar{q}\}, \quad (56.26)$$

or, if we take into account the fact that all the lines of $[\tilde{\mathcal{M}}]$ are the same:

$$\gamma = \sum_{j=1}^{M} \hat{p}_j q_j, \quad (56.27)$$

where $\hat{p}_j$, $j = 1, 2, ..., M$, are the elements of the lines of $[\tilde{\mathcal{M}}]$.

Of course, if the vector $\{v(0)\}$ is not zero in (56.10), it will enter into the calculation of $\{\bar{v}(n \to \infty)\}$, but in accordance with (56.26), it has no further influence on vector $\{\gamma\}$.

D. GENERAL CASE

Using (56.10),

$$\{\bar{v}^*(z)\} = \frac{z}{1-z} [\mathcal{C}(z)]^{-1}\{\bar{q}\} + [\mathcal{C}(z)]^{-1}\{v(0)\}. \quad (56.28)$$

For the generalization, we again decompose $[\mathcal{C}(z)]^{-1}$ into simple rational fractions,[1] but the expression obtained can have a different form

---

[1] Taking into account the fact that certain terms can be constants or powers of z (case where the matrix $[\mathcal{M}]$ possesses one or several values of its own which are null).

from (56.20); in any case, the decomposition and the search for the reciprocal z-transform will tell us the behavior of $\{\bar{v}(n)\}$ and $\{\gamma\}$ as $n \to \infty$.

First, let us examine an ergodic (but not fully ergodic) matrix, and then an example with one which is nonergodic.

E. CASE OF AN ERGODIC, BUT NOT COMPLETELY ERGODIC, MATRIX

If we consider the case of such a matrix, the components of the vector $\{\gamma\}$ are not identical, since all the lines of matrix $[\mathscr{M}]$ are not the same, so that we have

$$\gamma_i = \sum_{j=1}^{M} \tilde{p}_{ij}\bar{q}_j , \qquad i = 1, 2,..., M. \tag{56.29}$$

Hence, the average revenue will be a function of the initial state E_i . In the following example, let

$$[\mathscr{M}] = \begin{matrix} & \begin{matrix} (1) & (2) & (3) \end{matrix} \\ \begin{matrix} (1) \\ (2) \\ (3) \end{matrix} & \begin{bmatrix} 1 & 0 & 0 \\ 0 & 1 & 0 \\ 0.2 & 0.5 & 0.3 \end{bmatrix} \end{matrix}, \tag{56.30}$$

$$[R] = \begin{matrix} & \begin{matrix} (1) & (2) & (3) \end{matrix} \\ \begin{matrix} (1) \\ (2) \\ (3) \end{matrix} & \begin{bmatrix} 5 & 0 & 0 \\ 0 & 3 & 0 \\ -2 & 7 & 4 \end{bmatrix} \end{matrix}, \tag{56.31}$$

$$v(0) = \begin{Bmatrix} 0 \\ 0 \\ 0 \end{Bmatrix}. \tag{56.32}$$

Matrix (56.30) is ergodic, but not completely ergodic, since it contains two sets of recurring states, each set being in fact only one state.

Let us calculate $\{\bar{q}\}$:

$$\bar{q}_1 = (1)(5) = 5, \qquad \bar{q}_2 = (1)(3) = 3,$$
$$\bar{q}_3 = (0.2)(-2) + (0.5)(7) + (0.3)(4) = 4.3, \tag{56.33}$$

let

$$\{\bar{q}\} = \begin{Bmatrix} 5 \\ 3 \\ 4.3 \end{Bmatrix}. \tag{56.34}$$

Beginning with $[\mathscr{M}]$, we obtain $[\mathscr{C}(z)]^{-1}$ [see (57.8) and (57.9) for the details of the calculations]:

$$[\mathscr{C}(z)]^{-1} = \frac{1}{(1-z)(1-0.3z)} \begin{bmatrix} 1-0.3z & 0 & 0 \\ 0 & 1-0.3z & 0 \\ 0.2z & 0.5z & 1-z \end{bmatrix}, \quad (56.35)$$

which we can decompose thus:

$$[\mathscr{C}(z)]^{-1} = \frac{1}{1-z} \begin{bmatrix} 1 & 0 & 0 \\ 0 & 1 & 0 \\ \frac{2}{7} & \frac{5}{7} & 0 \end{bmatrix} + \frac{1}{1-0.3z} \begin{bmatrix} 0 & 0 & 0 \\ 0 & 0 & 0 \\ -\frac{2}{7} & -\frac{5}{7} & 1 \end{bmatrix}, \quad (56.36)$$

whence

$$\{\bar{v}^*(z)\} = \left(\frac{z}{(1-z)^2} \begin{bmatrix} 1 & 0 & 0 \\ 0 & 1 & 0 \\ \frac{2}{7} & \frac{5}{7} & 0 \end{bmatrix} \right.$$
$$\left. + \frac{z}{(1-z)(1-0.3z)} \begin{bmatrix} 0 & 0 & 0 \\ 0 & 0 & 0 \\ -\frac{2}{7} & -\frac{5}{7} & 1 \end{bmatrix} \right) \begin{Bmatrix} 5 \\ 3 \\ 4.3 \end{Bmatrix}, \quad (56.37)$$

$$\{\bar{v}^*(z)\} = \left(\frac{z}{(1-z)^2} \begin{bmatrix} 1 & 0 & 0 \\ 0 & 1 & 0 \\ \frac{2}{7} & \frac{5}{7} & 0 \end{bmatrix} + \frac{10/7}{1-z} \begin{bmatrix} 0 & 0 & 0 \\ 0 & 0 & 0 \\ -\frac{2}{7} & -\frac{5}{7} & 1 \end{bmatrix} \right.$$
$$\left. - \frac{10/7}{1-0.3z} \begin{bmatrix} 0 & 0 & 0 \\ 0 & 0 & 0 \\ -\frac{2}{7} & -\frac{5}{7} & 1 \end{bmatrix} \right) \begin{Bmatrix} 5 \\ 3 \\ 4.3 \end{Bmatrix}, \quad (56.38)$$

$$\{\bar{v}(n)\} = \left(n \begin{bmatrix} 1 & 0 & 0 \\ 0 & 1 & 0 \\ \frac{2}{7} & \frac{5}{7} & 0 \end{bmatrix} + \frac{10}{7} \begin{bmatrix} 0 & 0 & 0 \\ 0 & 0 & 0 \\ -\frac{2}{7} & -\frac{5}{7} & 1 \end{bmatrix} \right.$$
$$\left. - \frac{10}{7}(0.3)^n \begin{bmatrix} 0 & 0 & 0 \\ 0 & 0 & 0 \\ -\frac{2}{7} & -\frac{5}{7} & 1 \end{bmatrix} \right) \begin{Bmatrix} 5 \\ 3 \\ 4.3 \end{Bmatrix}. \quad (56.39)$$

If n is sufficiently large, the last term in the brackets disappears:

$$\{\bar{v}(n)\} = \left(n \begin{bmatrix} 1 & 0 & 0 \\ 0 & 1 & 0 \\ \frac{2}{7} & \frac{5}{7} & 0 \end{bmatrix} + \frac{10}{7} \begin{bmatrix} 0 & 0 & 0 \\ 0 & 0 & 0 \\ -\frac{2}{7} & -\frac{5}{7} & 1 \end{bmatrix} \right) \begin{Bmatrix} 5 \\ 3 \\ 4.3 \end{Bmatrix}. \quad (56.40)$$

The average revenue is

$$\{\gamma\} = \lim_{n \to \infty} \frac{1}{n}\{\bar{v}(n)\} = \begin{bmatrix} 1 & 0 & 0 \\ 0 & 1 & 0 \\ \frac{2}{7} & \frac{5}{7} & 0 \end{bmatrix} \begin{Bmatrix} 5 \\ 3 \\ 4.3 \end{Bmatrix} = \begin{Bmatrix} 5 \\ 3 \\ 3.57 \end{Bmatrix}. \tag{56.41}$$

Thus

$$\gamma_1 = 5, \qquad \gamma_2 = 3, \qquad \gamma_3 = 3.57. \tag{56.42}$$

The average revenue depends on the initial state.

F. CASE OF A NONERGODIC MATRIX

We shall study the behavior of $\{\bar{v}(n)\}$ when $n \to \infty$; the nonergodic matrix considered here is periodic.

$$[\mathscr{M}] = \begin{bmatrix} 0 & 1 \\ 1 & 0 \end{bmatrix}, \tag{56.43}$$

$$[R] = \begin{bmatrix} 0 & 8 \\ -3 & 0 \end{bmatrix}, \tag{56.44}$$

$$\{v(0)\} = \begin{Bmatrix} 0 \\ 0 \end{Bmatrix}. \tag{56.45}$$

First, we have

$$\{\bar{q}\} = \begin{Bmatrix} 8 \\ -3 \end{Bmatrix}. \tag{56.46}$$

Let us calculate $[\mathscr{C}(z)]^{-1}$:

$$[\mathscr{C}(z)] = \begin{bmatrix} 1 & -z \\ -z & 1 \end{bmatrix}; \tag{56.47}$$

thence

$$
\begin{aligned}
[\mathscr{C}(z)]^{-1} &= \frac{1}{1-z^2}\begin{bmatrix} 1 & z \\ z & 1 \end{bmatrix} \\
&= \frac{1}{1-z}\begin{bmatrix} \frac{1}{2} & \frac{1}{2} \\ \frac{1}{2} & \frac{1}{2} \end{bmatrix} + \frac{1}{1+z}\begin{bmatrix} \frac{1}{2} & -\frac{1}{2} \\ -\frac{1}{2} & \frac{1}{2} \end{bmatrix}.
\end{aligned} \tag{56.48}
$$

$$
\begin{aligned}
\{\bar{v}^*(z)\} &= \left(\frac{z}{(1-z)^2}\begin{bmatrix} \frac{1}{2} & \frac{1}{2} \\ \frac{1}{2} & \frac{1}{2} \end{bmatrix} \right. \\
&\quad \left. + \frac{z}{(1-z)(1+z)}\begin{bmatrix} \frac{1}{2} & -\frac{1}{2} \\ -\frac{1}{2} & \frac{1}{2} \end{bmatrix} \right)\begin{Bmatrix} 8 \\ -3 \end{Bmatrix} \\
&= \left(\frac{z}{(1-z)^2}\begin{bmatrix} \frac{1}{2} & \frac{1}{2} \\ \frac{1}{2} & \frac{1}{2} \end{bmatrix} + \frac{1}{1-z}\begin{bmatrix} \frac{1}{4} & -\frac{1}{4} \\ -\frac{1}{4} & \frac{1}{4} \end{bmatrix} \right. \\
&\quad \left. + \frac{1}{1+z}\begin{bmatrix} -\frac{1}{4} & \frac{1}{4} \\ \frac{1}{4} & -\frac{1}{4} \end{bmatrix} \right)\begin{Bmatrix} 8 \\ -3 \end{Bmatrix}.
\end{aligned} \tag{56.49}
$$

$$\{\bar{v}(n)\} = \left(n \begin{bmatrix} \frac{1}{2} & \frac{1}{2} \\ \frac{1}{2} & \frac{1}{2} \end{bmatrix} + \begin{bmatrix} \frac{1}{4} & -\frac{1}{4} \\ -\frac{1}{4} & \frac{1}{4} \end{bmatrix} \right.$$
$$\left. + (-1)^n \begin{bmatrix} -\frac{1}{4} & \frac{1}{4} \\ \frac{1}{4} & -\frac{1}{4} \end{bmatrix} \right) \begin{Bmatrix} 8 \\ -3 \end{Bmatrix}. \qquad (56.50)$$

Hence we establish

$$\{\gamma\} = \lim_{n \to \infty} \frac{1}{n} \{\bar{v}(n)\} = \begin{bmatrix} \frac{1}{2} & \frac{1}{2} \\ \frac{1}{2} & \frac{1}{2} \end{bmatrix} \begin{Bmatrix} 8 \\ -3 \end{Bmatrix} = \begin{Bmatrix} 5 \\ 5 \end{Bmatrix}, \qquad (56.51)$$

and does not depend on the initial state.

Nevertheless, we should not assume as a generalization that the average value over a large number of periods does not depend on the initial state when the matrix is periodic; this independence is subject to the properties of the graph of transition. Naturally, if the matrix is decomposable, except in the special case for certain values of the vector $\{\bar{q}\}$, the elements of the vector $\{\gamma\}$ are dependent on the initial state or on the separable set of states to which this belongs.

Observation. The preceding calculation may seem very complicated, but in practice, in many cases with numerical values, it is sufficient to begin calculating the limits by iteration based on (56.6), especially if the stochastic matrix is of a high order. The merit of Howard's method is that it shows the manner in which the process is convergent (if it is) and the changes that occur from stage to stage.

57. Study of Some Important Special Cases

We shall now show how to apply the z-transform to certain special cases such as a chain with one or more transitory states, a decomposable matrix, and a periodic matrix, before considering other developments connected with Markovian chains bearing a revenue.

A. CHAIN WITH A TRANSITORY STATE BUT A FULLY ERGODIC MATRIX

Let us consider the stochastic matrix with a transitory state E_2 (Fig. 57.1).

$$[\mathscr{M}] = \begin{matrix} & \begin{matrix} E_1 & E_2 & E_3 \end{matrix} \\ \begin{matrix} E_1 \\ E_2 \\ E_3 \end{matrix} & \begin{bmatrix} 0.2 & 0 & 0.8 \\ 0.2 & 0.3 & 0.5 \\ 0.1 & 0 & 0.9 \end{bmatrix} \end{matrix}. \qquad (57.1)$$

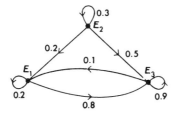

FIG. 57.1.

Let us calculate $[\mathscr{C}(z)]^{-1}$ for this matrix; we first have

$$|\mathscr{C}(z)| = \begin{bmatrix} 1-0.2z & 0 & -0.8z \\ -0.2z & 1-0.3z & -0.5z \\ -0.1z & 0 & 1-0.9z \end{bmatrix}$$

$$= (1-z)(1-0.1z)(1-0.3z). \qquad (57.2)$$

$$[\mathscr{C}(z)]^{-1} = \frac{1}{|\mathscr{C}(z)|} \begin{bmatrix} 1-1.2z+0.27z^2 & 0 & 0.8z-0.24z^2 \\ 0.2z-0.13z^2 & 1-1.1z+0.1z^2 & 0.5z+0.06z^2 \\ 0.1z-0.03z^2 & 0 & 1-0.5z+0.06z^2 \end{bmatrix}.$$
$$(57.3)$$

Let us identify it with

$$[\mathscr{C}(z)]^{-1} = \frac{1}{1-z}[A] + \frac{1}{1-0.1z}[A_1] + \frac{1}{1-0.3z}[A_2], \qquad (57.4)$$

it follows:

$$[A] = \begin{bmatrix} \frac{1}{9} & 0 & \frac{8}{9} \\ \frac{1}{9} & 0 & \frac{8}{9} \\ \frac{1}{9} & 0 & \frac{8}{9} \end{bmatrix}, \qquad [A_1] = \begin{bmatrix} \frac{16}{18} & 0 & -\frac{16}{18} \\ -\frac{11}{18} & 0 & \frac{11}{18} \\ -\frac{2}{18} & 0 & \frac{2}{18} \end{bmatrix},$$
$$(57.5)$$

$$[A_2] = \begin{bmatrix} 0 & 0 & 0 \\ -\frac{7}{9} & -\frac{14}{9} & \frac{21}{9} \\ 0 & 0 & 0 \end{bmatrix}.$$

Lastly:

$$[P(n)] = [P(0)]\left(\begin{bmatrix} \frac{1}{9} & 0 & \frac{8}{9} \\ \frac{1}{9} & 0 & \frac{8}{9} \\ \frac{1}{9} & 0 & \frac{8}{9} \end{bmatrix} + (0.1)^n \begin{bmatrix} \frac{16}{18} & 0 & -\frac{16}{18} \\ -\frac{11}{18} & 0 & \frac{11}{18} \\ -\frac{2}{18} & 0 & \frac{2}{18} \end{bmatrix} \right.$$

$$\left. + (0.3)^n \begin{bmatrix} 0 & 0 & 0 \\ -\frac{7}{9} & -\frac{14}{9} & \frac{21}{9} \\ 0 & 0 & 0 \end{bmatrix} \right). \qquad (57.6)$$

When $n \to \infty$:

$$[P(n)] = [P(0)] \begin{bmatrix} \frac{1}{9} & 0 & \frac{8}{9} \\ \frac{1}{9} & 0 & \frac{8}{9} \\ \frac{1}{9} & 0 & \frac{8}{9} \end{bmatrix} = [\frac{1}{9} \quad 0 \quad \frac{8}{9}].$$

(57.6)′

B. CHAIN WITH A TRANSITORY STATE BUT WITH A MATRIX WHICH
IS NOT FULLY ERGODIC (see Fig. 57.2)

$$[\mathcal{M}] = \begin{array}{c} \\ E_1 \\ E_2 \\ E_3 \end{array} \begin{matrix} E_1 & E_2 & E_3 \\ \begin{bmatrix} 1 & 0 & 0 \\ 0 & 1 & 0 \\ 0.2 & 0.5 & 0.3 \end{bmatrix} \end{matrix},$$

(57.7)

$$|\mathscr{C}(z)| = \begin{bmatrix} 1-z & 0 & 0 \\ 0 & 1-z & 0 \\ -0.2z & -0.5z & 1-0.3z \end{bmatrix}$$

$$= (1-z)^2(1-0.30z)$$

(57.8)

$$[\mathscr{C}(z)]^{-1} = \frac{1}{|\mathscr{C}(z)|} \begin{bmatrix} (1-z)(1-0.3z) & 0 & 0 \\ 0 & (1-z)(1-0.3z) & 0 \\ 0.20z(1-z) & 0.50z(1-z) & (1-z)^2 \end{bmatrix}$$

$$= \frac{1}{(1-z)(1-0.3z)} \begin{bmatrix} 1-0.3z & 0 & 0 \\ 0 & 1-0.3z & 0 \\ 0.2z & 0.5z & 1-z \end{bmatrix}.$$

(57.9)

$$[\mathscr{C}(z)]^{-1} = \frac{1}{1-z} \begin{bmatrix} 1 & 0 & 0 \\ 0 & 1 & 0 \\ \frac{2}{7} & \frac{5}{7} & 0 \end{bmatrix} + \frac{1}{1-0.3z} \begin{bmatrix} 0 & 0 & 0 \\ 0 & 0 & 0 \\ -\frac{2}{7} & -\frac{5}{7} & 1 \end{bmatrix}.$$

(57.10)

$$[P(n)] = [P(0)] \left(\begin{bmatrix} 1 & 0 & 0 \\ 0 & 1 & 0 \\ \frac{2}{7} & \frac{5}{7} & 0 \end{bmatrix} + (0.3)^n \begin{bmatrix} 0 & 0 & 0 \\ 0 & 0 & 0 \\ -\frac{2}{7} & -\frac{5}{7} & 1 \end{bmatrix} \right).$$

(57.11)

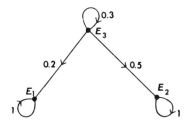

FIG. 57.2.

When $n \to \infty$:

$$[P(n)] = [P(0)] \begin{bmatrix} 1 & 0 & 0 \\ 0 & 1 & 0 \\ \frac{2}{7} & \frac{5}{7} & 0 \end{bmatrix}, \tag{57.12}$$

but this time $[P(n)]$ depends on $[P(0)]$; hence, starting successively from E_1, E_2, and E_3, it follows that leaving from E_1 :

$$[P(n)] = [1 \quad 0 \quad 0] \begin{bmatrix} 1 & 0 & 0 \\ 0 & 1 & 0 \\ \frac{2}{7} & \frac{5}{7} & 0 \end{bmatrix} = [1 \quad 0 \quad 0], \tag{57.13}$$

leaving from E_2 :

$$[P(n)] = [0 \quad 1 \quad 0] \begin{bmatrix} 1 & 0 & 0 \\ 0 & 1 & 0 \\ \frac{2}{7} & \frac{5}{7} & 0 \end{bmatrix} = [0 \quad 1 \quad 0], \tag{57.14}$$

leaving from E_3 :

$$[P(n)] = [0 \quad 0 \quad 1] \begin{bmatrix} 1 & 0 & 0 \\ 0 & 1 & 0 \\ \frac{2}{7} & \frac{5}{7} & 0 \end{bmatrix} = [\frac{2}{7} \quad \frac{5}{7} \quad 0]. \tag{57.15}$$

C. Decomposable Matrix (Fig. 57.3)

$$[\mathcal{M}] = \begin{array}{c} \\ E_1 \\ E_2 \\ \\ E_3 \\ E_4 \end{array} \begin{array}{cccc} E_1 & E_2 & E_3 & E_4 \\ \begin{bmatrix} 0.1 & 0.9 & 0 & 0 \\ 0.5 & 0.5 & 0 & 0 \\ \hline 0 & 0 & 0.5 & 0.5 \\ 0 & 0 & 0.3 & 0.7 \end{bmatrix} \end{array}. \tag{57.16}$$

It is sufficient to treat the two matrices $[\mathcal{M}_1]$ separately in relation to states E_1 and E_2, then $[\mathcal{M}_2]$ in relation to E_3 and E_4.

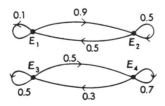

Fig. 57.3.

D. MATRIX WITH ITS OWN COMPLEX VALUES

The procedure for treating a stochastic matrix with complex values is as follows (Fig. 57.4). Let

$$[\mathcal{M}] = \begin{array}{c} \\ E_1 \\ E_2 \\ E_3 \end{array} \begin{array}{c} E_1 \quad E_2 \quad E_3 \\ \begin{bmatrix} 0.7 & 0.3 & 0 \\ 0 & 0.2 & 0.8 \\ 0.5 & 0 & 0.5 \end{bmatrix} \end{array}, \tag{57.17}$$

$$[\mathcal{C}(z)] = \begin{bmatrix} 1 - 0.7z & -0.3z & 0 \\ 0 & 1 - 0.2z & -0.8z \\ -0.5z & 0 & 1 - 0.5z \end{bmatrix}, \tag{57.18}$$

$$|\mathcal{C}(z)| = 1 - 1.4z + 0.59z^2 - 0.19z^3$$
$$= (1 - z)[1 - (0.2 - j\sqrt{0.15})z][1 - (0.2 + j\sqrt{0.15})z] \tag{57.19}$$

$$[\mathcal{C}(z)]^{-1} = \frac{1}{|\mathcal{C}(z)|} \begin{bmatrix} 1 - 0.7z + 0.1z^2 & 0.3z - 0.15z^2 & 0.24z^2 \\ 0.4z^2 & 1 - 1.2z + 0.35z^2 & 0.8z - 0.56z^2 \\ 0.5z - 0.1z^2 & 0.15z^2 & 1 - 0.9z + 0.14z^2 \end{bmatrix} \tag{57.20}$$

$$[\mathcal{C}(z)] = \frac{1}{1 - z}[A] + \frac{1}{1 - (0.2 - j\sqrt{0.15})z}[A_1]$$

$$+ \frac{1}{1 - (0.2 + j\sqrt{0.15})z}[\bar{A}_1]. \tag{57.21}$$

Multiplying (57.21) by $(1 - z)$, and letting $z = 1$, it follows that

$$[A] = \begin{bmatrix} 40/79 & 15/79 & 24/79 \\ 40/79 & 15/79 & 24/79 \\ 40/79 & 15/79 & 24/79 \end{bmatrix}. \tag{57.22}$$

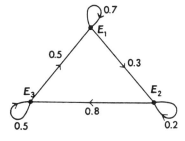

FIG. 57.4.

We have

$$[P(n)] = [P(0)]([A] + \bar{\alpha}^n[A_1] + \alpha^n[\bar{A}_1]), \qquad (57.23)$$

where

$$\alpha = 0.2 + j\sqrt{0.15}.$$

If he wishes, the reader can calculate the real matrix

$$\bar{\alpha}^n[A_1] + \alpha^n[\bar{A}_1],$$

but he should note that if $n \to \infty$, the moduli of α^{-n} and α^n approach zero.

Some authors give the name *pseudoperiodic stochastic matrices* to those in which certain of their own values are unreal.

E. PERIODIC MATRIX (Fig. 57.5)

$$[\mathscr{M}] = \begin{array}{c} \\ \begin{array}{c} E_1 \\ E_2 \\ \\ E_3 \\ E_4 \end{array} \begin{array}{cccc} E_1 & E_2 & E_3 & E_4 \\ \left[\begin{array}{cc|cc} 0 & 0 & 0.3 & 0.7 \\ 0 & 0 & 0.2 & 0.8 \\ \hline 0.1 & 0.9 & 0 & 0 \\ 0.4 & 0.6 & 0 & 0 \end{array}\right] \end{array} \end{array} \qquad (57.24)$$

$$[\mathscr{C}(z)] = \begin{bmatrix} 1 & 0 & -0.3z & -0.7z \\ 0 & 1 & -0.2z & -0.8z \\ -0.1z & -0.9z & 1 & 0 \\ -0.4z & -0.6z & 0 & 1 \end{bmatrix}, \qquad (57.25)$$

$$|\mathscr{C}(z)| = 1 - 0.97z^2 - 0.03z^4$$

$$= (1 - z)(1 + z)(1 + 0.03z^2). \qquad (57.26)$$

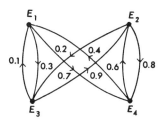

FIG. 57.5.

$$[\mathscr{C}(z)] = \frac{1}{1-z}[A] + \frac{1}{1+z}[A_1] + \frac{1}{1+j\sqrt{0.03}z}[\bar{A}_2]$$

$$+ \frac{1}{1-j\sqrt{0.03}z}[A_2]. \tag{57.27}$$

$$[P(n)] = [A] + (-1)^n[A_1] + (-j\sqrt{0.03})^n[\bar{A}_2] + (j\sqrt{0.03})[A_2], \tag{57.28}$$

where[1]

$$[A] = \begin{bmatrix} 34/206 & 69/206 & 24/206 & 79/206 \\ 34/206 & 69/206 & 24/206 & 79/206 \\ 34/206 & 69/206 & 24/206 & 79/206 \\ 34/206 & 69/206 & 24/206 & 79/206 \end{bmatrix}. \tag{57.29}$$

$$[A_1] = \begin{bmatrix} -34/206 & -69/206 & 24/206 & 79/206 \\ -34/206 & -69/206 & 24/206 & 79/206 \\ 34/206 & 69/206 & -24/206 & -79/206 \\ 34/206 & 69/206 & -24/206 & -79/206 \end{bmatrix}. \tag{57.30}$$

$$[A] + [A_1] = \begin{bmatrix} 0 & 0 & 24/103 & 79/103 \\ 0 & 0 & 24/103 & 79/103 \\ 34/103 & 69/103 & 0 & 0 \\ 34/103 & 69/103 & 0 & 0 \end{bmatrix}. \tag{57.31}$$

$$[A] - [A_1] = \begin{bmatrix} 34/103 & 69/103 & 0 & 0 \\ 34/103 & 69/103 & 0 & 0 \\ 0 & 0 & 24/103 & 24/103 \\ 0 & 0 & 24/103 & 24/103 \end{bmatrix}. \tag{57.32}$$

As $n \to \infty$, the terms containing powers of $(j\sqrt{0.03})$ disappear, but the term containing $(-1)^n$ does not disappear, so that matrix (57.24) is nonergodic

$$\lim_{n\to\infty}[\mathscr{M}]^n = [A] + [A_1] \quad \text{if } n \text{ is even,}$$

$$= [A] - [A_1] \quad \text{if } n \text{ is uneven.} \tag{57.33}$$

---

[1] Let us recall the matrix which we used to calculate $[\mathscr{M}]$ and $[A_1]$:

$$[\mathscr{C}(z)]^{-1} = \frac{1}{|\mathscr{C}(z)|}$$

$$\begin{bmatrix} 1 - 0.66z^2 & 0.69z^2 & 0.30z - 0.06z^3 & 0.07z + 0.09z^3 \\ 0.34z^2 & 1 - 0.31z^2 & 0.20z + 0.04z^3 & 0.80z - 0.01z^3 \\ 0.10z + 0.24z^3 & 0.90z - 0.21z^3 & 1 - 0.76z^2 & 0.79z^2 \\ 0.40z - 0.06z^3 & 0.60z + 0.09z^3 & 0.24z^2 & 1 - 0.21z^2 \end{bmatrix}$$

with

$$|\mathscr{C}(z)| = (1-z)(1+z)(1+0.03z^2).$$

58. Dynamic Programming in a Case of Uncertainty

A. D.P. PROCESS[1]

Let us first consider the discrete and stationary case of uncertainty shown in Fig. 58.1. At each time n, the system may be in a state $E_i \in \mathbf{E}$, and can move to $E_r' \in \Gamma_D E_i$ at time $n < n' < n + 1$ by a decision $E_i \to E_r'$ to which the value a_{ir} is attached. At time n' uncertainty intervenes and the system returns to one of the states $E_j \in \Gamma_H E_r' \subset \mathbf{E}$ with a probability p_{rj}, after a transition with a value b_{rj}. We shall call the interval between times n and $n + 1$, *stage n.*

If $\bar{f}_i(N - n, N)$ represents the probable optimal value (expected value) of the set of stages $n, n + 1,..., N - 1, N$, when the system is in state E_i at time n, we can state, as a result of the theorem of optimality and by the use of the probable optimal value as the criterion,

$$\bar{f}_i(N - n, N) = \underset{E_r' \in \Gamma_D E_i}{\text{OPT}} \Big\{ a_{ir} + \sum_{E_j \in \Gamma_H E_r'} p_{rj}[b_{rj} + \bar{f}_j(N - n + 1, N)] \Big\},$$

$$i = 1, 2,..., M,$$

$$n = 1, 2,..., N. \quad (58.1)$$

with

$$\bar{f}_i(N, N) = f_{0,i}, \quad \text{given.} \quad (58.2)$$

A system in which, at each stage, an uncertain influence follows a decision will be called a *D.P. process.*[2]

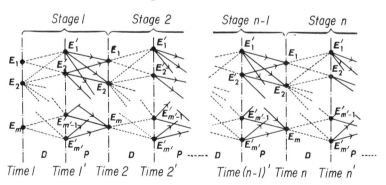

Stage 1   *Stage 2*   *Stage n-1*   *Stage n*

FIG. 58.1.   Sequential process of D.P. type.

[1] According to the term introduced by Companys in his thesis *Contribution à la théorie mathématique des processus de décisions séquentielles.*

[2] As the French word *hasard* is here translated as *probability* in accordance with common American usage, the French D.H. process (*Décision-hasard*) becomes the D.P. process and the H.D. process becomes the P.D. (Translator's note).

The formulas for multistage optimization (58.1 and 58.2) can be given as generalizations for nonstationary cases, and we then have

$$E_i(n) \in \mathbf{E}(n); \qquad E_r'(n') \in \Gamma_D(n)E_i(n); \qquad E_j(n+1) \in \Gamma_H(n')E_r'(n') \quad (58.3)$$

with $a_{ir}(n)$, $p_{rj}(n)$ and $b_{rj}(n)$. These formulas then become

$$\tilde{f}_i(N-n, N) = \underset{E'_r(n) \in \Gamma_D(n)E_i(n)}{\text{OPT}} \Big\{ a_{ir}(n)$$

$$+ \sum_{E_j(n) \in \Gamma_H(n)E_r'(n)} p_{rj}(n)[b_{rj}(n) + \tilde{f}_j(N-n+1, N)] \Big\},$$

$$n = 1, 2, ..., N \qquad (58.4)$$

and

$$\tilde{f}_i(N, N) = f_{0,i} . \qquad (58.5)$$

B. P.D. PROCESS

At each time n with the system in state $E_r' \in \mathbf{E}'$, uncertainty intervenes and the system moves into a state $E_j \in \Gamma_H E_r'$ with a probability p_{rj} and a value c_{rj}. At time n', such that $n < n' < n + 1$, a decision intervenes and allows the system to return to one of the states $E_s' \in \mathbf{E}'$, such that $E_s' \in \Gamma_D E_j$ with a value d_{js}. The times and stages are shown in Fig. 58.2.

With $\bar{g}_r(N-n, N)$ as the expected optimal value for the set of stages $n, n+1, ..., N-1, N$, when the system is in E_r' at time n, we can, as a result of

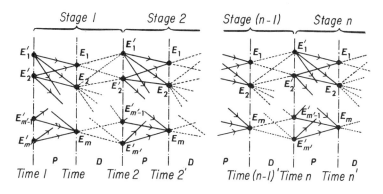

FIG. 58.2.   Sequential process of P.D. type.

the theorem of optimality, and by using the average value as our criterion, now state

$$\bar{g}_r(N-n, N) = \sum_{E_j \in \Gamma_H E_{r'}} \left\{ p_{rj} c_{rj} \right.$$

$$\left. + \operatorname*{OPT}_{E_s' \in \Gamma_D E_j} [d_{js} + \bar{g}_s(N-n+1, N)] \right\}, \qquad n = 1, 2, ..., N. \tag{58.6}$$

with

$$\bar{g}_r(N, N) = g_{0,r}, \tag{58.7}$$

as given quantities.

A system in which, at each stage, an influence of uncertainty precedes a decision is called a *P.D. process*.

In the case of a nonstationary process, and using the same notation as for a D.P. process, we have

$$\bar{g}_r(N-n, N) = \sum_{E_j(n) \in \Gamma_H(n) E_{r'}(n)} \left\{ p_{rj}(n) c_{rj}(n) \right.$$

$$\left. + \operatorname*{OPT}_{E_s'(n) \in \Gamma_D(n) E_j(n)} [d_{js}(n) + \bar{g}_s(N-n+1, N)] \right\},$$

$$i = 1, 2, ..., M; \qquad n = 1, 2, ..., N \tag{58.8}$$

with

$$\bar{g}_r(N, N) = g_{0,r}, \tag{58.9}$$

as given quantities.

C. MORE GENERAL PROCESSES

It is possible to consider processes where a succession of decisions and effects of probability occur within a stage, for example $P_1 \cdot D_1 \cdot D_2 \cdot P_2 \cdot P_3$. with given mappings of each set into the one which follows it. We shall not give the generalizations which can then be made for formulas (58.1), (58.4), (58.6), and (58.8), since they become very complicated and can, if required, be transformed for a given process. Instead, we shall begin the explanation of another and more important generalization.

D. REDUCTION TO MARKOVIAN DECISION CHAIN

All the discrete processes which we shall now examine are such that what happens at time n only depends on the state considered at time

$n - 1$, and they are therefore Markovian. If we recall the properties of a Markovian chain we can show that they form such chains when they are stationary. On the other hand, when the stages are decomposable into a succession of decisions and results of probability, the D. P. and P. D. processes and their generalization are as well suited to non-stationary as to stationary cases, and it is rare for the phenomena encountered in business administration to be stationary. In consequence of this, it is much more convenient and effective in practice to employ the D. P. or P. D. processes than concepts derived from Markovian decision chains.

Let us now show how these two processes can be converted into Markovian chains (a transformation which is very useful from an analytical standpoint), first examining how a stationary D. P. process (Fig. 58.3) can be reduced to a Markovian chain with transition values.

For the decision $D_1^{(1)}$ to move from E_1 into $E_1{}'$ there is a corresponding stochastic vector

$$p_1^{(1)} = [0.1 \quad 0.9 \quad 0 \quad 0],$$

which means that the move $E_1 \to E_1$ will be made with a probability of 0.1, the move from E_1 to E_2 with one of 0.9, while the other transitions will have a zero probability. To these transitions are attached values defined by the vector of revenue[1]:

$$R_1^{(1)} = [(2 + 4) \quad (2 - 5) \quad 0 \quad 0]$$
$$= [6 \quad -3 \quad 0 \quad 0]$$

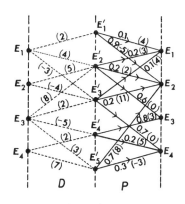

FIG. 58.3.

---

[1] We shall arbitrarily show a zero revenue when the corresponding transition is impossible; in point of fact, the revenue is not defined under these conditions.

It is thus that the following Table 58.1 has been drawn up.

TABLE 58.1

Decision	Stochastic vector	Vector of revenue
$D_1^{(1)} : E_1 \to E_1'$	$p_1^{(1)} = [0.1 \quad 0.9 \quad 0 \quad 0]$	$R_1^{(1)} = [6 \quad -3 \quad 0 \quad 0]$
$D_1^{(2)} : E_1 \to E_2'$	$p_1^{(2)} = [0.2 \quad 0.2 \quad 0.6 \quad 0]$	$R_1^{(2)} = [7 \quad 6 \quad 4 \quad 0]$
$D_1^{(3)} : E_1 \to E_3'$	$p_1^{(3)} = [0.1 \quad 0.2 \quad 0 \quad 0.7]$	$R_1^{(3)} = [1 \quad 8 \quad 0 \quad -3]$
$D_2^{(1)} : E_2 \to E_2'$	$p_2^{(1)} = [0.2 \quad 0.2 \quad 0.6 \quad 0]$	$R_2^{(1)} = [8 \quad 7 \quad 5 \quad 0]$
$D_2^{(2)} : E_2 \to E_3'$	$p_2^{(2)} = [0.1 \quad 0.2 \quad 0 \quad 0.7]$	$R_2^{(2)} = [0 \quad 7 \quad 0 \quad -4]$
$D_3^{(1)} : E_3 \to E_2'$	$p_3^{(1)} = [0.2 \quad 0.2 \quad 0.6 \quad 0]$	$R_3^{(1)} = [11 \quad 10 \quad 8 \quad 0]$
$D_3^{(2)} : E_3 \to E_3'$	$p_3^{(2)} = [0.1 \quad 0.2 \quad 0 \quad 0.7]$	$R_3^{(2)} = [6 \quad 13 \quad 0 \quad 2]$
$D_3^{(3)} : E_3 \to E_4'$	$p_3^{(3)} = [0 \quad 0 \quad 0.8 \quad 0.2]$	$R_3^{(3)} = [0 \quad 0 \quad -2 \quad 0]$
$D_3^{(4)} : E_3 \to E_5'$	$p_3^{(4)} = [0 \quad 0.7 \quad 0 \quad 0.3)$	$R_3^{(4)} = [0 \quad 11 \quad 0 \quad 0]$
$D_4^{(1)} : E_4 \to E_4'$	$p_1^{(1)} = [0 \quad 0 \quad 0.8 \quad 0.2]$	$R_4^{(1)} = [0 \quad 0 \quad 5 \quad 7]$
$D_4^{(2)} : E_4 \to E_5'$	$p_4^{(2)} = [0 \quad 0.7 \quad 0 \quad 0.3]$	$R_4^{(2)} = [0 \quad 15 \quad 0 \quad 4]$

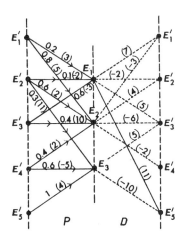

FIG. 58.4.

TABLE 58.2

Decision	Stochastic vector	Vector of revenue
$D_1^{(1)} : \begin{cases} E_1 \to E_1' \\ E_2 \to E_1' \end{cases}$	$p_1^{(1)} = [1 \quad 0 \quad 0 \quad 0 \quad 0]$	$R_1^{(1)} = [3.6 \quad 0 \quad 0 \quad 0 \quad 0]$
$D_1^{(2)} : \begin{cases} E_1 \to E_1' \\ E_2 \to E_2' \end{cases}$	$p_1^{(2)} = [0.2 \quad 0.8 \quad 0 \quad 0 \quad 0]$	$R_1^{(2)} = [10 \quad 9 \quad 0 \quad 0 \quad 0]$
$D_1^{(3)} : \begin{cases} E_1 \to E_1' \\ E_2 \to E_3' \end{cases}$	$p_1^{(3)} = [0.2 \quad 0 \quad 0.8 \quad 0 \quad 0]$	$R_1^{(3)} = [10 \quad 0 \quad -1 \quad 0 \quad 0]$
$D_1^{(4)} : \begin{cases} E_1 \to E_1' \\ E_2 \to E_4' \end{cases}$	$p_1^{(4)} = [0.2 \quad 0 \quad 0 \quad 0.8 \quad 0]$	$R_1^{(4)} = [10 \quad 0 \quad 0 \quad 3 \quad 0]$
$D_1^{(5)} : \begin{cases} E_1 \to E_2' \\ E_2 \to E_1' \end{cases}$	$p_1^{(5)} = [0.8 \quad 0.2 \quad 0 \quad 0 \quad 0]$	$R_1^{(5)} = [2 \quad 1 \quad 0 \quad 0 \quad 0]$
$D_1^{(6)} : \begin{cases} E_1 \to E_2' \\ E_2 \to E_2' \end{cases}$	$p_1^{(6)} = [0 \quad 1 \quad 0 \quad 0 \quad 0]$	$R_1^{(6)} = [0 \quad 7.4 \quad 0 \quad 0 \quad 0]$
$D_1^{(7)} : \begin{cases} E_1 \to E_2' \\ E_2 \to E_3' \end{cases}$	$p_1^{(7)} = [0 \quad 0.2 \quad 0.8 \quad 0 \quad 0]$	$R_1^{(7)} = [0 \quad 1 \quad -1 \quad 0 \quad 0]$
$D_1^{(8)} : \begin{cases} E_1 \to E_2' \\ E_2 \to E_4' \end{cases}$	$p_1^{(8)} = [0 \quad 0.2 \quad 0 \quad 0.8 \quad 0]$	$R_1^{(8)} = [0 \quad 1 \quad 0 \quad 3 \quad 0]$
$D_1^{(9)} : \begin{cases} E_1 \to E_3' \\ E_2 \to E_1' \end{cases}$	$p_1^{(9)} = [0.8 \quad 0 \quad 0.2 \quad 0 \quad 0]$	$R_1^{(9)} = [2 \quad 0 \quad 8 \quad 0 \quad 0]$
$D_1^{(10)} : \begin{cases} E_1 \to E_3' \\ E_2 \to E_2' \end{cases}$	$p_1^{(10)} = [0 \quad 0.8 \quad 0.2 \quad 0 \quad 0]$	$R_1^{(10)} = [0 \quad 9 \quad 8 \quad 0 \quad 0]$
$D_1^{(11)} : \begin{cases} E_1 \to E_3' \\ E_2 \to E_3' \end{cases}$	$p_1^{(11)} = [0 \quad 0 \quad 1 \quad 0 \quad 0]$	$R_1^{(11)} = [0 \quad 0 \quad 0.8 \quad 0 \quad 0]$
$D_1^{(12)} : \begin{cases} E_1 \to E_3' \\ E_2 \to E_4' \end{cases}$	$p_1^{(12)} = [0 \quad 0 \quad 0.2 \quad 0.8 \quad 0]$	$R_1^{(12)} = [0 \quad 0 \quad 8 \quad 3 \quad 0]$
$D_1^{(13)} : \begin{cases} E_1 \to E_5' \\ E_2 \to E_1' \end{cases}$	$p_1^{(13)} = [0.8 \quad 0 \quad 0 \quad 0 \quad 0.2]$	$R_1^{(13)} = [2 \quad 0 \quad 0 \quad 0 \quad 14]$
$D_1^{(14)} : \begin{cases} E_1 \to E_5' \\ E_2 \to E_2' \end{cases}$	$p_1^{(14)} = [0 \quad 0.8 \quad 0 \quad 0 \quad 0.2]$	$R_1^{(14)} = [0 \quad 9 \quad 0 \quad 0 \quad 14]$
$D_1^{(15)} : \begin{cases} E_1 \to E_5' \\ E_2 \to E_3' \end{cases}$	$p_1^{(15)} = [0 \quad 0 \quad 0.8 \quad 0 \quad 0.2]$	$R_1^{(15)} = [0 \quad 0 \quad -1 \quad 0 \quad 14]$
$D_1^{(16)} : \begin{cases} E_1 \to E_5' \\ E_2 \to E_4' \end{cases}$	$p_1^{(16)} = [0 \quad 0 \quad 0 \quad 0.8 \quad 0.2]$	$R_1^{(16)} = [0 \quad 0 \quad 0 \quad 3 \quad 14]$

TABLE 58.2   (Continued)

Decision	Stochastic vector	Vector of revenue
$D_2^{(1)}$: $\begin{cases} E_1 \to E_1' \\ E_2 \to E_1' \\ E_3 \to E_3' \end{cases}$	$p_2^{(1)} = [0.7 \quad 0 \quad 0.3 \quad 0 \quad 0]$	$R_2^{(1)} = [-0.143 \quad 0 \quad 16 \quad 0 \quad 0]$
$D_2^{(2)}$: $\begin{cases} E_1 \to E_1' \\ E_2 \to E_2' \\ E_3 \to E_3' \end{cases}$	$p_2^{(2)} = [0.1 \quad 0.6 \quad 0.3 \quad 0 \quad 0]$	$R_2^{(2)} = [5 \quad 6 \quad 16 \quad 0 \quad 0]$
$D_2^{(3)}$: $\begin{cases} E_1 \to E_1' \\ E_2 \to E_2' \\ E_3 \to E_5' \end{cases}$ $\vdots$	$p_2^{(3)} = [0.1 \quad 0.6 \quad 0 \quad 0 \quad 0.3]$	$R_2^{(3)} = [5 \quad 6 \quad 0 \quad 0 \quad 1]$
$\vdots$		
$\vdots$ $D_5^{(2)}$: $E_3 \to E_5'$	$p_5^{(2)} = [0 \quad 0 \quad 0 \quad 0 \quad 1]$	$R_5^{(2)} = [0 \quad 0 \quad 0 \quad 0 \quad -6]$

Let us now consider an example of a P. D. stationary process (Fig. 58.4), Taking $D_1^{(2)}$ as the decision to move from E_1 to E_1' and from E_2 to E_2', we have

$$p_1 = [0.2 \quad 0.8 \quad 0 \quad 0 \quad 0]$$
$$R_1 = [(3 + 7) \quad (5 + 4) \quad 0 \quad 0 \quad 0]$$
$$= [10 \quad 9 \quad 0 \quad 0 \quad 0].$$

It is in this way that Table 58.1 has been obtained. It will be observed how complicated the description of all the different decisions becomes; this is because, when faced with earlier probability, every possible situation must be defined (see Table 58.2).

59. Dynamic Program with Markovian Chain

A. DEFINITION

Given a discrete stochastic procedure with Markovian decisions such that, for each change of state $(E_i \to E_j)$, $i, j = 1, 2,..., M$, there are one or more probabilities $p_{ij}^{(r)}$. For each change of state E_i we consider a set of m_i stochastic vectors

$$[p_i^{(r)}] = [p_{i1}^{(r)} \quad p_{i2}^{(r)} \quad \cdots \quad p_{iM}^{(r)}];$$

the choice of such a vector of index r from among the m_i vectors is freely made, and this system constitutes a Markovian decision chain.

With each probability $p_{ij}^{(r)}$ we associate a value or revenue $R_{ij}^{(r)}$ which is a real number, and in this manner we form vectors of revenue

$$[R_i^{(r)}] = [R_{i1}^{(r)} \quad R_{i2}^{(r)} \quad \cdots \quad R_{iM}^{(r)}],$$

which will be associated with the corresponding stochastic vectors.

We call the *decision* at time n the choice of M stochastic vectors

$$[p_1^{(r_1)}], \quad [p_2^{(r_2)}],\ldots, [p_M^{(r_M)}]$$

corresponding to the transition probabilities chosen when we are in E_1, E_2,..., E_M at this time: this choice produces associated vectors of revenue.

A strategy is a sequence of decisions over a set of connected periods.

B. FINDING THE OPTIMAL STRATEGIES

In considering N successive transitions at times $0, 1, 2,..., N - 1$, we look for the optimal strategy which maximizes (or minimizes) the sum of the values of N.

If $\bar{q}_i^{(r)}$ is the expected value of a transition from E_i at time n when we have chosen the stochastic vector $[p_i^{(r)}]$, we have

$$\bar{q}_i^{(r)} = \sum_{j=1}^{M} p_{ij}^{(r)} R_{ij}^{(r)}, \qquad i = 1, 2,..., M. \tag{59.1}$$

If $\bar{v}_i(N - n, N)$ is the expected total value for n stages or periods from time $N - n$, when the system was in state E_i, to time N, we can state, as the result of the theorem of optimality:

$$\bar{v}_i(N - n - 1, N) = \max_{r=1,2,\ldots,m} \left[\bar{q}_i^{(r)} + \sum_{j=1}^{w} p_{ij}^{(r)} \bar{v}_j(N - n, N) \right] \tag{59.2}$$

$$i = 1, 2,..., M; \qquad n = 0, 1, 2,..., N - 1 \tag{59.3}$$

with

$$\bar{v}_j(N, N) = v_{0j}, \qquad j = 1, 2,..., M,$$

which is the value of the process when it is ended at time N in state j.

For simplification, let us assume

$$\bar{v}_i(n) = \bar{v}_i(N - n, N), \qquad n = 0, 1, 2,..., N, \qquad (59.4)$$

$$\bar{v}_i(0) = \bar{v}_i(N, N) = \bar{v}_{0i}, \qquad i = 1, 2,..., M. \qquad (59.5)$$

Let

$$\bar{v}_i(n + 1) = \max_{r=1,2,...,m} \left[\bar{q}_i^{(r)} + \sum_{j=1}^{M} p_{ij}^{(r)} \bar{v}_j(n) \right].$$

$$i = 1, 2,..., M; \qquad n = 0, 1, 2,..., N - 1. \qquad (59.6)$$

C. General Formulas

To find the optimal strategies we shall accordingly evaluate in succession:

$$\bar{v}_i(1) = \max_{r=1,2,...,m} \left[q_i^{(r)} + \sum_{j=1}^{M} \bar{p}_{ij}^{(r)} \bar{v}_{0j} \right], \qquad i = 1, 2,..., M \qquad (59.7)$$

$$\bar{v}_i(2) = \max_{r=1,2,...,m} \left[\bar{q}_i^{(r)} + \sum_{j=1}^{M} p_{ij}^{(r)} \bar{v}_j(1) \right]. \qquad (59.8)$$

$$\vdots$$

$$\bar{v}_i(N) = \max_{r=1,2,...,m} \left[\bar{q}_i^{(r)} + \sum_{j=1}^{M} p_{ij}^{(r)} \bar{v}_j(N - 1) \right]. \qquad (59.9)$$

These formulas have already been used in Section 16.

60. Long Term Situation (Case of a Fully Ergodic Matrix)

A. Calculating the Limits

Let us consider a type of problem where the matrix $[\mathcal{M}]$ is fully ergodic, and return to (56.24):

$$\{\bar{v}(n)\} = (n[\mathcal{M}] + [\mathcal{N}])\{\bar{q}\} + [\mathcal{M}]\{v(0)\}$$
$$= n[\mathcal{M}]\{\bar{q}\} + [\mathcal{N}]\{\bar{q}\} + [\mathcal{M}]\{v(0)\}. \qquad (60.1)$$

Let us assume:

$$\{w\} = [\mathcal{N}]\{\bar{q}\} + [\mathcal{M}]\{v(0)\}, \qquad (60.2)$$

whence

$$\{\bar{v}(n)\} = n[\mathcal{M}]\{\bar{q}\} + \{w\}. \qquad (60.3)$$

We know that $[\tilde{\mathcal{M}}]$ is formed of M identical lines with elements p_i, $i = 1, 2,..., M$, which form the long term probabilities, so that (60.3) can be expressed as

$$\bar{v}_i(n) = n \sum_{i=1}^{M} p_i \bar{q}_i + w_i, \qquad i = 1, 2,..., M, \tag{60.4}$$

or again,

$$\bar{v}_i(n) = n\gamma + w_i, \qquad i = 1, 2,..., M. \tag{60.5}$$

Using (56.6):

$$\bar{v}_i(n + 1) = \bar{q}_i + \sum_{j=1}^{M} p_{ij} \bar{v}_j(n), \qquad i = 1, 2,..., M. \tag{60.6}$$

Substituting (60.5) in (60.6):

$$(n + 1)\gamma + w_i = \bar{q}_i + \sum_{j=1}^{M} (n\gamma + w_j) p_{ij}, \qquad i = 1, 2,..., M. \tag{60.7}$$

Simplifying and taking into account the property:

$$\sum_{j=1}^{M} p_{ij} = 1, \qquad i = 1, 2,..., M,$$

we have

$$\gamma + w_i = \bar{q}_i + \sum_{j=1}^{M} p_{ij} w_j, \qquad i = 1, 2,..., M. \tag{60.8}$$

This forms a system of M equations with $M + 1$ unknowns, w_1, w_2,..., w_M and γ, which are indeterminate. Let us, for example, take

$$\begin{aligned} w_j &= w_1, & j &= 1 \\ &= w_1 + a_j, & j &= 2, 3,..., M, \end{aligned} \tag{60.9}$$

where the a_j terms are constants representing the separations between each w_j and w_1, and the system becomes

$$\gamma = \bar{q}_1 + \sum_{j=2}^{M} p_{ij} a_j \tag{60.10}$$

$$\gamma + a_i = \bar{q}_i + \sum_{j=2}^{M} p_{ij} a_j, \qquad i = 1, 2,..., M, \tag{60.11}$$

where we find M variables a_2, a_3,..., a_M, which are linearly independent, as we can easily prove, while the γ variable is determinate.

In practice, we consider system (60.8), in which w_1 is arbitrarily taken as 0, and solve the M equations with M unknowns, and hence obtain γ and the other w_j values.

B. EXAMPLE

As an example, let us rework the problem in Section 56.

$$[\mathcal{M}] = \begin{bmatrix} 0.5 & 0.5 \\ 0.7 & 0.3 \end{bmatrix}; \tag{60.12}$$

$$\{\bar{q}\} = \begin{Bmatrix} 325 \\ 20 \end{Bmatrix}. \tag{60.13}$$

It follows that

$$\gamma + w_1 = 325 + 0.5w_1 + 0.5w_2, \tag{60.14}$$

$$\gamma + w_2 = 20 + 0.7w_1 + 0.3w_2. \tag{60.15}$$

Let us assume $w_1 = 0$:

$$\gamma = 325 + 0.5w_2, \tag{60.16}$$

$$\gamma = 20 - 0.7w_2; \tag{60.17}$$

whence

$$\gamma = 197.92 \quad \text{and} \quad w_2 = -254.16, \tag{60.18}$$

the result already obtained in (56.19).

61. Iterative Optimization (Howard's Method [16])

A. DESCRIPTION OF THE METHOD

Let us return to the Markovian dynamic program explained in Section 59, and assume that all the matrices $[\mathcal{M}^{(r)}]$ formed by any M vectors $[p_i^{(r)}]$, as defined in Section 58, in which the subscript i corresponds to line i of this matrix, are fully ergodic.

We now have

$$\bar{v}_i(n + 1) = \underset{r=1,2,\dots,m}{\text{MAX}} \sum_{j=1}^{M} [\bar{q}_i^{(r)} + p_{ij}^{(r)} \bar{v}_j(n)] \qquad i = 1, 2, \dots, M, \tag{61.1}$$

where

$$\bar{q}_i^{(r)} = \sum_{j=1}^{M} p_{ij}^{(r)} R_{ij}^{(r)} \cdots, \qquad i = 1, 2, \dots, M. \tag{61.2}$$

Let us look for the long term optimal strategy; that is to say, for the high values of n. For each decision r when we are in state i at time $N - n - 1$($n + 1$ periods before N), if n is sufficiently large, we can use (60.5) and state

$$\bar{v}_i^{(r)}(n) = n\gamma + w_i^{(r)}, \qquad i = 1, 2,..., M. \tag{61.3}$$

Substituting (61.3) in (61.1), and following the same process as in (60.7) and (60.8), it follows that

$$\gamma + w_i^{(r)} = \underset{r=1,2,...,m}{\text{MAX}} \left[\bar{q}_i^{(r)} + \sum_{j=1}^{M} p_{ij}^{(r)} w_j^{(r)} \right], \qquad i = 1, 2,..., M. \tag{61.4}$$

One method consists of evaluating the best decision r in state i at time $N - n - 1$ ($n + 1$ periods before N), but this can lead to lengthy calculations where M and m are large numbers, since there are M^m quantities to be calculated and compared. We shall therefore use the following iterative method where

$$r = \{r_1, r_2,..., r_M\}$$

is the decision to choose the vector $p_1{}^r$ for line (1), $p_2{}^r$ for line (2), etc.

Stage 1.   Choose any strategy r and solve the following linear system in which each $\bar{q}_i$ and p_{ij} corresponds to this strategy:

$$\gamma + w_i = \bar{q}_i + \sum_{j=1}^{M} p_{ij} w_j, \qquad i = 1, 2,..., M \tag{61.5}$$

taking $w_1 = 0$ for example, which gives us $w_2, w_3,..., w_M$ and γ.

Stage 2.   For the w values obtained, and for each possible decision, evaluate the quantities:

$$z_i^{(r)} = \bar{q}_i^{(r)} + \sum_{j=1}^{M} p_{ij}^{(r)} w_j, \qquad \begin{cases} i = 1, 2,..., M, \\ r = 1, 2,..., m. \end{cases} \tag{61.6}$$

Stage 3.   For each state E_i choose r_i' such that

$$z_i^{(r_i')} = \underset{r=1,2,...,m}{\text{MAX}} (z_i^{(r_i)}). \tag{61.7}$$

Stage 4.   If we take r_i' as the decision for E_i, $\bar{q}_i^{(r_i')}$ becomes $\bar{q}_i$ and $p_{ij}^{(r_i')}$ becomes p_{ij}. With these data, let us recommence stage 1. If γ' is the new value of revenue per period thus obtained, then

(a)   If $\gamma' = \gamma$, the optimum has been found, and we stop;

(b)   If $\gamma' > \gamma$, we pass to Stage 2.

We continue, if necessary, calculating γ'', γ''',..., until we have found the same revenue per period twice. There may be several solutions; that is to say, several optimal strategies corresponding to the optimal value of γ.

B. EXAMPLE

Let us take the example given in Section 16 of Part I, and find the permanent optimal strategy by iterative numerical calculation for each month.

Let us take

$$[p_1^{(1)}] = [p_{11}^{(1)} \quad p_{12}^{(1)}] = [0.5 \quad 0.5],$$

$$[r_1^{(1)}] = [r_{11}^{(1)} \quad r_{12}^{(1)}] = [500 \quad 150],$$

$$[p_2^{(1)}] = [p_{21}^{(1)} \quad p_{22}^{(1)}] = [0.7 \quad 0.3],$$

$$[r_2^{(1)}] = [r_{21}^{(1)} \quad r_{22}^{(1)}] = [200 \quad -400],$$

$$[p_1^{(2)}] = [p_{11}^{(2)} \quad p_{12}^{(2)}] = [0.6 \quad 0.4], \tag{61.8}$$

$$[r_1^{(2)}] = [r_{11}^{(2)} \quad r_{12}^{(2)}] = [400 \quad 200],$$

$$[p_2^{(2)}] = [p_{21}^{(2)} \quad p_{22}^{(2)}] = [0.8 \quad 0.2],$$

$$[r_2^{(2)}] = [r_{21}^{(2)} \quad r_{22}^{(2)}] = [100 \quad -800].$$

whence, in accordance with (16.1),

$$\bar{q}_1^{(1)} = 325, \quad \bar{q}_1^{(2)} = 320, \quad \bar{q}_2^{(1)} = 20 \quad \text{and} \quad \bar{q}_2^{(2)} = -80. \tag{61.9}$$

Let us make an arbitrary choice of the strategy $r = 1$ when we are in E_1, and $r = 2$ when in E_2.

Stage 1. We have

$$\gamma + w_1 = 325 + 0.5w_1 + 0.5w_2,$$
$$\gamma + w_2 = -80 + 0.8w_1 + 0.2w_2. \tag{61.10}$$

taking $w_1 = 0$, it follows that

$$w_2 = -311.53, \quad \gamma = 169.24. \tag{61.11}$$

Stage 2.

$$z_1^{(1)} = \bar{q}_1^{(1)} + p_{11}^{(1)}w_1 + p_{12}^{(1)}w_2$$

$$= 325 + (0.5)(-311.53) = 169.24,$$

$$z_1^{(2)} = \bar{q}_1^{(2)} + p_{11}^{(2)}w_1 + p_{12}^{(2)}w_2$$

$$= 320 + (0.4)(-311.53) = 195.24,$$

(61.12)

$$z_2^{(1)} = \bar{q}_2^{(1)} + p_{21}^{(1)}w_1 + p_{22}^{(1)}w_2$$

$$= 20 + (0.3)(-311.53) = -73.459,$$

$$z_2^{(2)} = \bar{q}_2^{(2)} + p_{21}^{(2)}w_1 + p_{22}^{(2)}w_2$$

$$= -80 + (0.2)(-311.53) = -142.30.$$

(61.13)

Stage 3.

$$z_1^{(r_1')} = \text{MAX}[169.24;\ 195.24]$$

$$= 195.24,\quad \text{hence}\quad r_1' = 2,$$

$$z_2^{(r_2')} = \text{MAX}[-73.459;\ -142.30]$$

$$= -73.459,\quad \text{hence}\quad r_2' = 1.$$

(61.14)

Stage 4.

Hence, we now choose the strategy $r = 2$ if we are in E_1 and $r = 1$ if we are in E_2.

$$\gamma' + w_1 = 320 + 0.6w_1 + 0.4w_2,$$

$$\gamma' + w_2 = 20 + 0.7w_1 + 0.3w_2.$$

(61.15)

Making $w_1 = 0$, it follows that

$$w_2 = -272.72 \quad \text{and}\quad \gamma' = 210.92.$$

We have $\gamma' > \gamma$ and go back to stage 2.

Stage 2'.

$$z_1^{(1)} = 325 + (0.5)(-272.72) = 188.64,$$

$$z_1^{(2)} = 320 + (0.4)(-272.72) = 210.92,$$

$$z_2^{(1)} = 20 + (0.3)(-272.72) = -61.81,$$

$$z_2^{(2)} = -80 + (0.2)(-272.72) = -134.54.$$

(61.16)

Stage 3'.

$$z_1^{(r_1'')} = \text{MAX}[188.64; \ 210.92] = 210.92, \quad \text{hence} \quad r_1'' = 2,$$

$$z_2^{(r_2'')} = \text{MAX}[-61.81; \ -134.54] = -61.81, \quad \text{hence} \quad r_2'' = 1.$$

(61.17)

Stage 4'. We choose $r = 2$ if we are in E_1, and $r = 1$ if we are in E_2, the same strategy as in stage 4 of the preceding repetition; we then use (61.15) and find the optimal strategy with the corresponding value $\gamma = 210.92$. This is the same result as we obtained in Section 16 (16.14), with a different notation.

62. Long Term Strategy When the Chain Is Not Fully Ergodic

A. ANALYTIC STUDY

Let us consider the case where matrix $[\mathcal{M}]$, representing a certain strategy, is not fully ergodic (as in 57.7), and let us use Eqs. (60.7):

$$(n + 1)\gamma + w_i = \bar{q}_i + \sum_{j=1}^{M} p_{ij}(n\gamma + w_j), \qquad i = 1, 2, ..., M, \quad (62.1)$$

but γ is not now a single unknown, for there is a γ for each state, and each state which is a part of the same recurring chain has the same γ. Hence, every γ_i depends on the initial state E_i (time 0), and (62.1) becomes

$$(n + 1)\gamma_i + w_i = \bar{q}_i + \sum_{j=1}^{w} p_{ij}(n\gamma_j + w_j), \qquad i = 1, 2, ..., M. \quad (62.2)$$

We are dealing with a system with $2M$ variables, γ_i, and w_i and M equations, but we can obtain $2M$ equations, for (62.2) can be written as

$$\gamma_i + n\left(\gamma_i - \sum_{j=1}^{M} p_{ij}\gamma_j\right) + w_i - \bar{q}_i - \sum_{j=1}^{M} p_{ij}w_j = 0, \qquad i = 1, 2, ..., M. \quad (62.3)$$

If these equations are to be independent of n (hence, checked for $n \to \infty$), we must have:

$$\gamma_i = \sum_{j=1}^{M} p_{ij}\gamma_j, \qquad i = 1, 2, ..., M, \quad (62.4)$$

$$\gamma_i + w_i = \bar{q}_i + \sum_{j=1}^{M} p_{ij}w_j, \qquad i = 1, 2, ..., M. \quad (62.5)$$

But as the matrix $([1] - [\mathcal{M}])$ is single, the $2M$ equations are not independent; in order to obtain a solution we must make an arbitrary choice L of the w_i quantities, where L is the number of recurring chains.

B. EXAMPLE

Given matrix (57.7) which we studied in Section 56:

$$[\mathcal{M}] = \begin{bmatrix} 1 & 0 & 0 \\ 0 & 1 & 0 \\ 0.2 & 0.5 & 0.3 \end{bmatrix}. \tag{62.6}$$

Let us assume the revenues are given by the matrix:

$$[\mathcal{R}] = \begin{bmatrix} 300 & 200 & 50 \\ -100 & 100 & 0 \\ 200 & -200 & -300 \end{bmatrix}, \tag{62.7}$$

in which case:

$$\{\bar{q}\} = \begin{Bmatrix} 300 \\ 100 \\ -150 \end{Bmatrix}. \tag{62.8}$$

The system of Eqs. (62.4) and (62.5) becomes:

$$\begin{aligned} \gamma_1 &= \gamma_1, \\ \gamma_2 &= \gamma_2, \\ \gamma_3 &= 0.2\gamma_1 + 0.5\gamma_2 + 0.3\gamma_3; \end{aligned} \tag{62.9}$$

$$\begin{aligned} \gamma_1 + w_1 &= 300 + w_1, \\ \gamma_2 + w_2 &= 100 + w_2, \\ \gamma_3 + w_3 &= -150 + 0.2w_1 + 0.5w_2 + 0.3w_3. \end{aligned} \tag{62.10}$$

It follows:

$$\gamma_1 = 300, \qquad \gamma_2 = 100, \qquad \gamma_3 = 1100/7 \tag{62.11}$$

Let us take

$$w_1 = w_2 = 0, \qquad \text{then} \quad w_3 = -260. \tag{62.12}$$

Direct use can be made of formula (56.26), but matrix

$$[\tilde{\mathcal{M}}] = \lim_{n \to \infty} [\mathcal{M}]^n$$

is not formed of n identical lines; nevertheless, the reasoning given in Section 56 is still valid:

$$\{\gamma\} = [\tilde{\mathscr{M}}]\{q\} = \begin{bmatrix} 1 & 0 & 0 \\ 0 & 1 & 0 \\ \frac{2}{7} & \frac{5}{7} & 0 \end{bmatrix} \begin{pmatrix} 300 \\ 100 \\ -150 \end{pmatrix} = \begin{pmatrix} 2100/7 \\ 700/7 \\ 1100/7 \end{pmatrix}. \qquad (62.13)$$

C. ANOTHER EXAMPLE (Fig. 62.1)

$$[\mathscr{M}] = \begin{bmatrix} 1 & 0 & 0 & 0 \\ 0.3 & 0.4 & 0.3 & 0 \\ 0 & 0 & 0.7 & 0.3 \\ 0 & 0 & 0.2 & 0.8 \end{bmatrix}. \qquad (62.14)$$

There are two recurring chains: I, formed by E_1, and II, formed by E_3 and E_4.

Assuming that each q_i is given by

$$\{q\} = \begin{Bmatrix} 1 \\ 2 \\ 3 \\ -2 \end{Bmatrix}, \qquad (62.15)$$

Eqs. (62.4) and (62.5) become

$$\begin{aligned} \gamma_1 &= \gamma_1 \\ \gamma_2 &= 0.3\gamma_1 + 0.4\gamma_2 + 0.3\gamma_3 \\ \gamma_3 &= 0.7\gamma_3 + 0.3\gamma_4 \\ \gamma_4 &= 0.2\gamma_3 + 0.8\gamma_4 \end{aligned} \qquad (62.16)$$

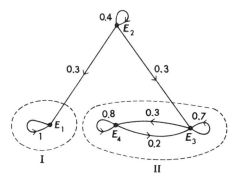

FIG. 62.1.

and

$$
\begin{aligned}
\gamma_1 + w_1 &= 1 + w_1 \\
\gamma_2 + w_2 &= 2 + 0.3w_1 + 0.4w_2 + 0.3w_3 \\
\gamma_3 + w_3 &= 3 + 0.7w_3 + 0.3w_4 \\
\gamma_4 + w_4 &= -2 + 0.2w_3 + 0.8w_4 .
\end{aligned}
\tag{62.17}
$$

There are two recurring chains, and with an arbitrary choice of $w_1 = w_2 = 0$; it follows that

$$
\begin{aligned}
\gamma_{\mathrm{I}} &= \quad \gamma_1 = 1, &\quad w_1 &= 0, \\
& \qquad \gamma_2 = 0.50, &\quad w_2 &= 0, \\
\gamma_{\mathrm{II}} &= \begin{cases} \gamma_3 = 0, \\ \gamma_4 = 0, \end{cases} &\quad \begin{aligned} w_3 &= -5, \\ w_4 &= -15. \end{aligned}
\end{aligned}
\tag{62.18}
$$

$\gamma_3 = \gamma_4$ is the result of belonging to the same chain; the fact that they are both zero is a special case due to the numbers chosen for (62.15).

In Howard's work [16, p. 63] the reader will find a variant of the repetitive method given in Section 60 for the case of a matrix which is not fully ergodic.

63. Discounted Value of the Revenue

Let us assume that a rate of interest α for each period is introduced into a stationary procedure under uncertainty; formulas (58.1, 58.6, and 59.1), which could easily be extended to nonstationary cases, would then become:

A. D.P. PROCESS

$$
f_i(N-n, N) = \operatorname*{OPT}_{E_r' \in \Gamma_D E_i} \left\{ a_{ir} + \sum_{E_j \in \Gamma_H E_r'} p_{rj} \left[b_{rj} + \frac{1}{1+\alpha} f_j(N-n+1, N) \right] \right\},
$$
$$
n = 1, 2, ..., N. \tag{63.1}
$$

B. P.D. PROCESS

$$
\bar{g}_r(N-n, N) = \sum_{E_j \in \Gamma_H E_r'} \left\{ p_{rj} c_{rj} + \operatorname*{OPT}_{E_j' \in \Gamma_D E_j} [d_{js} + \bar{g}_s(N-n+1, N)] \right\},
$$
$$
n = 1, 2, ..., N. \tag{63.2}
$$

C. DESCRIPTION BY MEANS OF A MARKOVIAN DECISION CHAIN

$$\bar{v}_i(N - n, N) = \operatorname*{OPT}_{r=1,2,\ldots,m} \left[\bar{q}_i^{(r)} + \frac{1}{1 + \alpha} \sum_{j=1}^{M} p_{ij}^{(r)} \bar{v}_j(N - n + 1, N) \right],$$

$$n = 1, 2, \ldots, N. \quad (63.3)$$

It is interesting to follow the development of the equations which rule the Markovian chains with revenue where discounting enters. If the process is not *decisional* (Section 56), Eqs. (56.6) become

$$\{\bar{v}(n + 1)\} = \{\bar{q}\} + \frac{1}{1 + \alpha}[M]\{\bar{v}(n)\}, \qquad n = 0, 1, 2, \ldots, N - 1 \quad (63.4)$$

or by taking $r = 1/(1 + \alpha)$:

$$\{\bar{v}(n + 1)\} = \{\bar{q}\} + r[M]\{\bar{v}(n)\}. \quad (63.5)$$

Thence

$$\frac{1}{z}\{\bar{v}^*(z) - \bar{v}(0)\} = \frac{1}{1 - z}\{\bar{q}\} + r[M]\{\bar{v}^*(z)\}. \quad (63.6)$$

From these equations we obtain

$$\{\bar{v}^*(z) - \bar{v}(0)\} = \frac{z}{1 - z}\{\bar{q}\} + rz[M]\{\bar{v}^*(z)\}. \quad (63.7)$$

$$([1] - rz[M])\{\bar{v}^*(z)\} = \frac{z}{1 - z}\{\bar{q}\} + \{v(0)\}. \quad (63.8)$$

Let us assume

$$[\mathscr{C}(rz)] = [1] - rz[M]. \quad (63.9)$$

It follows that

$$\{\bar{v}^*(z)\} = \frac{z}{1 - z}[\mathscr{C}(rz)]^{-1}\{\bar{q}\} + [\mathscr{C}(rz)]^{-1}\{\bar{v}(0)\}. \quad (63.10)$$

Hence, we have obtained exactly the same result as in (56.10) except that in $[\mathscr{C}(z)]$, z would be replaced by rz.

D. EXAMPLE

For this we shall use (56.11, 56.12, and 56.12′), taking $r = 0.9$. It follows:

$$[\mathscr{C}(rz)]^{-1} = \frac{1}{1-0.9z}\begin{bmatrix} \frac{7}{12} & \frac{5}{12} \\ \frac{7}{12} & \frac{5}{12} \end{bmatrix}$$

$$+ \frac{1}{1+0.18z}\begin{bmatrix} \frac{5}{12} & -\frac{5}{12} \\ -\frac{7}{12} & \frac{7}{12} \end{bmatrix}. \tag{63.11}$$

$$\{\bar{v}^*(z)\} = \left(\frac{z}{(1-z)(1-0.9z)}\begin{bmatrix} \frac{7}{12} & \frac{5}{12} \\ \frac{7}{12} & \frac{5}{12} \end{bmatrix}\right.$$

$$+ \frac{z}{(1-z)(1+0.18z)}\begin{bmatrix} \frac{5}{12} & -\frac{5}{12} \\ -\frac{7}{12} & \frac{7}{12} \end{bmatrix}\right)\begin{Bmatrix} 325 \\ 20 \end{Bmatrix}$$

$$= \left(\frac{10}{1-z}\begin{bmatrix} \frac{7}{12} & \frac{5}{12} \\ \frac{7}{12} & \frac{5}{12} \end{bmatrix} - \frac{10}{1-0.9z}\begin{bmatrix} \frac{7}{12} & \frac{5}{12} \\ \frac{7}{12} & \frac{5}{12} \end{bmatrix}\right.$$

$$+ \frac{1/1.18}{1-z}\begin{bmatrix} \frac{5}{12} & -\frac{5}{12} \\ -\frac{7}{12} & \frac{7}{12} \end{bmatrix}$$

$$\left. - \frac{1/1.18}{1+0.18z}\begin{bmatrix} \frac{5}{12} & -\frac{5}{12} \\ -\frac{7}{12} & \frac{7}{12} \end{bmatrix}\right)\begin{Bmatrix} 325 \\ 20 \end{Bmatrix}. \tag{63.12}$$

Thence:

$$\{\bar{v}(n)\} = \left(10\begin{bmatrix} \frac{7}{12} & \frac{5}{12} \\ \frac{7}{12} & \frac{5}{12} \end{bmatrix} + \frac{1}{1.18}\begin{bmatrix} \frac{5}{12} & -\frac{5}{12} \\ -\frac{7}{12} & \frac{7}{12} \end{bmatrix}\right.$$

$$\left. - 10(0.9)^n\begin{bmatrix} \frac{7}{12} & \frac{5}{12} \\ \frac{7}{12} & \frac{5}{12} \end{bmatrix} - \frac{1}{1.18}(-0.18)^n\begin{bmatrix} \frac{5}{12} & -\frac{5}{12} \\ -\frac{7}{12} & \frac{7}{12} \end{bmatrix}\right)\begin{Bmatrix} 325 \\ 20 \end{Bmatrix} \tag{63.13}$$

When $k \to \infty$, it follows that

$$\{\bar{v}(\infty)\} = \left(\begin{bmatrix} \frac{70}{12} & \frac{50}{12} \\ \frac{70}{12} & \frac{50}{12} \end{bmatrix} + \begin{bmatrix} 4.24/12 & -4.24/12 \\ -5.93/12 & 5.93/12 \end{bmatrix}\right)\begin{Bmatrix} 325 \\ 20 \end{Bmatrix}$$

$$= \frac{1}{12}\begin{bmatrix} 74.24 & 45.76 \\ 64.07 & 55.93 \end{bmatrix}\begin{Bmatrix} 325 \\ 20 \end{Bmatrix} = \begin{Bmatrix} 2086.9 \\ 1828.4 \end{Bmatrix}. \tag{63.14}$$

We conclude that, if $r < 1$, $\{\bar{v}(n)\}$ has a limit as $n \to \infty$.

Returning to the general case, we decided in Section 56 that a matrix such as $[\mathscr{C}(z)]^{-1}$ has at least one value of its own, $z = 1$, where a rate of interest is not introduced. Hence, in a case of discounting where the coefficient $r = 1/(1 + \alpha)$, there must be at least one value $z = 1/r$, since rz has been substituted for r, whereas there can be no value $z = 1$. Hence, all the values of the matrix have a modulus less than 1, and can be single or multiple. If we return to (63.10) we see that the first part of the right member contains a denominator with a single factor $(1 - z)$, the second part contains no such factor. By consulting the table of z-transforms, and in particular (54.15, 54.19, and 54.20), we see that the reciprocal transform of (63.10) contains a constant term, whereas all the other terms will approach 0 as $n \to \infty$. Hence, if $r < 1$ (in other words, if $\alpha > 0$), there will always be a limit, and the process with an introduced rate of interest is always convergent, whatever the nature of $[M]$; in every case, the limit of $\{\bar{v}(n)\}$ will depend on the initial state.[1]

If we now consider a decision process it can be proved [18] that, for any strategy, there is a limit $\{\bar{v}(\infty)\}$, and that the system is always convergent.

We have not sufficient space to give Howard's method of repetitive optimization for cases of an introduced rate, but the reader will find it in Reference [16].

It should be noted that the sequences of strategy may be different if we compare the same case with and without discounting.

E. Observation Regarding the Numerical Calculation of Sequential Procedures

While the reduction of a discrete Markovian decision process into a decision chain is very useful as an analytic exercise, both as regards the numerical calculations and the search for optimal strategies, it is more convenient to reduce it to a D. P. procedure. In the latter, the treatment is very simple and the programming considerably easier.

---

[1] Since $\{\bar{v}(\infty)\}$ is always finite if $r < 1$,

$$\lim_{n \to \infty} \frac{v(n)}{n} = 0,$$

and the criterion of average value has no meaning in a case of an introduced rate; hence, it is replaced by that of the actual value.

64. Applying Dynamic Programming to a Real Problem[1]

A. MANAGEMENT OF STOCK AND CALCULATION OF
   OPTIMAL STORAGE CAPACITY

Basically, the problem to be studied was whether there was sufficient warehouse space at a port used for its exports by a firm which had undertaken a limited expansion program, a program which would reach its peak in 1961. If not, new warehouses were to be built, and the amount of stock which they would absorb had to be calculated.

Very quickly, the study revealed that the problem could be reduced to the scientific management of stocking, whereby the storage capacity would be used to optimal advantage, and investments in buildings and handling would be reduced to a minimum.

The study of the model confirmed that the rules for stocking which had been found rendered any increase of warehouse space unnecessary.

The model which was used had these characteristics:

(1)   The process was nonstationary;

(2)   An entire absence of assumption as to the type of decision with regard to the replenishment of the stock.

The dynamic program which was used produced an optimal policy for restocking, thus demonstrating the effectiveness of this method for nonstationary procedures.

B. THE PROBLEM

A well-known manufacturer plans to have a total warehouse capacity S at the port from which his goods are shipped overseas. The transhipments X_i, which will be exported during a given montly period i, are uncertain, though the normal law for the exports is known from the past statistics, and is expected to be valid for the next five or six years.

The goods are sent to the warehouses from a central depot several hundred miles from the port, and two methods of transportation are used. The normal method, which is cheap but subject to delays, is by water, and the other, which is more expensive but at the same time quicker and more dependable, and is therefore used in cases of urgency, is by rail. The latter is taken as being twice as expensive as the former.

---

[1] An article by M. E. Venture, Director of S.E.P.R.O. in France, which won the American Lanchester Prize in 1961 and which he has kindly given us permission to reproduce. It appeared in *Operational Research Quarterly*, Vol. 2, No. 1.

Transportation by water has a number of uncertain factors which are mainly due to the varying tonnages shipped at different seasons of the year, and when there is a shortage of barges the surplus merchandise has to be transported by rail in spite of the extra cost.

The problem to be solved is whether an increase of storage space at the docks, which would be filled up when there were plenty of barges available, would obviate the whole or a part of the greater cost of transport by rail.

What has to be found, in fact, is the optimal storage space which would minimize the over-all cost of transportation and warehousing in the course of a year.

C. THE ECONOMIC FUNCTION

For the expected value over a yearly period, the following terms will be used:

M the annual profit due to the increased storage space;

R the annual cost of shipment by water;

F the annual cost of shipment by rail:

D the eventual extra cost of handling due to greater stocking and unstocking;

A the interest on the capital represented by the value of the stock.

There is a final term to be added: the penalty which is exacted when the warehouses are filled, and new goods arriving by water have to be kept in the barges. The cost of such extra storage for a year will be called P.

We must now examine these functions of $\hat{S}$ (the optimal capacity), adding S_i as the stock at the beginning of the period; unless otherwise stated, periods i will be numbered in the direction of the past.

D. BUILDING NEW WAREHOUSES (M)

The importance of adequate warehouses, as well as considerations of topography, led the company to decide that the only satisfactory plan for increasing the storage space would be to add either one or two new buildings of approximately the same size and cost. The term $M(\hat{S})$ is therefore shown as a noncontinuous function with three stair-heads (Fig. 64.1). There is, in fact, no difficulty in calculating the annual allowance for the supplementary capacity alone.

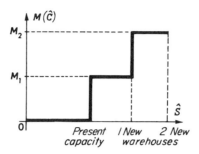

FIG. 64.1.

Hence, we have been able to adopt the simple approach of calculating the over-all annual cost of management for the three hypotheses as to the value $\hat{S}$ (0, 1, or 2 extra warehouses) so as to choose the one with the minimal cost.

E. VALUE OF THE STOCK (A)

To calculate this value the firm was not being judicious in valuing the stock at its cost price at the port. The merchandise could, in fact, be stored either at the port or at the central warehouse. Being stored at the port merely increased the capital which was tied up, since the value of a ton delivered to the port was augmented by the cost of transportation. Hence, it is the annual allowance with a given charge θ of the unit cost r of the normal transportation by water which has been taken as the value of immobilization in the warehouse. We have made the classic assumption that the volume of stock S varied in a linear manner over a monthly period; for the period i, where it varied from S_i to S_{i-1}, the cost of immobilization in the warehouse will therefore amount to

$$\tfrac{1}{2}\theta r_i(S_i + S_{i-1}). \tag{64.1}$$

F. COST OF HANDLING

We make the assumption, which is confirmed by the facts, that if over the period i, the supplies Q_i brought to the port exceed the shipments from it X_i, the difference is added to the stock, with the excess d in the unit cost of handling which this causes; whereas the portion of the incoming supplies which is equal to the transhipments is loaded directly from the barges onto the freighters. It is the same with regard to removals from the stock, if the incoming merchandise is less than the outgoing.

Hence, the expected cost of the handling for period i is

$$D = d \left[\int_{-\infty}^{0} (Q_i - X_i) f_i(X_i - Q_i) \, d(X_i - Q_i) \right.$$

$$\left. + \int_{0}^{\infty} (X_i - Q_i) f_i(X_i - Q_i) \, d(X_i - Q_i) \right], \qquad (64.2)$$

where $f_i(X_i - Q_i)$ is the density of the probability of the difference $X_i - Q_i$ for the period i.

A rough estimate shows that 10–15% of the merchandise is subject to this extra handling, which amounts to some 5–6% of the cost of transportation by water. The effect of factor D will therefore amount to less than 1% of the overall cost of management.

G. Transportation by Rail (F)

Taking into account its cost and its rapidity, we consider this as a *penalty* which only intervenes when the stock at the port has fallen to zero and freighters are awating shipments, in other words, when, during period i,

$$X_i > Q_i + S_i \quad (S_i \text{ being the stock at the beginning of the period}). \quad (64.3)$$

This is equivalent to transporting by rail the difference $X_i - (Q_i + S_i)$ with a unit cost f.

Hence, the expected value of the cost for a period is

$$F = f \int_{S_i}^{\infty} (X_i - Q_i - S_i) f_i(X_i - Q_i) \, d(X_i - Q_i). \qquad (64.4)$$

H. Transportation by Water (R)

We have mentioned that this involves uncertainties of an important kind, and an enquiry has revealed that the principal causes of a periodic shortage of barges are as follows:

(a) The shallow water which prevents the barges being loaded to capacity;
(b) Ice and fog which slow down the turn-round of the barges;
(c) Long winter nights, for the barges only operate by day;
(d) Traffic jams.

An extended study of these factors, particularly of the first, would have caused much delay and might have involved considerable difficulties. Instead, therefore, of studying the physical *causes* of the problem, the *consequences* of these external conditions were considered.

The chartering of barges is carried out by the firm giving the barge owners a statement of the tonnage which it will require at the beginning of each month. Statistics show that in the summer these requirements are met, but that in the winter there is often a more or less serious margin between the needs and the tonnage available. Let us take C_i for the required tonnage for month i, and Q_i as the tonnage available for this month.

A *retrogressive* comparison between the deficit $C - Q$ and the *depth of water* proved very rewarding, since a certain critical depth H_0 in the river has to be reached if the deficit is to be zero. In this retrogressive study, the daily heights of the river were reduced to the level H_0, all those above it being given this symbol, and the monthly average $\bar{H}_x$ of these reduced heights H_x was then calculated. The coefficient correlating the deficit for month $i(C_i - Q_i)$ and the difference $H_0 - \bar{H}_x = \Delta H$ is more than 0.70, which accounts for more than half of the variance of the deficit, the other factors (traffic jams, ice, etc.) being responsible for the remainder.

In addition, it is a well-known fact that the average height of a river one month is generally related to its average height the previous one. The regressive relation between the variable ΔH for any given month and that for the preceding month was accordingly calculated, using the statistics for the last 60 years, and provided 12 regressions of the type

$$\Delta H_{\text{April}} = a_{\text{April}} \, \Delta H_{\text{March}} + \epsilon_{\text{April}} t, \tag{64.5}$$

where

  a is the regressive coefficient for April;

  ϵ^2 the residual variance for this month;

  t a reduced Laplace-Gauss distributed variable, having regard to the many different causes which can affect the height of a river.

Armed with these equations, the company was in a position to determine from the height of water the previous month, the deficit that could be anticipated for the current one.

If its transportation requirements were $\bar{Q}_i$, all it need do was to fix the desired tonnage at the level:

$$C_i - \bar{Q}_i = \alpha_i \, \Delta H_{i-1} + \beta_i, \tag{64.6}$$

where the coefficients α_i, β_i have been determined by the above-mentioned considerations. The real value of Q_i will then differ from $\bar{Q}_i$ by an uncertain value U_i with a zero average.

The term representing transportation by water in the economic function is therefore expressed as $r_i \bar{Q}_i$, where $\bar{Q}_i$ is the average quantity and r_i is the rate for freight for the period considered, this rate being subject to certain complicated regulations. Hence, for the period i, R could be given the expected value:

$$R = r_i[C_i - (\alpha_i \Delta H_{i-1} + \beta_i)]. \tag{64.7}$$

I. Storage of Surplus (F)

The penalty to be paid when the stock at the port exceeds the capacity $\hat{S}$ of the warehouses is to be forced to rent surplus space at a rate per ton-day for a minimum period. With the given conditions, the following approximation seems fully justified:

If the stock at the beginning of a period is $S_i > \hat{S}$, the surplus $S_i - \hat{S}$ is stored on the barges and costs $c(S_i - \hat{S})$ over a period, whether or not the stock falls below $\hat{S}$ during this period.

J. Over-All Cost for a Period (L_i)

The introduction of two new variables helps us to simplify the formulas.

(1) Y_i, the available funds for a period when the requirements have been fully satisfied; that is, $C_i + S_i$;

(2) Z_i, the fictitious removals (exports + deficit). Hence:

$$Y_i = C_i + S_i, \tag{64.8}$$

$$Z_i = X_i + C_i - Q_i = X_i + \alpha_i \Delta H_{i-1} + \beta_i + U_i. \tag{64.9}$$

This makes it possible to combine the two uncertain variables X and U as a single uncertain variable Z. The latter will be treated as a Laplace-Gauss variable, since it is the sum of two uncertain independent variables, one of which is strictly of this type and the other almost so, owing to the large number of unconnected causes which contribute to the shortage of barges. $g_i(z_i)$ and $G_i(z_i)$ will be, respectively, the density and cumulative probability of this variable z_i during period i, with an average $\bar{Z}_i$, and with standard deviation σ_i.

With these new variables, $S_i - (X_i - Q_i)$, which represents the stock at the end of a period, becomes $Y_i - Z_i$, and we obtain the following economic function for period i:

$$L_i(S_i) = r_i[C_i - (\alpha_i \Delta H_{i+1} + \beta_i)]$$

$$+ f \int_{y_i}^{\infty} (z_i - y_i) g_i(z_i) \, dz_i$$

$$+ c \int_0^{y_i - \hat{S}} (y_i - \hat{S} - z_i) g_i(z_i) \, dz_i$$

$$+ d \left[\int_0^{y_i - S_i} (y_i - S_i - z_i) g_i(z_i) \, dz_i \right.$$

$$\left. + \int_{y_i - S_i}^{\infty} (S_i - y_i + z_i) g_i(z_i) \, dz_i \right] + \tfrac{1}{2}\theta\bar{r}(S_i + y_i - z_i). \quad (64.10)$$

The precise significance of the three variables is clear. S is the state variable representing the passage from one period to another; Z is the uncertain variable which depends on the orders and the uncertainties of navigation on the waterways; and Y is the decision variable which provides the person who is responsible for chartering the barges, and who knows the stock S_i at the beginning of the period, with the value C_i which enables him to charter the optimal tonnage from the barge owners.

K. CHOICE OF THE PERIOD. NONSTATIONARY CHARACTER OF THE PROCESS

The fact that the company estimates its sales X_i for a monthly period, and charters the barges at the beginning of each month, has led us to choose a month as the elementary period. Hence, it is necessary to calculate the optimal value of the variable Y_i each month, and this, together with the state of the stock S_i at the beginning of the month, makes it possible to calculate C_i for each monthly period.

L. THE SEQUENTIAL PROGRAM

Since $L_i(y_i, S_i)$ is the value of the economic function for a period i, we shall take $F_i(S_i)$ for its cumulative value for the periods, $i, i - 1$, $i - 2,..., 3, 2, 1$ (the values being discounted by the coefficient α). In accordance with the fundamental theorem of dynamic programming for a case of uncertainty, the minimum for F_i is obtained provided the optimum management is realized for the set of subsequent periods. Hence we have

$$F_i(S_i) = \underset{y_i}{\text{MIN}}[L_i(y_i, S_i) + \alpha\overline{F_{i-1}(S_{i-1})}], \quad (64.11)$$

where $\bar{F}_{i-1}$ is the expected value of F_{i-1} for all possible values of S_{i-1}. Since $S_{i-1} = y_i - z_i$, this can be expressed as

$$F_i(S_i) = \underset{y_i}{\mathrm{MIN}} \left[L_i(y_i, S_i) + \alpha \int_0^{y_i} F_{i-1}(y_i - z_i) g_i(z_i) \, dz_i \right.$$

$$\left. + \alpha \int_{y_i}^{\infty} F_{i-1}(0) g_i(z_i) \, dz_i \right], \qquad (64.12)$$

this minimum satisfying the relation $C_i \geqslant 0$ if $y_i \geqslant S_i$.

The minimum will be obtained for a value y_i^* such that $\partial F_i / \partial y_i = 0$. Let

$$\frac{\partial L_i}{\partial y_i} + \alpha \int_0^{y_i} \frac{\partial F_{i-1}(y_i - z_i)}{\partial y_i} g_i(z_i) \, dz_i + \alpha F_{i-1}(0) g_i(z_i) - \alpha F_{i-1}(0) g_i(z_i) = 0,$$
$$(64.13)$$

or

$$\frac{\partial L_i}{\partial y_i} + \alpha \int_0^{y_i} \frac{\partial}{\partial y_i} [F_{i-1}(y_i - z_i)] g_i(z_i) \, dz_i = 0. \qquad (64.14)$$

When the process reaches the last period to be considered, and the residual stock is S_0, it is reasonable to assume that the administrative cost for the following period, which will be confined to disposing of S_0, will contain a term proportional to S_0, and eventually a constant.

With this premise, let us find the conditions under which functions $F_i(S_i)$, which are linear for S_i, are the solution of Eq. (64.12). Let

$$F_i(S_i) = p_i S_i + F_i(0), \qquad (64.15)$$

whence:

$$F_{i-1}(y_i - z_i) = p_{i-1}(y_i - z_i) + F_{i-1}(0). \qquad (64.16)$$

The derivative becomes

$$\frac{\partial L_i}{\partial y_i} + \alpha p_{i-1} \int_0^{y_i} g_i(z_i) \, dz_i = \frac{\partial L_i}{\partial y_i} + \alpha p_{i-1} G_i(y_i). \qquad (64.17)$$

The expression (64.10) of $L_i(y_i, S_i)$ gives, by replacing C_i by $y_i - S_i$:

$$\frac{\partial L_i}{\partial y_i} = r_i - f[1 - G_i(y_i)] + c G_i(y_i - \hat{S}) + 2d \, G_i(y_i - S_i) - d + \tfrac{1}{2}\theta\bar{r}, \quad (64.18)$$

which enables us to calculate y_i, taking into account the relation (64.14):

$$(r_i - d + \tfrac{1}{2}\theta\bar{r} - f) + f G_i(y_i) + c G_i(y_i - \hat{S}) + 2d \, G_i(y_i - S_i)$$
$$+ \alpha p_{i-1} G_i(y_i) = 0. \qquad (64.19)$$

This relation has a solution y^* if

$$r_i - d + \tfrac{1}{2}\theta\bar{r} - f < 0,$$

which is certainly the case, given the importance of f.

The second derivative is positive:

$$fg_i(y_i) + cg_i(y_i - \hat{S}) + 2d\,g_i(y_i - S_i) + \alpha p_{i-1} g_i(y_i) > 0, \qquad (64.20)$$

since all these terms are positive; also, as the G_i terms are monotone increasing functions, we have only one *minimum*, and *one only*.

Hence, for the optimum, we can state

$$F_i(S_i) = L_i(y_i^*, S_i) + \alpha F_{i-1}(0) + \alpha p_{i-1} \int_0^{y_i^*} (y_i^* - z_i)\,g_i(z_i)\,dz_i, \qquad (64.21)$$

the two last terms being independent of S_i.

On the other hand, $L_i(y_i^*, S_i)$ is expressed as

$$L_i(y_i^*, S_i) = -r_i S_i + d \int_0^{y_i^* - S_i} (y_i^* - z_i)\,g_i(z_i)\,dz_i$$

$$+ d \int_{y_i^* - S_i}^{\infty} (S_i - y_i^* + z_i)\,g_i(z_i)\,dz_i + L_i(y_i^*, 0) \qquad (64.22)$$

and with this form of L_i, $F_i(S_i)$ is not linear in S_i. Indeed, we conclude that the degree of F_i in S_i is that of L_i.

Calculation shows that the sensitiveness of y^* to d is very weak and that the cost of handling is comparatively low. This allows us to suppress d in L_i with the great advantage that we now obtain a linear expression in S_i for L_i, and in consequence for F_i.

$$L_i(y_i^*, S_i) = -r_i S_i + L_i(y_i^*, 0), \qquad (64.23)$$

and

$$F_i^*(S_i) = -r_i S_i + F_i^*(0), \qquad (64.24)$$

which enables us to say $p_i = -r_i$.

Hence, the equation which gives y_i^* is

$$(r_i + \theta\bar{r}/2 - f) + (f - \alpha r_{i-1})G_i(y_i) + cG_i(y_i - \hat{S}) = 0 \qquad (64.25)$$

or

$$G_i(y_i) + \frac{c}{f - \alpha r_{i-1}} G_i(y_i - \hat{S}) = \frac{f - r_i - \tfrac{1}{2}\theta\bar{r}}{f - \alpha r_{i-1}}, \qquad (64.26)$$

again, using the reduced variable,

$$v_i = \frac{z_i - \bar{z}_i}{\sigma_i} \qquad (64.27)$$

and by assuming

$$\mu_i = \frac{c}{f - \alpha r_{i-1}} \qquad (64.28)$$

and

$$\lambda_i = \frac{f - r_i - \theta \bar{r}/2}{f - \alpha r_{i-1}} , \qquad (64.29)$$

$$G(v_i) + \mu G(v_i - \hat{S}/\sigma_i) = \lambda_i . \qquad (64.30)$$

The solution can be obtained graphically by constructing the curve $G(v_i)$, then the same displaced curve of S/σ_i towards the right of the v axis and reduced by the coefficient $-\mu_i$; all that we now need is the abscissa $v_i{}^*$, where the distance between the two curves is equal to λ_i (Fig. 64.2).

Each month the person responsible for the management has to calculate the value of $\bar{z}_i$, adding to his estimate of exports $\bar{X}_i$, the probable shortage of barges:

$$C_i - Q_i = \alpha_i \Delta H_{i+1} + \beta_i ,$$

where α_i and β_i are found from the regressions, and ΔH_{i+1} from the average for the previous month $(i + 1)$ of the heights of water reduced to H_0 .

M. Annual Cost of Administration

Since we have obtained the optimal value $y_i{}^*$ for the variable y_i , we can now calculate the expected value of the firm's operating costs for a period, and hence for a year. If this is worked out for the different

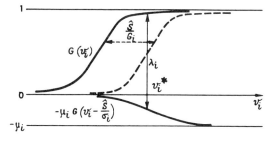

FIG. 64.2.

assumptions as to the value $\hat{S}$ of the storage capacity at the port, we shall be in a position to decide between these assumptions and determine whether it is advisable to build one or two new warehouses.

This annual cost will be $\Gamma(\hat{S})$:

$$\Gamma(\hat{S}) = \sum_{i=\text{January}}^{\text{December}} \left\{ r_i[C_i - (\alpha_i \Delta H_{i+1} + \beta_i)] \right.$$

$$+ f \int_{y_i^*}^{\infty} (z_i - y_i^*)\, g(z_i)\, dz_i + c \int_0^{y_i^* - \hat{S}} (y_i^* - z_i - \hat{S})\, g_i(z_i)\, dz_i$$

$$\left. + \tfrac{1}{2}\theta\bar{r}(S_i + y_i^* - z_i) \right\}. \tag{64.31}$$

To this annual cost which can be expressed as

$$\Gamma(\hat{S}) = \sum_{i=\text{January}}^{\text{December}} \left\{ r_i C_i - r_i \alpha_i \Delta H_{i+1} - r_i \beta_i + f\sigma_i\, g(v_i^*) \right.$$

$$+ f(\bar{z}_i - y_i^*)[1 - G(v_i^*)] + c\sigma_i\, g(v_i^* - \hat{S}/\sigma_i) - c\sigma_i\, g(-\bar{z}/\sigma_i)$$

$$\left. + c(y_i^* - z_i - \hat{S})[G(v_i^* - \hat{S}/\sigma_i) - G(-\bar{z}/\sigma_i)] + \tfrac{1}{2}\theta\bar{r}(S_i + y_i^* - z_i) \right\}, \tag{64.32}$$

we must eventually add M_1 or M_2, representing the annual allowance for the construction of one or two new warehouses.

As documentation for our case, we now give graphic representations of the level of stock for the three possible assumptions as to the value of $\hat{S}$, and of the total operating costs which would result from taking a monthly average for the depth of water based on the statistics for the last 50 years (Figs. 64.3 and 64.4).

The graph of the operating costs provides striking evidence. The existing situation $\hat{S}_0$ is definitely more advantageous than $\hat{S}_1$ or $\hat{S}_2$, under which one or two new warehouses would be built. It is, of course, obvious that this is only true when the optimal management y_i^* is carried out. We notice, in particular, that in $\hat{S}_0$ the tonnage transported by rail is extremely low.

N. STABILITY OF THE SOLUTION

The solution $\hat{S}_0$, which avoids the contruction of new warehouses, results in much lower *average* costs. Nevertheless, the possibility remains that under exceptional conditions, such as persistently low water in

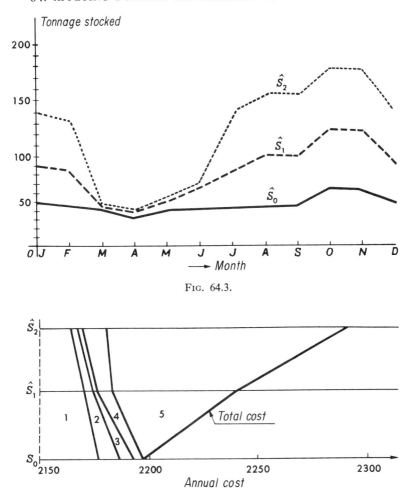

FIG. 64.3.

FIG. 64.4. 1, Cost of transportation by water; 2, cost of rail transportation; 3, cost of storing surplus; 4, cost of stock; 5, annuities for additional warehouses.

the river, a high expenditure on rail transportation might be incurred which could have been avoided if extra storage space had been available.

To test this possibility, a simulation was carried out based on the heights of water during a year when they were exceptionally low, whereas the export figures were retained at the estimated monthly average. In this simulation, the rail costs were obviously much greater, leading to a considerably higher cost of management. But even under these exceptional conditions, the building of new warehouses would not have proved economic, and the solution is therefore stable under all circumstances.

O. THE EFFECT OF PARAMETERS

The optimal management is represented by the monthly value of the decision variable $y_i{}^*$, which was the solution of Eq. (64.25).

$$G_i(y_i) + \frac{C}{f - \alpha r_{i-1}} G_i(y_i - \hat{S}) = \frac{f - r_i - \theta \bar{r}/2}{f - \alpha r_{i-1}}. \tag{64.33}$$

It is interesting to notice how the value $y_i{}^*$, the solution of (64.33), is affected by variations of the different parameters C, r_i and r_{i-1}, f and σ_i (implicit in the function G_i).

Let us turn back to Eq. (64.30), which gives $v_i{}^*$ and $y_i{}^*$,

$$G(v_i{}^*) + \mu_i G(v_i - \hat{S}/\sigma_i) = \lambda_i. \tag{64.34}$$

The total differential is expressed as

$$g_i(v_i)\, dv_i + \mu_i\, g(v_i - \hat{S}/\sigma_i)\, d(v_i - \hat{S}/\sigma_i) + G(v_i - \hat{S}/\sigma_i)\, d\mu_i = d\lambda_i. \tag{64.35}$$

We note that
$$d(v - \hat{S}/\sigma) = dv - d\hat{S}/\sigma + \hat{S}\, d\sigma/\sigma^2. \tag{64.36}$$

Knowing that

$$\mu_i \approx \frac{c}{f - r_{i-1}} \tag{64.37}$$

and

$$\lambda_i \approx \frac{f - r_i}{f - r_{i-1}} = \frac{f - r_{i-1}(1 + K)}{f - r_{i-1}}, \tag{64.38}$$

with

$$K = \frac{r_i}{r_{i-1}} - 1 \tag{64.39}$$

$$\begin{aligned} d\mu &= \frac{dc}{f - r_{i-1}} - \frac{c\, df}{(f - r_{i-1})^2} + \frac{c\, dr_{i-1}}{(f - r_{i-1})^2} \\ &= \frac{2}{C}\left(\frac{dC}{\mu} - df + dr_{i-1}\right) \end{aligned} \tag{64.40}$$

$$d\lambda = K \frac{r_{i-1}\, df - f\, dr_{i-1}}{(f - r_{i-1})^2} = K \frac{2}{C^2}(r_{i-1}\, df - f\, dr_{i-1}), \tag{64.41}$$

whence

$$g(v_i)\, dv_i + \mu_i\, g(v_i - \hat{S}/\sigma_i)(dv_i - d\hat{S}/\sigma_i + \hat{S}/\sigma_i{}^2\, d\sigma_i)$$
$$+ G(v_i - S/\sigma_i)\mu^2/C(dC/\mu - df + dr_{i-1}) = K\mu^2/C^2(r_{i-1}\, df - f\, dr_{i-1}). \tag{64.42}$$

We can easily deduce

$$\frac{dv_i}{dC} = -\frac{\mu_i}{C}\frac{C(v_i - \hat{S}/\sigma_i)}{(gv_i) + \mu_i g(v_i - \hat{S}/\sigma_i)}$$

$$
\begin{aligned}
&c \nearrow 10\% \\
&v \searrow 0.01 \text{ to } 0.05 \\
&y \searrow 0.3 \text{ to } 1.5 \text{ (thousand tons)}
\end{aligned}
$$
(64.43)

varies from -0.0003 to -0.0015;

$$\frac{dv}{dS} = \frac{\mu_i}{\sigma_i}\frac{g(v_i - \hat{S}/\sigma_i)}{g(v_i) + \mu_i g(v_i - \hat{S}/\sigma_i)}$$

$$
\begin{aligned}
&\hat{S} \nearrow 10\% \\
&v \nearrow 0.3 \text{ to } 0.05 \\
&y \nearrow 8 \text{ to } 1.5 \text{ (thousand tons)}
\end{aligned}
$$
(64.44)

varies from 0.03 to 0.005;

$$\frac{dv}{d\sigma} = \frac{\hat{S}\mu_i}{\sigma_i^2}\frac{g(v_i - \hat{S}/\sigma_i)}{g(v_i) + \mu_i g(v_i - \hat{S}/\sigma_i)}$$

$$
\begin{aligned}
&\sigma \nearrow 10\% \\
&v \searrow 0.06 \text{ to } 0.25 \\
&y \searrow 2 \text{ to } 8
\end{aligned}
$$
(64.45)

varies from -0.02 to -0.08 for S_0;

$$\frac{dv}{df} = \frac{\mu^2}{C}\frac{G(v_i - \hat{S}/\sigma_i) + Kr_{i-1}/C}{g(v_i) + \mu_i g(v_i - \hat{S}/\sigma_i)}$$

$$
\begin{aligned}
&f \nearrow 10\% \\
&v \nearrow 0.07 \\
&y \nearrow 0.02 \text{ (max)}
\end{aligned}
$$
(64.46)

varies from 0.00023 to 0.000005; and

$$\frac{dv}{dr_{i-1}} = -\frac{\mu^2}{C}\frac{G(v_i - \hat{S}/\sigma_i) + Kf/C}{g(v_i) + \mu_i g(v_i - \hat{S}/\sigma_i)}$$

$$
\begin{aligned}
&r_{i-1} \nearrow 10\% \\
&v \genfrac{}{}{0pt}{}{\nearrow 0.006}{\searrow 0.06} \\
&y \genfrac{}{}{0pt}{}{\nearrow 0.18}{\searrow 0.8.}
\end{aligned}
$$
(64.47)

varies from -0.00003 to $+0.0004$

We conclude that the variations of v^* and hence of y^* have the same sign as those of $\hat{S}$ and f; the opposite sign to those of C; and a variable sign, according to the period compared, to those of r_{i-1}.

We notice in particular that the value of the angular coefficients is usually low, which points to the stability of the solutions y_i^* when faced with the eventual fluctuations of the different parameters.

It will be noted that there has been no *a priori* choice of a policy for the management of the stock, and that the monthly decision depends solely on one criterion: the minimization of the function of over-all cost as an expected value. It has been proved, by the method of dynamic programming, that the policy of completing an initial stock S_i to a predetermined level y_i^* by means of a simple relation is the optimal solution, provided that the decision for the increase is made at regular intervals.

CHAPTER VI

MATHEMATICAL PROPERTIES OF
GAMES OF STRATEGY

---

65. Introduction

We shall now give the main theorems and proofs for rectangular games with two players and a zero value, and our treatment will, to a considerable extent, rely on that of MacKinsey [J16]. Other sections will deal with the numerical calculation of optimal strategies in games with continuous functions in which there are n players and a nonzero sum. Our treatment of the latter may appear somewhat summary, but we must emphasize that our aim throughout this book has been to raise the reader's knowledge to the level from which he will be able to undertake more advanced studies.

66. Various Properties Connected with the Maximal and Minimal Values of Functions

The proofs given in the later sections will be based on the various lemmas which we shall now state.

A. LEMMA I

If $f(x)$ is a real function of the real variable x, then

$$\operatorname*{MAX}_{x}[-f(x)] = -\operatorname*{MIN}_{x}[f(x)] \tag{66.1}$$

and

$$\operatorname*{MIN}_{x}[-f(x)] = -\operatorname*{MAX}_{x}[f(x)]. \tag{66.2}$$

In particular, if x is the discrete variable, $i = 1, 2,..., m$, and $f(i) = a_i$

$$\operatorname*{MAX}_{i}(-a_i) = -\operatorname*{MIN}_{i} a_i, \tag{66.3}$$

$$\operatorname*{MIN}_{i}(-a_i) = -\operatorname*{MAX}_{i} a_i. \tag{66.4}$$

406

Example.

$$\{\ a_1,\quad a_2,\quad a_3,\quad a_4\} = \{\ 2, -3, -5,\quad 1\},$$
$$\{-a_1,\ -a_2,\ -a_3,\ -a_4\} = \{-2,\quad 3,\quad 5, -1\}. \tag{66.5}$$

$$\mathop{\text{MAX}}_{i}(-a_i) = 5, \qquad \mathop{\text{MIN}}_{i} a_i = -5, \tag{66.6}$$

$$\mathop{\text{MIN}}_{i}(-a_i) = -2, \qquad \mathop{\text{MAX}}_{i} a_i = 2. \tag{66.7}$$

B. Lemma II

If $f(x, y)$ is a function with real values defined for any $x \in \mathbf{A}$ and $y \in \mathbf{B}$ (**A** and **B** being subsets of **R**), we assume that

$$\mathop{\text{MAX}}_{x \in A}[\mathop{\text{MIN}}_{y \in B} f(x, y)] \tag{66.8}$$

and

$$\mathop{\text{MIN}}_{y \in B}[\mathop{\text{MAX}}_{x \in A} f(x, y)] \tag{66.9}$$

exist; then

$$\mathop{\text{MAX}}_{x \in A}[\mathop{\text{MIN}}_{y \in B} f(x, y)] \leqslant \mathop{\text{MIN}}_{y \in B}[\mathop{\text{MAX}}_{x \in A} f(x, y)]. \tag{66.10}$$

Proof. For every x and y, from the definition of the minimum and maximum, we have

$$\mathop{\text{MIN}}_{y \in B} f(x, y) \leqslant f(x, y) \leqslant \mathop{\text{MAX}}_{x \in A} f(x, y). \tag{66.11}$$

Thus

$$\mathop{\text{MIN}}_{y \in B} f(x, y) \leqslant \mathop{\text{MAX}}_{x \in A} f(x, y). \tag{66.12}$$

Since the left member of (66.11) is independent of y, we can state

$$\mathop{\text{MIN}}_{y \in B} f(x, y) \leqslant \mathop{\text{MIN}}_{y \in B}[\mathop{\text{MAX}}_{x \in B} f(x, y)]. \tag{66.13}$$

Since the right member of (66.6) is independent of x, we can state

$$\mathop{\text{MAX}}_{x \in A}[\mathop{\text{MIN}}_{y \in B} f(x, y)] \leqslant \mathop{\text{MIN}}_{y \in B}[\mathop{\text{MAX}}_{x \in A} f(x, y)]. \tag{66.14}$$

In particular, if a is a matrix with elements

$$a_{ij}, \qquad i = 1, 2, ..., m; \qquad j = 1, 2, ..., n,$$

which belong to **R**: the above theorem applies. We then have $f(i, j) = a_{ij}$:

$$\text{MAX}_i[\text{MIN}_j a_{ij}] \leqslant \text{MIN}_j[\text{MAX}_i a_{ij}].^1 \tag{66.15}$$

C. SADDLE POINT

If $f(x, y)$ is a real function defined wherever $x \in \mathbf{A}$, $y \in \mathbf{B}$ (**A** and **B** being subsets of **R**), then $[x_0 y_0]$ is a *saddle point* if

$$f(x, y_0) \leqslant f(x_0, y_0) \leqslant f(x_0, y) \tag{66.16}$$

or

$$f(x, y_0) \geqslant f(x_0, y_0) \geqslant f(x_0, y). \tag{66.17}$$

In a case where $f(x, y)$ is continuous in the neighborhood of $[x_0 y_0]$, we prove that the necessary and sufficient condition for it to be a saddle point is that

$$\left.\frac{\partial f}{\partial x}\right|_{\substack{x=x_0\\y=y_0}} = 0, \qquad \left.\frac{\partial f}{\partial y}\right|_{\substack{x=x_0\\y=y_0}} = 0, \tag{66.18} \tag{66.19}$$

$$\left[\left(\frac{\partial^2 f}{\partial x\, \partial y}\right)^2 - \frac{\partial^2 f}{\partial x^2}\frac{\partial^2 f}{\partial y^2}\right]_{\substack{x=x_0\\y=y_0}} > 0. \tag{66.20}$$

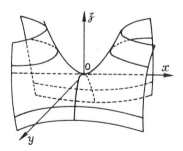

FIG. 66.1.

[1] In this special case, a direct, elementary proof of the property can be given. Let

$$a = \text{MIN}_j[\text{MAX}_i a_{ij}] \quad \text{and} \quad \beta = \text{MAX}_i[\text{MIN}_j a_{ij}].$$

Let us consider the element γ at the intersection of the column of a and the line of β

$$\begin{bmatrix} a & \\ \vdots & \\ \gamma & \cdots & \beta \end{bmatrix}.$$

Since γ is in the column of a: $\gamma \leqslant a$ and since γ is in the line of β; $\gamma \geqslant \beta$, then $\beta \leqslant \alpha$.

Example. Given the hyperbolic paraboloid of Fig. 66.1 with the general equation:

$$z = \frac{1}{2c}\left(\frac{x^2}{a^2} - \frac{y^2}{b^2}\right). \tag{66.21}$$

$$\frac{\partial f}{\partial x} = \frac{x}{a^2 c}, \quad \frac{\partial f}{\partial y} = -\frac{y}{b^2 c}, \quad \frac{\partial^2 f}{\partial x \, \partial y} = 0, \quad \frac{\partial^2 f}{\partial x^2} = \frac{1}{a^2 c}, \quad \frac{\partial^2 f}{\partial y^2} = -\frac{1}{b^2 c} \tag{66.22}$$

$$\left.\frac{\partial f}{\partial x}\right|_{x=0} = 0, \quad \left.\frac{\partial f}{\partial y}\right|_{y=0} = 0, \quad \left[\left(\frac{\partial^2 f}{\partial x \, \partial y}\right)^2 - \frac{\partial^2 f}{\partial x^2}\frac{\partial^2 f}{\partial y^2}\right]_{\substack{x=0 \\ y=0}} = \frac{1}{a^2 b^2 c^2} > 0 \tag{66.23}$$

if a and b have the same sign.

Hence the hyperbolic paraboloid has a saddle point or collar.

A necessary and sufficient condition if a function $f(x, y)$ is to possess a saddle point, and is not to be restricted to a case of continuous functions, is given in the following lemma.

D. LEMMA III

If $f(x, y)$ is a real function defined for $x \in \mathbf{A}$, $y \in \mathbf{B}$ (with $\mathbf{A}$ and $\mathbf{B}$ subsets of $\mathbf{R}$), we assume that

$$\underset{x \in \mathbf{A}}{\text{MAX}}\left[\underset{y \in \mathbf{B}}{\text{MIN}} f(x, y)\right] \quad \text{and} \quad \underset{y \in \mathbf{B}}{\text{MIN}}\left[\underset{x \in \mathbf{A}}{\text{MAX}} f(x, y)\right] \tag{66.24}$$

exist; then a necessary and sufficient condition for the function $f(x, y)$ to have a saddle point is that

$$\underset{x \in \mathbf{A}}{\text{MAX}}\left[\underset{y \in \mathbf{B}}{\text{MIN}} f(x, y)\right] = \underset{y \in \mathbf{B}}{\text{MIN}}\left[\underset{x \in \mathbf{A}}{\text{MAX}} f(x, y)\right]. \tag{66.25}$$

If the coordinates of this point are $[x_0 y_0]$, then

$$\underset{x \in \mathbf{A}}{\text{MAX}}\left[\underset{y \in \mathbf{B}}{\text{MIN}} f(x, y)\right] = \underset{y \in \mathbf{B}}{\text{MIN}}\left[\underset{x \in \mathbf{A}}{\text{MAX}} f(x, y)\right] = f(x_0, y_0). \tag{66.26}$$

Proof. Let us show first that the condition is sufficient. Let us assume that $[x_0, y_0]$ is a saddle point of $f(x, y)$. Then we must have[1]:

$$\left.\begin{cases} f(x, y_0) \leqslant f(x_0, y_0) \\ f(x_0, y_0) \leqslant f(x_0, y) \end{cases}\right\} \quad \forall x \in \mathbf{A} \quad \text{and} \quad \forall y \in \mathbf{B}. \qquad \begin{matrix} (66.27) \\ (66.28) \end{matrix}$$

---

[1] We are considering (66.17) in this proof, which would also be valid for (66.18) if ·we replace MAX by MIN, and reciprocally, wherever required.

From (66.21), we deduce that

$$\operatorname*{MAX}_{x\in\mathbf{A}} f(x, y_0) \leqslant f(x_0, y_0),\tag{66.29}$$

and from (66.22):

$$f(x_0, y_0) \leqslant \operatorname*{MIN}_{y\in\mathbf{B}} f(x_0, y).\tag{66.30}$$

Let

$$\operatorname*{MAX}_{x\in\mathbf{A}} f(x, y_0) \leqslant f(x_0, y_0) \leqslant \operatorname*{MIN}_{y\in\mathbf{B}} f(x_0, y)\tag{66.31}$$

and *a fortiori*:

$$\operatorname*{MIN}_{y\in\mathbf{B}}[\operatorname*{MAX}_{x\in\mathbf{A}} f(x, y)] \leqslant f(x_0, y_0) \leqslant \operatorname*{MAX}_{x\in\mathbf{A}}[\operatorname*{MIN}_{y\in\mathbf{B}} f(x, y)].\tag{66.32}$$

But in accordance with (66.17) the first term of (66.32) must be greater than or equal to the third, hence they are equal and the condition is sufficient.

Let us now prove that it is necessary. Let $x_0 \in \mathbf{A}$ be a value which makes $\operatorname*{MIN}_{y\in\mathbf{B}} f(x, y)$ maximal and $y_0 \in \mathbf{B}$ a value which makes $\operatorname*{MAX}_{x\in\mathbf{A}} f(x, y)$ minimal, that is to say,

$$x_0 \in \mathbf{A} \quad \text{satisfies} \quad \operatorname*{MIN}_{y\in\mathbf{B}} f(x_0, y) = \operatorname*{MAX}_{x\in\mathbf{A}}[\operatorname*{MIN}_{y\in\mathbf{B}} f(x, y)],\tag{66.33}$$

$$y_0 \in \mathbf{B} \quad \text{satisfies} \quad \operatorname*{MAX}_{x\in\mathbf{A}} f(x, y_0) = \operatorname*{MIN}_{y\in\mathbf{B}}[\operatorname*{MAX}_{x\in\mathbf{A}} f(x, y)].\tag{66.34}$$

Let us show that $[x_0, y_0]$ is a saddle point of $f(x, y)$. We have assumed that

$$\operatorname*{MAX}_{x\in\mathbf{A}}[\operatorname*{MIN}_{y\in\mathbf{B}} f(x, y)] = \operatorname*{MIN}_{y\in\mathbf{B}}[\operatorname*{MAX}_{x\in\mathbf{A}} f(x, y)].\tag{66.35}$$

We conclude that

$$\operatorname*{MIN}_{y\in\mathbf{B}} f(x_0, y) = \operatorname*{MAX}_{x\in\mathbf{A}} f(x, y_0).\tag{66.36}$$

But, by the definition of the minimum:

$$\operatorname*{MIN}_{y\in\mathbf{B}} f(x_0, y) \leqslant f(x_0, y_0),\tag{66.37}$$

and hence, by (66.30):

$$\operatorname*{MAX}_{x\in\mathbf{A}} f(x, y_0) \leqslant f(x_0, y_0).\tag{66.38}$$

If we refer to (66.38) and to the definition of the maximum of a function, we conclude that

$$\forall x \in \mathbf{A}, \quad f(x, y_0) \leqslant f(x_0, y_0). \tag{66.39}$$

In the same way, from the definition of the maximum:

$$\max_{x \in B} f(x, y_0) \geqslant f(x_0, y_0), \tag{66.40}$$

we obtain

$$\forall y \in \mathbf{B}, \quad f(x_0, y_0) \leqslant f(x_0, y). \tag{66.41}$$

E. NOTE ON THE USE OF THE TERM "SADDLE POINT"

In the theory of games of strategy we say that a rectangular game with two players and a zero sum, which is defined on a rectangular matrix

$$a_{ij}, \quad i = 1, 2, ..., m; \quad j = 1, 2, ..., n,$$

contains a "saddle point" or "point of equilibrium" if

$$\max_{x \in A}[\min_{y \in B} a_{ij}] = \min_{y \in B}[\max_{x \in A} a_{ij}]. \tag{66.42}$$

As we have just seen, this definition can be given a much more general sense, and we shall show further on that if a rectangular game does not have a saddle point, the function

$$E(X, Y) = \sum_{i=1}^{m} \sum_{j=1}^{n} a_{ij} x_i y_j$$

$$\tag{66.43}$$

$$x_i, y_j \geqslant 0, \quad \forall x_i \text{ and } \forall y_i \quad \sum_{i=1}^{m} x_i = 1 \text{ and } \sum_{j=1}^{n} y_j = 1$$

always possesses one.

In the terminology of games of strategy, this means that if there is no pure optimal strategy corresponding to the saddle point, there is always a mixed optimal strategy for each of the players.

67. Fundamental Theorem for Games of Strategy

Given a rectangular game with a matrix of which the elements are

$$a_{ij}, \quad i = 1, 2, ..., m; \quad j = 1, 2, ..., n.$$

(1) For player A there is an optimal strategy,[1] in other words, relative frequencies

$$X^* = [x_1{}^*, x_2{}^*, ..., x_m{}^*]$$

such that if he plays line (1) with the frequency $x_1{}^*$, line (2) with that of $x_2{}^*$,..., line (m) with that of $x_m{}^*$, he is certain of *at least making a profit* equal to the value of the game.

For player B there is an optimal strategy or relative frequencies

$$Y^* = [y_1{}^*, y_2{}^*, ..., y_n{}^*]$$

such that if he plays column (1) with the frequency $y_1{}^*$, column (2) with that of $y_2{}^*$,..., column (n) with that of $y_n{}^*$, he is certain of *not losing more* than the value v of the game.

Hence, an optimal strategy for each player is formed from the solution of $m + n + 2$ equations or inequalities, which are not strict, with $m + n + 1$ unknowns $x_1, x_2, ..., x_m, y_1, y_2, ..., y_n, v$. The x_i and y_i terms are positive, but v can be any real number.

$$
\begin{aligned}
x_1 + x_2 + \cdots + x_m &= 1, \\
a_{11}x_1 + a_{21}x_2 + \cdots + a_{m1}x_m &\geqslant v, \\
a_{12}x_1 + a_{22}x_2 + \cdots + a_{m2}x_m &\geqslant v, \\
&\;\;\vdots \\
a_{1n}x_1 + a_{2n}x_2 + \cdots + a_{mn}x_m &\geqslant v, \\
0 \leqslant x_i, \qquad i = 1, 2, ..., m.
\end{aligned}
\tag{67.1}
$$

$$
\begin{aligned}
y_1 + y_2 + \cdots + y_n &= 1, \\
a_{11}y_1 + a_{12}y_2 + \cdots + a_{1n}y_n &\leqslant v, \\
a_{21}y_1 + a_{22}y_2 + \cdots + a_{2n}y_n &\leqslant v, \\
&\;\;\vdots \\
a_{m1}y_1 + a_{m2}y_2 + \cdots + a_{mn}y_n &\leqslant v, \\
0 \leqslant y_j, \qquad j = 1, 2, ..., n.
\end{aligned}
\tag{67.2}
$$

(2) There is always a solution to the system formed by (67.1) and (67.2); let

$$X^* = [x_1{}^* \quad x_2{}^* \quad \cdots \quad x_m{}^*] \quad \text{and} \quad Y^* = [y_1{}^* \quad y_2{}^* \quad \cdots \quad y_n{}^*]$$

be one solution. If there are several, any linear form produced by the X^* terms, on the one hand, and by the Y^* terms, on the other, of which the

---

[1] When no distinction is made, "strategy" is to be read as "mixed or pure strategy," with the understanding that a pure strategy is only a special case of a set of mixed strategies.

coefficients are nonnegative and of a sum equal to 1 (convex weighting), is also a solution. Hence for A, there are, in this case, r optimal strategies with any two of which, for example $X^{*\prime}$ and $X^{*\prime\prime}$, we form an infinity:

$$X^{*\prime\prime\prime} = \lambda_1 X^{*\prime} + \lambda_2 X^{*\prime\prime}, \qquad \text{where} \quad \lambda_1, \lambda_2 \geqslant 0, \quad \lambda_1 + \lambda_2 = 1. \quad (67.3)$$

Similarly, if there are s optimal strategies for B.[1]

(3) There is only one value for v.

(4) If $X^* = [x_1^* x_2^* \cdots x_m^*]$ is an optimal strategy for A and

$$a_{1j}x_1^* + a_{2j}x_2^* + \cdots + a_{mj}x_m^* > v, \qquad (67.4)$$

then

$$y_j^* = 0; \qquad (67.5)$$

but the converse is not always true.

Similarly, if $Y^* = [y_1^* y_2^* \cdots y_n^*]$ is an optimal strategy of B and

$$a_{i1}y_1^* + a_{i2}y_2^* + \cdots + a_{in}y_n^* < v, \qquad (67.6)$$

then

$$x_i^* = 0; \qquad (67.7)$$

but the converse is not always true.

The proof of this theorem is carried out in successive stages; these are given in the next section and closely follow the treatment in MacKinsey's work [J16].

68. Various Proofs

A. NOTATIONS

We shall use the following notations:

$\mathbf{S}_m$,	the set of all A's strategies,
$\mathbf{S}_n$,	the set of all B's strategies,
$X = [x_1 \ x_2 \cdots x_m]$,	a strategy for A,
$X^* = [x_1^* \ x_2^* \cdots x_m^*]$,	an optimal strategy for A,
$Y = [y_1 \ y_2 \cdots y_n]$,	a strategy for B,
$Y^* = [y_1^* \ y_2^* \cdots y_n^*]$,	an optimal strategy for B.

---

[1] What has been shown above is also true for any solution, whether it is optimal or not, in accordance with the properties of convex bodies.

Let us call

$$E(X, Y) = \sum_{j=1}^{n} \sum_{i=1}^{m} a_{ij} x_i y_j , \tag{68.1}$$

$$E(X^*, Y) = \sum_{j=1}^{n} \sum_{i=1}^{m} a_{ij} x_i{}^* y_j , \tag{68.2}$$

$$E(X, Y^*) = \sum_{j=1}^{n} \sum_{i=1}^{m} a_{ij} x_i y_j{}^*, \tag{68.3}$$

$$E(X^*, Y^*) = \sum_{j=1}^{n} \sum_{i=1}^{m} a_{ij} x_i{}^* y_j{}^*. \tag{68.4}$$

$X \in \mathbf{S}_m$ means strategy X belongs to set $\mathbf{S}_m$ of all A's strategies [the set of all the solutions of (67.1)];

$Y \in \mathbf{S}_n$ has the same meaning for B's strategies [the set of all the solutions of (67.2)];

$\mathbf{S}_{m+n}$ means the set of all the solutions of (67.1) and (67.2) which belongs [1] to $\mathbf{E}_{m+n}$.

The following proofs depend on various theorems about convex bodies, of which the reader will find an explanation in [J16] and [J33].

B. THEOREM I

Given a matrix $[a]$ with elements a_{ij} on which a game of strategy is to be played, one or other of the following propositions is true:

(a)   There is a strategy $X \in \mathbf{S}_m$ such that:

$$a_{1j} x_1 + a_{2j} x_2 + \cdots + a_{mj} x_m \geqslant 0, \qquad \text{for} \quad j = 1, 2, ..., n; \tag{68.5}$$

(b)   There is a strategy $Y \in \mathbf{S}_n$ such that:

$$a_{1i} y_1 + a_{2i} y_2 + \cdots + a_{ni} y_n \leqslant 0 \qquad \text{for} \quad i = 1, 2, ..., m. \tag{68.6}$$

Proof.   Let us use Kronecker's notation

$$\begin{aligned} \delta_{ij} &= 0 \qquad \text{if} \quad i \neq j \\ &= 1 \qquad \text{if} \quad i = j, \end{aligned} \tag{68.7}$$

[1] $\mathbf{E}_n$ signifies "Euclidean space" and is the set of all the n-tuples $[x_1 x_2 \cdots x_n]$ where the x_i terms are real numbers and in which the distance between $X^{(1)}$ and $X^{(2)}$ is given by

$$d(X^{(1)}, X^{(2)}) = [(x_1^{(1)} - x_1^{(2)})^2 + (x_2^{(1)} - x_2^{(2)})^2 + \cdots + (x_n^{(1)} - x_n^{(2)})^2]^{1/2}$$

and let us assume

$$\delta^{(1)} = [\delta_{11} \quad \delta_{21} \quad \cdots \quad \delta_{m1}],$$
$$\delta^{(2)} = [\delta_{12} \quad \delta_{22} \quad \cdots \quad \delta_{m2}],$$
$$\vdots$$
$$\delta^{(m)} = [\delta_{1m} \quad \delta_{2m} \quad \cdots \quad \delta_{mm}].$$

(68.8)

Thus $\delta^{(j)}$, $j = 1, 2, ..., m$, is a point of $\mathbf{E}_n$ of which the j coordinate is equal to 1 and the others are equal to 0. Let us also assume

$$a^{(1)} = [a_{11} \quad a_{21} \quad \cdots \quad a_{m1}],$$
$$a^{(2)} = [a_{12} \quad a_{22} \quad \cdots \quad a_{m2}],$$
$$\vdots$$
$$a^{(n)} = [a_{1n} \quad a_{2n} \quad \cdots \quad a_{mn}].$$

(68.9)

Thus $a^{(j)}$, $j = 1, 2, ..., n$ is the point of which the coordinates are formed by the jth column of a. Let $\mathbf{C}$ be the convex body of set $\mathbf{A}$ of $m + n$ points:

$$\delta^{(1)}, \delta^{(2)}, ..., \delta^{(m)}, \quad a^{(1)}, a^{(2)}, ..., a^{(n)}$$

(68.10)

and let us call the origin of $\mathbf{E}_n$

$$Z = [0 \quad 0 \quad \cdots \quad 0],$$

(68.11)

For an actual example, let us consider the following matrix $[a]$:

$$[a] = [a^{(1)} \quad a^{(2)} \quad a^{(3)} \quad a^{(4)} \quad a^{(5)}]$$

$$= \begin{bmatrix} a_{11} & a_{12} & a_{13} & a_{14} & a_{15} \\ a_{21} & a_{22} & a_{23} & a_{24} & a_{25} \end{bmatrix}$$

(68.12)

$$= \begin{bmatrix} 0.6 & 1.7 & 1.5 & 1.8 & 1 \\ -0.4 & 0.4 & 0.7 & 0.9 & 1.3 \end{bmatrix},$$

then the convex body $\mathbf{C}$ of set $\mathbf{A}$ of the $m + n = 2 + 5 = 7$ points, formed by the columns of the following matrix:

$$\begin{bmatrix} \delta_{11} & \delta_{12} & a_{11} & a_{12} & a_{13} & a_{14} & a_{15} \\ \delta_{21} & \delta_{22} & a_{21} & a_{22} & a_{23} & a_{24} & a_{25} \end{bmatrix} = \begin{bmatrix} 1 & 0 & 0.6 & 1.7 & 1.5 & 1.8 & 1 \\ 0 & 1 & -0.4 & 0.4 & 0.7 & 0.9 & 1.3 \end{bmatrix}$$

(68.13)

will be given by the polygon shown in Fig. 68.1.

Let us now examine two cases:

$$\text{(a)} \quad Z \in \mathbf{C}, \tag{68.14}$$

$$\text{(b)} \quad Z \notin \mathbf{C}. \tag{68.15}$$

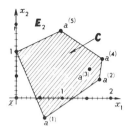

FIG. 68.1.

(a)   If $Z \in \mathbf{C}$, then Z is a convex linear combination of the points of the set $\mathbf{A}$. Hence, there is an element $[u_1 u_2 \cdots u_m \quad v_1 v_2 \cdots v_n]$ of $\mathbf{S}_{m+n}$ such that

$$u_1 \delta^{(1)} + u_2 \delta^{(2)} + \cdots + u_m \delta^{(m)} + v_1 a^{(1)} + v_2 a^{(2)} + \cdots + v_n a^{(n)} = Z, \quad (68.16)$$

that is to say,

$$u_1 \delta_{i1} + u_2 \delta_{i2} + \cdots + u_m \delta_{im} + v_1 a_{i1} + v_2 a_{i2} + \cdots$$
$$+ v_n a_{in} = 0, \quad i = 1, 2, \ldots, m, \quad (68.17)$$

from (68.7), let

$$u_i + v_1 a_{i1} + v_2 a_{i2} + \cdots + v_n a_{in} = 0, \quad i = 1, 2, \ldots, m. \quad (68.18)$$

Since

$$[u_1 \quad u_2 \quad \cdots \quad u_m \quad v_1 \quad v_2 \quad \cdots \quad v_n] \in \mathbf{S}_{m+n}, \quad (68.19)$$

we conclude that u_i is positive, and hence

$$v_1 a_{i1} + v_2 a_{i2} + \cdots + v_n a_{in} \leqslant 0, \quad i = 1, 2, \ldots, m. \quad (68.20)$$

But

$$v_1 + v_2 + \cdots + v_n > 0 \quad (68.21)$$

since, on the one hand, we have $v_i \geqslant 0$ for $i = 1, 2, \ldots, n$; we therefore find

$$v_1 = v_2 = \cdots = v_n = 0. \quad (68.22)$$

From (68.18), it follows that

$$u_1 = 0, \quad i = 1, 2, \ldots, m, \quad (68.23)$$

which contradicts (68.19). Hence we can say

$$y_1 = \frac{v_1}{v_1 + v_2 + \cdots + v_n},$$

$$y_2 = \frac{v_2}{v_1 + v_2 + \cdots + v_n}, \qquad (68.24)$$

$$\vdots$$

$$y_n = \frac{v_n}{v_1 + v_2 + \cdots + v_n};$$

and, going back to (68.20) and (68.21):

$$y_1 a_{i1} + y_2 a_{i2} + \cdots + y_n a_{in} \leqslant 0, \qquad i = 1, 2,\ldots, m, \qquad (68.25)$$

since, from (68.24):

$$[y_1 \quad y_2 \quad \cdots \quad y_n] \in \mathbf{S}_n .$$

Thus if $Z \in \mathbf{C}$, (68.6) is true.

(b)   Let us examine what happens if $Z \notin \mathbf{C}$ (Fig. 68.1), and refer to a well-known theorem.[1]

If $\mathbf{X}$ is any closed subset of $\mathbf{E}_n$ and there is a point $Z \in \mathbf{E}_n$, $Z \notin \mathbf{X}$, then there is a *hyperplane P* which contains Z and such that $\mathbf{X}$ is a subset of one of the half spaces determined by P.

Let us assume that the equation of this hyperplane is

$$b_1 t_1 + b_2 t_2 + \cdots + b_m t_m = b_{m+1} . \qquad (68.26)$$

As Z belongs to the plane, we have

$$b_{m+1} = 0. \qquad (68.27)$$

Hence the equation becomes

$$b_1 t_1 + b_2 t_2 + \cdots + b_m t_m = 0, \qquad (68.28)$$

and we can assume that any point

$$[t_1 \quad t_2 \quad \cdots \quad t_m] \in \mathbf{C}$$

satisfies the inequality:

$$b_1 t_1 + b_2 t_2 + \cdots + b_m t_m > 0. \qquad (68.29)$$

---

[1] See Reference [J16]. This theorem has a very simple proof: showing that the two half spaces mutually exclude each other and are combined in $\mathbf{E}_n$. This property is obvious in $\mathbf{E}_2$ or $\mathbf{E}_3$.

If the points of **C** satisfy the inequality:

$$c_1 t_1 + c_2 t_2 + \cdots + c_n t_n > 0, \tag{68.30}$$

then, by multiplying this inequality by (-1), it follows that

$$(-c_1)t_1 + (-c_2)t_2 + \cdots + (-c_n)t_n > 0, \tag{68.31}$$

and we obtain (68.29) by substituting b_i, $i = 1, 2, ..., m$ for $-c_i$.

But the inequality (68.29) is also valid for the points $\delta^{(1)}, \delta^{(2)}, ..., \delta^{(m)}$ of **C**, and hence

$$b_1 \delta_{1i} + b_2 \delta_{2i} + \cdots + b_m \delta_{mi} > 0, \qquad i = 1, 2, ..., m, \tag{68.32}$$

which means, given property (68.7) of the δ_{ij} terms:

$$b_i > 0, \qquad i = 1, 2, ..., m. \tag{68.33}$$

But the inequalities (68.29) must satisfy all the points $a^{(1)}, a^{(2)}, ..., a^{(n)}$; hence

$$b_1 a_{1j} + b_2 a_{2j} + \cdots + b_m a_{mj} > 0, \qquad j = 1, 2, ..., n. \tag{68.34}$$

Considering (68.33), it is clear that

$$b_1 + b_2 + \cdots + b_m > 0,$$

and we can therefore state that

$$x_1 = \frac{b_1}{b_1 + b_2 + \cdots + b_m},$$

$$x_2 = \frac{b_2}{b_1 + b_2 + \cdots + b_m}, \tag{68.35}$$

$$\vdots$$

$$x_m = \frac{b_m}{b_1 + b_2 + \cdots + b_m}.$$

A consideration of (68.34) and (68.35) gives us

$$x_1 a_{1j} + x_2 a_{2j} + \cdots + x_m a_{mj} > 0, \qquad j = 1, 2, ..., n \tag{68.36}$$

and *a fortiori*:

$$x_1 a_{1j} + x_2 a_{2j} + \cdots + x_m a_{mj} \geqslant 0, \qquad j = 1, 2, ..., n. \tag{68.37}$$

From (68.33) and (68.35), we can see that

$$[x_1 \quad x_2 \cdots x_m] \in \mathbf{S}_m,$$

so that (68.5) is satisfied when $Z \notin \mathbf{C}$. Theorem I is therefore fully proved.

C. THEOREM II

Let $[a]$ be the matrix used in Theorem I, and $E(X, Y)$ the expected value defined by (68.1), where

$$X = [x_1 \quad x_2 \cdots x_m] \in \mathbf{S}_m \quad \text{and} \quad Y = [y_1 \quad y_2 \cdots y_n] \in \mathbf{S}_n,$$

then

$$\underset{X\in\mathbf{S}_m}{\text{MAX}}[\underset{Y\in\mathbf{S}_n}{\text{MIN}}\, E(X, Y)] \quad \text{and} \quad \underset{Y\in\mathbf{S}_n}{\text{MIN}}[\underset{X\in\mathbf{S}_m}{\text{MAX}}\, E(X, Y)] \tag{68.38}$$

exist and are equal. This common value is referred to as the *value v of the game*.

Proof. For each $Y = [y_1 y_2 \cdots y_n]$, the function:

$$E(X, Y) = \sum_{j=1}^{n} \sum_{i=1}^{m} a_{ij} x_i y_j$$

$$= x_1 \sum_{j=1}^{n} a_{1j} y_j + x_2 \sum_{j=1}^{n} a_{2j} y_j + \cdots$$

$$+ x_m \sum_{j=1}^{n} a_{mj} y_j \tag{68.39}$$

is continuous, for it is a linear form of X and is defined by the closed subset $\mathbf{S}_m$ of $\mathbf{E}_m$, so that we can state that

$$\underset{X\in\mathbf{S}_m}{\text{MAX}}\, E(X, Y) \tag{68.40}$$

exists for every $Y \in \mathbf{S}_n$. We can easily verify that (68.40) is a continuous function in parts and linear for $[y_1 y_2 \cdots y_n]$. Hence $\mathbf{S}_n$ is a closed subset of $\mathbf{E}_n$, so that we conclude that

$$\underset{Y\in\mathbf{S}_n}{\text{MIN}}[\text{MAX}\, E(X, Y)] \tag{68.41}$$

exists. But, in accordance with (68.5), there is an element

$$X = [x_1 \quad x_2 \cdots x_m] \in \mathbf{S}_m$$

such that

$$a_{1j}x_1 + a_{2j}x_2 + \cdots + a_{mj}x_m \geqslant 0, \quad j = 1, 2, \ldots, n \tag{68.42}$$

and we can thus state for every $Y \in \mathbf{S}_n$:

$$E(X, Y) = \sum_{j=1}^{n} (a_{1j}x_1 + a_{2j}x_2 + \cdots + a_{mj}x_m) \geqslant 0. \tag{68.43}$$

Since (68.15) is satisfied for each $Y \in \mathbf{S}_n$, we conclude that

$$\underset{Y \in \mathbf{S}_n}{\text{MIN}}\, E(X, Y) \geqslant 0, \qquad (68.44)$$

whence

$$\underset{X \in \mathbf{S}_m}{\text{MAX}}[\underset{Y \in \mathbf{S}_n}{\text{MIN}}\, E(X, Y)] \geqslant 0. \qquad (68.45)$$

The same result is obtained from the property expressed in (68.6):

$$\underset{Y \in \mathbf{S}_n}{\text{MIN}}[\underset{X \in \mathbf{S}_m}{\text{MAX}}\, E(X, Y)] \leqslant 0. \qquad (68.46)$$

Since one or other of properties (a) and (b) of Theorem I [the relations (68.5) and (68.6)] must be satisfied, we conclude that at least one of those in (68.43) and (68.45) must also be satisfied, and hence that we cannot have

$$\underset{X \in \mathbf{S}_m}{\text{MAX}}[\underset{Y \in \mathbf{S}_n}{\text{MIN}}\, E(X, Y)] < 0 < \underset{Y \in \mathbf{S}_n}{\text{MIN}}[\underset{X \in \mathbf{S}_m}{\text{MAX}}\, E(X, Y)]. \qquad (68.47)$$

Let us now use $E_k(X, Y)$ for the expression:

$$E_k(X, Y) = \sum_{i=1}^{m} \sum_{j=1}^{n} (a_{ij} - k)x_i y_j \qquad (68.48)$$

where k is a real number,

$$X = [x_1 \ x_2 \cdots x_m] \in \mathbf{S}_m \quad \text{and} \quad Y = [y_1 \ y_2 \cdots y_n] \in \mathbf{S}_n .$$

Since the condition (68.47) cannot be satisfied for the matrix with elements a_{ij}, it cannot equally be satisfied for the one with elements $a_{ij} - k$; hence we cannot have

$$\underset{X \in \mathbf{S}_m}{\text{MAX}}[\underset{Y \in \mathbf{S}_n}{\text{MIN}}\, E_k(X, Y)] < 0 < \underset{Y \in \mathbf{S}_n}{\text{MIN}}[\underset{X \in \mathbf{S}_m}{\text{MAX}}\, E_k(X, Y)]. \qquad (68.49)$$

It is easy to reduce (68.48) as follows:

$$E_k(X, Y) = \left(\sum_{i=1}^{m} \sum_{j=1}^{n} a_{ij}x_i y_j\right) - \sum_{i=1}^{m} \sum_{j=1}^{n} kx_i y_j ; \qquad (68.50)$$

let

$$E_k(X, Y) = E(X, Y) - k. \qquad (68.51)$$

Comparing (68.49) and (68.51), we conclude that the following condition cannot be satisfied:

$$\underset{X \in \mathbf{S}_m}{\text{MAX}}[\underset{Y \in \mathbf{S}_n}{\text{MIN}}\, E(X, Y)] < k < \underset{Y \in \mathbf{S}_n}{\text{MIN}}[\underset{X \in \mathbf{S}_m}{\text{MAX}}\, E(X, Y)]. \qquad (68.52)$$

Since (68.52) is false for every k, we conclude that the following inequality is equally false:

$$\underset{X\in\mathbf{S}_m}{\text{MAX}}[\underset{Y\in\mathbf{S}_n}{\text{MIN}}\,E(X,\,Y)] < \underset{Y\in\mathbf{S}_n}{\text{MIN}}[\underset{X\in\mathbf{S}_m}{\text{MAX}}\,E(X,\,Y)]; \qquad (68.53)$$

and finally, conclude that the following is true:

$$\underset{X\in\mathbf{S}_m}{\text{MAX}}[\underset{Y\in\mathbf{S}_n}{\text{MIN}}\,E(X,\,Y)] \geqslant \underset{Y\in\mathbf{S}_n}{\text{MIN}}[\underset{X\in\mathbf{S}_m}{\text{MAX}}\,E(X,\,Y)]. \qquad (68.54)$$

But the fundamental theorem expressed by (66.7) is clearly valid for the bilinear form $E(X,\,Y)$, so that we have

$$\underset{X\in\mathbf{S}_m}{\text{MAX}}[\underset{Y\in\mathbf{S}_n}{\text{MIN}}\,E(X,\,Y)] \leqslant \underset{Y\in\mathbf{S}_n}{\text{MIN}}[\underset{X\in\mathbf{S}_m}{\text{MAX}}\,E(X,\,Y)]. \qquad (68.55)$$

Comparing (68.52) and (68.53), and taking v as the value of the two equal members, it follows that

$$v = \underset{X\in\mathbf{S}_m}{\text{MAX}}[\underset{Y\in\mathbf{S}_n}{\text{MIN}}\,E(X,\,Y)] = \underset{Y\in\mathbf{S}_n}{\text{MIN}}[\underset{X\in\mathbf{S}_m}{\text{MAX}}\,E(X,\,Y)]. \qquad (68.56)$$

Lastly, using Theorems I and II, we can conclude that if a is the matrix of any rectangular game with elements a_{ij}, every such game has a value v and each player has at least one optimal strategy.

D. THEOREM III

If $X \in \mathbf{S}_m$ is a strategy of A and $Y \in \mathbf{S}_n$ a strategy of B, then a necessary and sufficient condition for $X^* \in \mathbf{S}_m$ to be an optimal strategy for A is that

$$E(X^*,\,Y) \geqslant v, \qquad \forall Y \in \mathbf{S}_n\,; \qquad (68.57)$$

in the same way, $Y^* \in \mathbf{S}_n$ will be an optimal strategy for B if

$$E(X,\,Y^*) \leqslant v, \qquad \forall X \in \mathbf{S}_m\,. \qquad (68.58)$$

This theorem is often a very useful method of finding the optimal strategies when v can be found by inspection or intuition.

Proof. If X^* is an optimal strategy for A, there must be a $Y^* \in \mathbf{S}_n$ such that $[X^*Y^*]$ is a saddle point of $E(X,\,Y)$, and hence for every $Y \in \mathbf{S}_n$:

$$v = E(X^*,\,Y^*) \leqslant E(X^*,\,Y) \qquad \text{(by the definition of a minimum).} \quad (68.59)$$

Let us now assume that $X^* \in \mathbf{S}_m$, so that for each $Y \in \mathbf{S}_n$:

$$v \leqslant E(X^*, Y); \tag{68.60}$$

in accordance with Theorem II, there is a point $[X'Y']$ such that, for each $X \in \mathbf{S}_m$ and each $Y \in \mathbf{S}_n$

$$E(X, Y') \leqslant E(X', Y') \leqslant E(X', Y); \tag{68.61}$$

and thus, as v is the assumed value of the game, we have

$$E(X', Y') = v. \tag{68.62}$$

Using (68.60) and (68.61), we conclude that

$$E(X', Y') \leqslant E(X^*, Y). \tag{68.63}$$

Replacing Y by Y' in (68.63) and X by X^* in the first member of (68.61), we have

$$E(X^*, Y') \leqslant E(X', Y') \leqslant E(X^*, Y'); \tag{68.64}$$

thus

$$E(X', Y') = E(X^*, Y'). \tag{68.65}$$

Refering to (68.61), (68.63), and (68.65), we can now conclude that

$$E(X, Y') \leqslant E(X^*, Y') \leqslant E(X^*, Y); \tag{68.66}$$

hence $[X^*, Y']$ is a saddle point of $E(X, Y)$, and thus X^* is an optimal strategy for A.

A similar proof can be used for (68.58).   (68.67)

E. THEOREM IV

Still following MacKinsey's treatment, let us employ the following notation:

$X_i \in \mathbf{S}_m$, X_i means that all the components are zero except for i which is 1;

$Y_j \in \mathbf{S}_n$ with a similar meaning to the above;

$E(i, Y)$ means $E(X_i, Y)$;

$E(X, j)$ means $E(X, Y_j)$.

Thus

$$E(i, Y) = \sum_{j=1}^{n} a_{ij} y_j, \qquad (68.68)$$

$$E(X, j) = \sum_{i=1}^{m} a_{ij} x_i \qquad (68.69)$$

and

$$E(X, Y) = \sum_{i=1}^{m} E(i, Y) \cdot x_i = \sum_{j=1}^{n} E(X, j) \cdot y_j. \qquad (68.70)$$

Let us now enunciate this theorem:

Given $E(X, Y)$, as defined by (68.1), v a real number, $X^* \in \mathbf{S}_m$ and $Y^* \in \mathbf{S}_n$, then if v is to be the value of the game, and X^*, Y^* are to be the optimal strategies of A and B, a necessary and sufficient condition is that $1 \leqslant i \leqslant m$ and $1 \leqslant j \leqslant n$:

$$E(i, Y^*) \leqslant v \leqslant E(X^*, j). \qquad (68.71)$$

Proof. To prove that the condition is necessary, it is sufficient to refer to the definition of a saddle point of $E(X, Y)$ and to replace X by X_i or Y by Y_j.

If the condition is satisfied, we obtain

$$\forall(X = [x_1 \quad x_2 \cdots x_m]) \in \mathbf{S}_m : \sum_{i=1}^{m} E(i, Y^*) x_i \leqslant \sum_{i=1}^{m} v x_i = v; \qquad (68.72)$$

let

$$E(X, Y^*) \leqslant v. \qquad (68.73)$$

In the same way,

$$\forall(Y = [y_1 \quad y_2 \cdots y_n]) \in \mathbf{S}_n : \quad E(X^*, Y) \geqslant v. \qquad (68.74)$$

Replacing X by X^* in (68.73) and Y by Y^* in (68.74), we obtain

$$E(X^*, Y^*) \leqslant v \qquad (68.75)$$

and

$$v \leqslant E(X^*, Y^*). \qquad (68.76)$$

Hence

$$v = E(X^*, Y^*). \qquad (68.77)$$

Lastly, from (68.73), (68.75), and (68.77), we see that

$$E(X, Y^*) \leqslant E(X^*, Y^*) \leqslant E(X^*, Y), \qquad (68.78)$$

hence $[X^* Y^*]$ is a saddle point of $E(X, Y)$ and v is the value of the game.

Note. The relation (68.71) is the same as (67.1) and (67.2):

$E(1, Y^*) \leqslant v$ can be expressed $a_{11} y_1{}^* + a_{12} y_2{}^* + \cdots + a_{1n} y_n{}^* \leqslant v;$

$E(2, Y^*) \leqslant v$ $\qquad\qquad\qquad$ $a_{21} y_1{}^* + a_{22} y_2{}^* + \cdots + a_{2n} y_n{}^* \leqslant v;$

$\qquad\vdots$ $\qquad\qquad\qquad\qquad\qquad\qquad\qquad\qquad\qquad\qquad \vdots$

$E(m, Y^*) \leqslant v$ $\qquad\qquad\qquad$ $a_{m1} y_1{}^* + a_{m2} y_2{}^* + \cdots + a_{mn} y_n{}^* \leqslant v;$

$$(68.79)$$

and

$E(X^*, 1) \geqslant v$ can be expressed $a_{11} x_1{}^* + a_{21} x_2{}^* + \cdots + a_{m1} x_m{}^* \geqslant v;$

$E(X^*, 2) \geqslant v$ $\qquad\qquad\qquad$ $a_{12} x_1{}^* + a_{22} x_2{}^* + \cdots + a_{m2} x_m{}^* \geqslant v;$

$\qquad\vdots$ $\qquad\qquad\qquad\qquad\qquad\qquad\qquad\qquad\qquad\qquad \vdots$

$E(X^*, n) \geqslant v$ $\qquad\qquad\qquad$ $a_{1n} x_1{}^* + a_{2n} x_2{}^* + \cdots + a_{mn} x_m{}^* \geqslant v.$

$$(68.80)$$

This note has been given for the benefit of those readers who may have found some difficulty over the notation used by MacKinsey in (68.68) and (68.69). This symbolism is very elegant, but it is somewhat difficult to handle for people unfamiliar with bilinear forms.

F. THEOREM V

This theorem depends on Theorem III.

Given $E(X, Y)$ as defined in (68.1) and with v the value of the game, then a necessary and sufficient condition for $X^* \in \mathbf{S}_m$ to be an optimal strategy for A is that, for each $1 \leqslant j \leqslant n$:

$$E(X^*, j) \geqslant v; \qquad (68.81)$$

in the same way, if $Y^* \in \mathbf{S}_n$ is to be an optimal strategy for B, the required condition is that, for $1 \leqslant i \leqslant m$:

$$E(i, Y^*) \leqslant v. \qquad (68.82)$$

G. THEOREM VI

Given $E(X, Y)$ defined by (68.1), v the value of the game, and $[X^* Y^*]$ a solution of the game, then

$$\operatorname*{MAX}_{1 \leqslant i \leqslant m} E(i, Y^*) = \operatorname*{MIN}_{1 \leqslant j \leqslant n} E(X^*, j). \qquad (68.83)$$

Proof.   In accordance with Theorem V, for every $1 \leqslant j \leqslant n$:

$$E(X^*, j) \geqslant v; \qquad (68.84)$$

thus

$$\underset{1 \leqslant j \leqslant n}{\text{MIN}}\ E(X^*, j) \geqslant v. \qquad (68.85)$$

But if we had

$$\underset{1 \leqslant j \leqslant n}{\text{MIN}}\ E(X^*, j) > v, \qquad (68.86)$$

then, for each $1 \leqslant j \leqslant n$:

$$E(X^*, j) > v; \qquad (68.87)$$

and thus

$$\sum_{j=1}^{n} E(X^*, j) y_j{}^* > \sum_{j=1}^{n} v y_j{}^*; \qquad (68.88)$$

in other words,

$$E(X^*, Y^*) > v, \qquad (68.89)$$

which contradicts the assumption that v is the value of the game. Hence

$$\underset{1 \leqslant j \leqslant n}{\text{MIN}}\ E(X^*, j) = v. \qquad (68.90)$$

In the same way we can prove that

$$\underset{1 \leqslant i \leqslant m}{\text{MAX}}\ E(i, Y^*) = v. \qquad (68.91)$$

This theorem enables us to state that the optimal solution $[X^* Y^*]$ reduces certain inequalities of (68.79) and (68.80) which are not strict to equations.

H. THEOREM VII

Given $E(X, Y)$ defined by (68.1), v the value of the game, and $X^* \in \mathbf{S}_m$ and $Y^* \in \mathbf{S}_n$ the optimal strategies for A and B, respectively, then

$$\forall i \quad \text{such that} \quad E(i, Y^*) < v, \qquad \text{we have:} \quad x_i{}^* = 0, \qquad (68.92)$$

and

$$\forall j \quad \text{such that} \quad E(X^*, j) > v, \qquad \text{we have:} \quad y_j{}^* = 0. \qquad (68.93)$$

Proof.   Let us assume that for a certain k it is possible that

$$E(k, Y^*) < v \quad \text{and} \quad x_k{}^* \neq 0. \qquad (68.94)$$

We should conclude that

$$E(k, Y^*)x_k^* < vx_k^*. \tag{68.95}$$

Since, for $r = 1, 2,..., k - 1, k + 1,... m$ we have

$$E(r, Y^*) \leqslant v \tag{68.96a}$$

and from there,

$$E(r, Y^*)x_r^* \leqslant vx_r^*, \tag{68.96b}$$

we conclude

$$\sum_{i=1}^{m} E(i, Y^*)x_i^* < \sum_{i=1}^{m} vx_i, \tag{68.97}$$

that is to say,

$$E(X^*, Y^*) < v, \tag{68.98}$$

which contradicts the assumption. We can prove (68.93) in the same way. Hence, with the notation of Section 67, we have

$$a_{1j}x_1^* + a_{2j}x_2^* + \cdots + a_{mj}x_m^* > v \Rightarrow y_j^* = 0, \tag{68.99}$$

and

$$a_{i1}y_1^* + a_{i2}y_2^* + \cdots + a_{in}y_n^* < v \Rightarrow x_i^* = 0. \tag{68.100}$$

It must be noted that the implication is not reciprocal.

Finally, the seven theorems which we have just proved can be summed up in the form of the fundamental theorem enunciated in Section 67 which is thereby itself proved.

69. Calculating the Numerical Solution of a Rectangular Game

There are a number of methods for calculating the numerical value of a rectangular game, and we shall describe the best known of them in the following sections.

A. USE OF THE FUNDAMENTAL THEOREM (Section 67) TO FIND THE OPTIMAL STRATEGIES

The method consists of taking the relations in (67.1) and (67.2) and in solving the systems which they form, in the assumption that certain inequalities which are not strict will become equations, and that the others will become strict inequalities, the rank of the matrix naturally being taken into account. This method can lead to lengthy calculations, since it requires a combinatorial enumeration, but in some cases they can be shortened by suppressing lines or columns which are strictly dominated.

Example 1.   Let us consider the game of Table 69.1, the solution of which will be given by the strategies:

TABLE 69.1

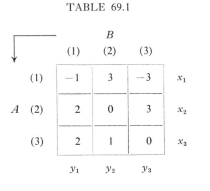

$$[x^*] = [x_1^* \quad x_2^* \quad x_3^*] \tag{69.1}$$

and

$$[y^*] = [y_1^* \quad y_2^* \quad y_3^*], \tag{69.2}$$

which should respectively satisfy the systems:

$$\begin{aligned}
-x_1 + 2x_2 + 2x_3 &\geqslant v, \\
3x_1 + x_3 &\geqslant v, \\
-3x_1 + 3x_2 &\geqslant v, \\
x_1 + x_2 + x_3 &= 1, \\
x_i &\geqslant 0, \qquad i = 1, 2, 3,
\end{aligned} \tag{69.3}$$

and

$$\begin{aligned}
-y_1 + 3y_2 - 3y_3 &\leqslant v, \\
2y_1 + 3y_3 &\leqslant v, \\
2y_1 + y_2 &\leqslant v, \\
y_1 + y_2 + y_3 &= 1, \\
y_j &\geqslant 0, \qquad j = 1, 2, 3.
\end{aligned} \tag{69.4}$$

To begin with, we are going to find whether the system of (69.3), in which all the inequalities become equations, has a nonnegative solution for the x_i terms.

Let

$$-x_1 + 2x_2 + 2x_3 - v = 0,$$
$$3x_1 + x_3 - v = 0,$$
$$-3x_1 + 3x_2 - v = 0,$$
$$x_1 + x_2 + x_3 = 1.$$

(69.5)

The determinant of the coefficients of the left member is not zero, and this system has one, and only one, solution; it is found that the x_i terms are all nonnegative. This solution is

$$[x^*] = [\tfrac{1}{3} \quad \tfrac{2}{3} \quad 0], \qquad v = 1.$$

(69.6)

Let us now seek B's optimal strategies; to do so we must solve the system:

$$-y_1 + 3y_2 - 3y_3 \leqslant 1 \qquad (1)$$
$$2y_1 + 3y_3 \leqslant 1 \qquad (2)$$
$$2y_1 + y_2 \leqslant 1 \qquad (3)$$
$$y_1 + y_2 + y_3 = 1 \qquad (4).$$

(69.7)

The four equations obtained from the three inequalities together with (4) are compatible and have a single solution:

$$[y^*] = [\tfrac{1}{5} \quad \tfrac{3}{5} \quad \tfrac{1}{5}].$$

(69.8a)

But if there is only one value of the game, which is equal to 1, this is not the case for the optimal strategies.

We must find out what occurs when we transform certain of the inequalities into equations; that is to say, which relations become inequalities or equalities. This is a fairly long process, and the details will not be given, for these calculations will be carried out further on by a systematized method.

We find, in fact, that there is another solution:

$$[x'^*] = [\tfrac{1}{3} \quad \tfrac{2}{3} \quad 0], \qquad [y'^*] = [0 \quad \tfrac{2}{3} \quad \tfrac{1}{3}],$$

(69.8b)

and lastly the more general solution where $\lambda_1 + \lambda_2 = 1$ and $\mu_1 + \mu_2 = 1$ is of the form:

$$[X^*] = \lambda_1[x^*] + \lambda_2[x'^*] = [x^*] = [\tfrac{1}{3} \quad \tfrac{2}{3} \quad 0]$$
$$[Y^*] = \mu_1[y^*] + \mu_2[y'^*]$$
$$= \mu_1[\tfrac{1}{5} \quad \tfrac{3}{5} \quad \tfrac{1}{5}] + \mu_2[0 \quad \tfrac{2}{3} \quad \tfrac{1}{3}]$$
$$= [\tfrac{1}{5}\mu_1 \quad \tfrac{2}{3} - \tfrac{1}{15}\mu_1 \quad \tfrac{1}{3} - \tfrac{2}{15}\mu_1],$$

(69.8c)

with $0 \leqslant \mu_1 \leqslant 1$.

Example 2 (Table 69.2).

TABLE 69.2

		B		
	(1)	(2)	(3)	(4)
(1)	6	0	3	8
A (2)	8	−2	3	9
(3)	4	6	5	7

TABLE 69.3

		B		
	(1)	(2)	(3)	
(1)	6	0	3	x_1
A (2)	8	−2	3	x_2
(3)	4	6	5	x_3
	y_1	y_2	y_3	

To begin, we eliminate column (4) which strictly dominates the other columns, thereby obtaining Table 69.3. This game lacks a saddle point and must therefore contain mixed strategies and a value which satisfy the following system:

$$
\begin{aligned}
6x_1 + 8x_2 + 4x_3 &\geqslant v \qquad &(1), \\
-2x_2 + 6x_3 &\geqslant v \qquad &(2), \\
3x_1 + 3x_2 + 5x_3 &\geqslant v \qquad &(3), \\
x_1 + x_2 + x_3 &= 1 \qquad &(4), \\
x_i \geqslant 0, \qquad i &= 1, 2, 3 &
\end{aligned}
\qquad (69.9)
$$

$$
\begin{aligned}
6y_1 \qquad + 3y_3 &\leqslant v \qquad &(1), \\
8y_1 - 2y_2 + 3y_3 &\leqslant v \qquad &(2), \\
4y_1 + 6y_2 + 5y_3 &\leqslant v \qquad &(3), \\
y_1 + y_2 + y_3 &= 1 \qquad &(4), \\
y_j \geqslant 0, \qquad j &= 1, 2, 3. &
\end{aligned}
\qquad (69.10)
$$

Let us first consider system (69.10)[1]:

$$
\begin{aligned}
6y_1 \qquad + 3y_3 - v &= 0, \\
8y_1 - 2y_2 + 3y_3 - v &= 0, \\
4y_1 + 6y_2 + 5y_3 - v &= 0, \\
y_1 + y_2 + y_3 &= 1.
\end{aligned}
\qquad (69.11)
$$

[1] If we had begun with (69.9) we should have immediately found A's mixed strategy and the value of the game, but the present example was intended to show how to overcome various difficulties which may arise.

The determinant of the coefficients of the left member is zero and it is of the third rank; since system (69.11) has no solution, we successively consider whether the nonstrict inequalities of (69.10) can be taken with the following signs for the first three lines:

$$
\begin{array}{cccccc}
= & < & < & < & = & = \\
< & = & < & = & < & = \\
< & < & = & = & = & < \,.
\end{array}
$$

If we assume that the following signs have been checked: $\genfrac{}{}{0pt}{}{(=}{\genfrac{}{}{0pt}{}{<}{<)}}$, it follows that x_2 and x_3 are zero in accordance with (67.4) and (67.5). By making $x_2 = x_3 = 0$ in Eq. (69.9) we can see that the system is not satisfied by these values, and the same is true for $\genfrac{}{}{0pt}{}{(\leq}{\genfrac{}{}{0pt}{}{=}{<)}}$ and $\genfrac{}{}{0pt}{}{(\leq}{\genfrac{}{}{0pt}{}{<}{=)}}$. Let us take $\genfrac{}{}{0pt}{}{(\leq}{\genfrac{}{}{0pt}{}{<}{=)}}$; hence we assume that $x_1 = 0$. It follows that

$$8x_2 + 4x_3 \geqslant v,$$
$$-2x_2 + 6x_3 \geqslant v,$$
$$3x_2 + 5x_3 \geqslant v,$$
$$x_2 + x_3 = 1.$$

(69.12)

It is at once apparent that this system has a single solution which is $x_2 = \frac{1}{6}$, $x_3 = \frac{5}{6}$ and $v = \frac{14}{3}$.

If we now examine the remaining signs to be considered: $\genfrac{}{}{0pt}{}{(=}{\genfrac{}{}{0pt}{}{\geq}{=)}}$ and $\genfrac{}{}{0pt}{}{(=}{\genfrac{}{}{0pt}{}{=}{>)}}$, we can check that they are not suitable. Hence:

$$[x^*] = [0 \quad \tfrac{1}{6} \quad \tfrac{5}{6}] \quad \text{and} \quad v = \tfrac{14}{3}.$$

Knowing that $v = \frac{14}{3}$ we can obtain the mixed strategy for B which must satisfy

$$6y_1 \quad\quad + 3y_3 < \tfrac{14}{3},$$
$$8y_1 - 2y_2 + 3y_3 = \tfrac{14}{3},$$
$$4y_1 + 6y_2 + 5y_3 = \tfrac{14}{3},$$
$$y_1 + y_2 + y_3 = 1.$$

(69.13)

We find from this

$$y_1 = \tfrac{2}{3}, \quad y_2 = \tfrac{1}{3}, \quad y_3 = 0;$$

in other words,

$$[y^*] = [\tfrac{2}{3} \quad \tfrac{1}{3} \quad 0].$$

These examples show that it is necessary to determine whether the system, which contains $n + m + 2$ relations formed by (69.1) and (69.2) and includes $n + m$ inequalities which are not strict, will yield a solution if these inequalities are considered in a certain order; this order cannot be laid down, except in some special cases.

$n + m$	equations,			
$n + m - 1$	equations,	and	1	strict inequality,
$n + m - 2$	equations,	and	2	strict inequalities,
$\vdots$			$\vdots$	
3	equations,	and	$n + m - 3$	strict inequalities,
2	equations,	and	$n + m - 2$	strict inequalities,

hence all the permutations must be tried.

This process of enumeration is no longer suitable if we are dealing with larger matrices, but an attempt has been made to systematize it by matricial calculus.

B. Use of Matricial Calculus to Find the Optimal Strategies; Examples

To save space, we shall give this procedure without accompanying proofs, which will be found in [J16]. This method does not avoid the difficulties of a combinatorial kind referred to above, but owing to its systematized character it can be fairly easily programmed to a computer. Hence, it is to some extent possible to use it to find the solution of games with an appreciably larger number of lines and columns; nevertheless, as we shall discover later on, the best method, in the case of large matrices, is to seek the solution of the linear program which is associated with the game.

Let us introduce the following notation. Let

$$[a] = \begin{bmatrix} a_{11} & a_{12} & \cdots & a_{1n} \\ a_{21} & a_{22} & \cdots & a_{2n} \\ & & \vdots & \\ a_{m1} & a_{m2} & \cdots & a_{mn} \end{bmatrix} \tag{69.14}$$

be the matrix of a given rectangular game; $[\alpha]$ a square submatrix of order r included in $[a]$;

$$[j_r] = [1 \quad 1 \; \cdots \; 1], \text{ matrix } 1 \times r; \quad 2 \leqslant r \leqslant \min(m, n);$$

$$[\alpha]^\Gamma, \quad \text{associated matrix of } [\alpha]; \tag{69.15}$$

$$[M]', \quad \text{transposed matrix of a given matrix } [M]; \tag{69.16}$$

$$[x] = [x_1 \quad x_2 \; \cdots \; x_m], \qquad \text{vector line } 1 \times m; \tag{69.17}$$

$$[x^*] = [x_1^* \quad x_2^* \; \cdots \; x_m^*], \quad \text{optimal strategy for } A; \tag{69.18}$$

$$[y] = [y_1 \quad y_2 \; \cdots \; y_n], \qquad \text{vector line } 1 \times n; \tag{69.19}$$

$$\lfloor y^* \rfloor = [y_1^* \quad y_2^* \; \cdots \; y_n^*], \quad \text{optimal strategy for } B; \tag{69.20}$$

$[\tilde{x}]$ vector $[x]$ in which we have suppressed the elements corresponding to the *lines* removed from $[a]$ to obtain $[\alpha]$; $\quad$ **(69.21)**

$[\tilde{y}]$ vector $[y]$ in which we have suppressed the elements corresponding to the *columns* removed from $[a]$ to obtain $[\alpha]$. $\quad$ **(69.22)**

We are considering a matrix $[a]$ which is without a point of equilibrium (otherwise, of course, the solution could at once be found), and the procedure to be employed is as follows:

(1) Choose a square matrix $[\alpha]$ of the type $r \geqslant 2$ in $[a]$, beginning with $r = \min(m, n)$ where m and n are the number of lines and columns in $[a]$, respectively.

(2) Calculate:

$$[\tilde{x}] = \frac{[j_r][\alpha]^\Gamma}{[j_r][\alpha]^\Gamma[j_r]'} = [x_{r_1} \quad x_{r_2} \; \cdots \; x_{r_r}], \tag{69.23}$$

$$[\tilde{y}] = \frac{[j_r]([\alpha]')^\Gamma}{[j_r]([\alpha]')^\Gamma[j_r]'} = [y_{r_1} \quad y_{r_2} \; \cdots \; y_{r_r}]. \tag{69.24}$$

(3) If some x_{r_i} or y_{r_j} are negative we discard this choice of $[\alpha]$ and choose another $[\alpha]$.

(4) If all $x_{r_i} \geqslant 0$, $i = 1, 2, \ldots, r$, and all $y_{r_j} \geqslant 0$, $j = 1, 2, \ldots, r$, calculate

$$w = \frac{|\alpha|}{[j_r][\alpha]^\Gamma[j_r]'}. \tag{69.25}$$

(5) Construct $[x]$ and $[y]$ corresponding to $[\tilde{x}]$ and $[\tilde{y}]$ by adding zeros in the required places.

(6)   Check whether $[x]$ and $[y]$ satisfy the relations:

$$\sum_{i=1}^{m} a_{ij}x_i \geq w \qquad \text{for every} \quad j = 1, 2,..., n, \qquad (69.26)$$

$$\sum_{j=1}^{n} a_{ij}y_j \leq w \qquad \text{for every} \quad i = 1, 2,..., m. \qquad (69.27)$$

(7)   Then, if these relations are satisfied,

$$ {}^{\cdot}[x] = [x^*], \qquad [y] = [y^*] \qquad (69.28)$$

and $v = w$ is the value of the game.

(8)   Next, look for the other optimal strategies, if there are any, and last, provide the most general ones by means of the special strategies which have been found, and by convex linear weighting.

TABLE 69.4

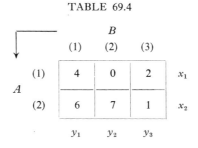

Example 1.   Given the game of Table 69.4, let us arbitrarily select:

$$[\alpha] = \begin{bmatrix} 4 & 0 \\ 6 & 7 \end{bmatrix}. \qquad (69.29)$$

It follows, that

$$[\alpha]^\Gamma = \begin{bmatrix} 7 & 0 \\ -6 & 4 \end{bmatrix}, \qquad (69.30)$$

$$([\alpha]')^\Gamma = \begin{bmatrix} 7 & -6 \\ 0 & 4 \end{bmatrix}, \qquad (69.31)$$

$$[j_r][\alpha]^\Gamma = [1 \quad 1] \begin{bmatrix} 7 & 0 \\ -6 & 4 \end{bmatrix} = [1 \quad 4], \qquad (69.32)$$

$$[j_r][\alpha]^\Gamma[j_r]' = [1 \quad 4] \begin{bmatrix} 1 \\ 1 \end{bmatrix} = 5, \qquad (69.33)$$

$$[j_r]([\alpha]')^\Gamma = [1 \quad 1] \begin{bmatrix} 7 & -6 \\ 0 & 4 \end{bmatrix} = [7 \quad -2], \qquad (69.34)$$

$$[j_r]([\alpha]')^\Gamma[j_r]' = [7 \quad -2]\begin{bmatrix}1\\1\end{bmatrix} = 5, \tag{69.35}$$

$$[\tilde{x}] = \frac{[1 \quad 4]}{5} = [\tfrac{1}{5} \quad \tfrac{4}{5}], \tag{69.36}$$

$$[\tilde{y}] = \frac{[7 \quad -2]}{5} = [\tfrac{7}{5} \quad -\tfrac{2}{5}]. \tag{69.37}$$

These vectors cannot be components of strategies.
Let us now choose:

$$[\alpha] = \begin{bmatrix}0 & 2\\7 & 1\end{bmatrix}. \tag{69.38}$$

$$[\alpha]^\Gamma = \begin{bmatrix}1 & -2\\-7 & 0\end{bmatrix}, \tag{69.39}$$

$$([\alpha]')^\Gamma = \begin{bmatrix}1 & -7\\-2 & 0\end{bmatrix}, \tag{69.40}$$

$$[j_r][\alpha]^\Gamma = [1 \quad 1]\begin{bmatrix}1 & -2\\-7 & 0\end{bmatrix} = [-6 \quad -2], \tag{69.41}$$

$$[j_r][\alpha]^\Gamma[j_r]' = [-6 \quad -2]\begin{bmatrix}1\\1\end{bmatrix} = -8, \tag{69.42}$$

$$[j_r]([\alpha]')^\Gamma = [1 \quad 1]\begin{bmatrix}1 & -7\\-2 & 0\end{bmatrix} = [-1 \quad -7], \tag{69.43}$$

$$[j_r]([\alpha'])^\Gamma[j_r]' = [-1 \quad -7]\begin{bmatrix}1\\1\end{bmatrix} = -8, \tag{69.44}$$

$$[\tilde{x}] = \frac{[-6 \quad -2]}{-8} = [\tfrac{3}{4} \quad \tfrac{1}{4}], \tag{69.45}$$

$$[\tilde{y}] = \frac{[-1 \quad -7]}{-8} = [\tfrac{1}{8} \quad \tfrac{7}{8}]. \tag{69.46}$$

These vectors form components of strategies. Let us proceed, taking:

$$[\alpha] = \begin{bmatrix}4 & 2\\6 & 1\end{bmatrix}; \tag{69.47}$$

we perceive that the only possible vectors are (69.45) and (69.46). Hence, taking into account the position of the submatrix (69.38) in $[a]$, we can state:

$$[x] = [\tfrac{3}{4} \quad \tfrac{1}{4}], \qquad [y] = [0 \quad \tfrac{1}{8} \quad \tfrac{7}{8}]. \tag{69.48}$$

Let us calculate w:

$$w = \frac{\begin{vmatrix}0 & 2\\7 & 1\end{vmatrix}}{-8} = \frac{-14}{-8} = \frac{7}{4}. \tag{69.49}$$

Checking

$$\begin{bmatrix} \frac{3}{4} & \frac{1}{4} \end{bmatrix} \begin{bmatrix} 4 & 0 & 2 \\ 6 & 7 & 1 \end{bmatrix} = \begin{bmatrix} \frac{11}{2} & \frac{7}{4} & \frac{7}{4} \end{bmatrix} \geqslant \begin{bmatrix} \frac{7}{4} & \frac{7}{4} & \frac{7}{4} \end{bmatrix},$$ (69.50)

$$\begin{bmatrix} 4 & 0 & 2 \\ 6 & 7 & 1 \end{bmatrix} \begin{Bmatrix} 0 \\ \frac{1}{8} \\ \frac{7}{8} \end{Bmatrix} = \begin{Bmatrix} \frac{7}{4} \\ \frac{4}{7} \end{Bmatrix}.$$ (69.51)

We have found the solution:

$$[x^*] = \begin{bmatrix} \frac{3}{4} & \frac{1}{4} \end{bmatrix}, \qquad [y^*] = \begin{bmatrix} 0 & \frac{1}{8} & \frac{7}{8} \end{bmatrix}.$$ (69.52)

and $v = w = \frac{7}{4}$, which is the value of the game.

Example 2.   Let us refer to Table 69.1:

$$[a] = \begin{bmatrix} -1 & 3 & -3 \\ 2 & 0 & 3 \\ 2 & 1 & 0 \end{bmatrix}.$$ (69.53)

To begin, let us take $[\alpha] = [a]$.

$$[\alpha]^\Gamma = \begin{bmatrix} -3 & -3 & 9 \\ 6 & 6 & -3 \\ 2 & 7 & -6 \end{bmatrix}, \qquad ([\alpha]')^\Gamma = \begin{bmatrix} -3 & 6 & 2 \\ -3 & 6 & 7 \\ 9 & -3 & -6 \end{bmatrix}.$$ (69.54)

$$[j_r][\alpha]^\Gamma = \begin{bmatrix} 1 & 1 & 1 \end{bmatrix}[\alpha]^\Gamma = \begin{bmatrix} 5 & 10 & 0 \end{bmatrix},$$

$$[j_r][\alpha]^\Gamma[j_r]' = \begin{bmatrix} 5 & 10 & 0 \end{bmatrix}\begin{bmatrix} 1 \\ 1 \\ 1 \end{bmatrix} = 15,$$ (69.55)

$$[j_r]([\alpha]')^\Gamma = \begin{bmatrix} 1 & 1 & 1 \end{bmatrix}([\alpha]')^\Gamma = \begin{bmatrix} 3 & 9 & 3 \end{bmatrix},$$

$$[\tilde{x}] = \frac{\begin{bmatrix} 5 & 10 & 0 \end{bmatrix}}{15} = \begin{bmatrix} \frac{1}{3} & \frac{2}{3} & 0 \end{bmatrix},$$ (69.56)

$$[\tilde{y}] = \frac{\begin{bmatrix} 3 & 9 & 3 \end{bmatrix}}{15} = \begin{bmatrix} \frac{1}{5} & \frac{3}{5} & \frac{1}{5} \end{bmatrix}.$$

Let us calculate the corresponding value of w:

$$w = \frac{\begin{vmatrix} -1 & 3 & -3 \\ 2 & 0 & 3 \\ 2 & 1 & 0 \end{vmatrix}}{\begin{bmatrix} 1 & 1 & 1 \end{bmatrix}\begin{bmatrix} -3 & -3 & 9 \\ 6 & 6 & -3 \\ 2 & 7 & -6 \end{bmatrix}\begin{bmatrix} 1 \\ 1 \\ 1 \end{bmatrix}} = \frac{15}{15} = 1.$$ (69.57)

Let us check the relations (69.26) and (69.27):

$$\begin{bmatrix} \frac{1}{3} & \frac{2}{3} & 0 \end{bmatrix} \begin{bmatrix} -1 & 3 & -3 \\ 2 & 0 & 3 \\ 2 & 1 & 0 \end{bmatrix} = \begin{bmatrix} 1 & 1 & 1 \end{bmatrix},$$

(69.58)

$$\begin{bmatrix} -1 & 3 & -3 \\ 2 & 0 & 3 \\ 2 & 1 & 0 \end{bmatrix} \begin{Bmatrix} \frac{1}{5} \\ \frac{3}{5} \\ \frac{1}{5} \end{Bmatrix} = \begin{Bmatrix} 1 \\ 1 \\ 1 \end{Bmatrix}.$$

Hence

$$[x^*] = \begin{bmatrix} \frac{1}{3} & \frac{2}{3} & 0 \end{bmatrix}, \qquad [y^*] = \begin{bmatrix} \frac{1}{5} & \frac{3}{5} & \frac{1}{5} \end{bmatrix}, \qquad v = 1,$$

constitutes the solution, but is it the only one? We must examine all the minors of (69.53); there are 9 submatrices of type 2 in $[a]$, and we shall take $[\alpha_{ij}]$ for the matrix corresponding to the minor of the element a_{ij} in $[a]$.

$$[\alpha_{11}] = \begin{bmatrix} 0 & 3 \\ 1 & 0 \end{bmatrix}, \qquad [\alpha_{11}]^\Gamma = \begin{bmatrix} 0 & -3 \\ -1 & 0 \end{bmatrix}$$

is provisionally satisfactory and gives

$$[x]_{11} = \begin{bmatrix} 0 & \frac{1}{4} & \frac{3}{4} \end{bmatrix}, \qquad [y]_{11} = \begin{bmatrix} 0 & \frac{3}{4} & \frac{1}{4} \end{bmatrix};$$

$$[\alpha_{12}] = \begin{bmatrix} 2 & 3 \\ 2 & 0 \end{bmatrix}, \qquad [\alpha_{12}]^\Gamma = \begin{bmatrix} 0 & -3 \\ -2 & 2 \end{bmatrix}$$

satisfies it provisionally and gives

$$[x]_{12} = \begin{bmatrix} 0 & \frac{2}{3} & \frac{1}{3} \end{bmatrix}, \qquad [y]_{12} = \begin{bmatrix} 1 & 0 & 0 \end{bmatrix};$$

$$[\alpha_{13}] = \begin{bmatrix} 2 & 0 \\ 2 & 1 \end{bmatrix}, \qquad [\alpha_{13}]^\Gamma = \begin{bmatrix} 1 & 0 \\ -2 & 2 \end{bmatrix} \text{ is not satisfactory;}$$

$$[\alpha_{21}] = \begin{bmatrix} 3 & -3 \\ 1 & 0 \end{bmatrix}, \qquad [\alpha_{21}]^\Gamma = \begin{bmatrix} 0 & 3 \\ -1 & 3 \end{bmatrix} \text{ is not satisfactory;}$$

(69.59)

$$[\alpha_{22}] = \begin{bmatrix} -1 & -3 \\ 2 & 0 \end{bmatrix}, \qquad [\alpha_{22}]^\Gamma = \begin{bmatrix} 0 & 3 \\ -2 & -1 \end{bmatrix} \text{ is not satisfactory;}$$

$$[\alpha_{23}] = \begin{bmatrix} -1 & 3 \\ 2 & 1 \end{bmatrix}, \qquad [\alpha_{23}]^\Gamma = \begin{bmatrix} 1 & -3 \\ -2 & -1 \end{bmatrix}$$

satisfies it provisionally and gives

$$[x]_{23} = \begin{bmatrix} \frac{1}{5} & 0 & \frac{4}{5} \end{bmatrix}, \qquad [y]_{23} = \begin{bmatrix} \frac{2}{5} & \frac{3}{5} & 0 \end{bmatrix};$$

$$[\alpha_{31}] = \begin{bmatrix} 3 & -3 \\ 0 & 3 \end{bmatrix}, \qquad [\alpha_{31}]^{\Gamma} = \begin{bmatrix} 3 & 3 \\ 0 & 3 \end{bmatrix}$$

is provisionally satisfactory and gives

$$[x]_{31} = \begin{bmatrix} \frac{1}{3} & \frac{2}{3} & 0 \end{bmatrix}, \qquad [y]_{31} = \begin{bmatrix} 0 & \frac{2}{3} & \frac{1}{3} \end{bmatrix};$$

$$[\alpha_{32}] = \begin{bmatrix} -1 & -3 \\ 2 & 3 \end{bmatrix}, \qquad [\alpha_{32}]^{\Gamma} = \begin{bmatrix} 3 & 3 \\ -2 & -1 \end{bmatrix} \text{ is not satisfactory;}$$

$$[\alpha_{33}] = \begin{bmatrix} -1 & 3 \\ 2 & 0 \end{bmatrix}, \qquad [\alpha_{33}]^{\Gamma} = \begin{bmatrix} 0 & -3 \\ -2 & -1 \end{bmatrix}$$

is provisionally satisfactory and gives

$$[x]_{33} = \begin{bmatrix} \frac{1}{3} & \frac{2}{3} & 0 \end{bmatrix}, \qquad [y]_{33} = \begin{bmatrix} \frac{1}{2} & \frac{1}{2} & 0 \end{bmatrix}.$$

For the $[\alpha_{ij}]$ terms which are provisionally satisfactory, let us calculate the values of w (which we shall call w_{ij}), for each $w_{ij} \neq 1$; this submatrix will be separated since, as we already know, the value of the game is 1.

$$w_{11} = \frac{\begin{vmatrix} 0 & 3 \\ 1 & 0 \end{vmatrix}}{\begin{bmatrix} 1 & 1 \end{bmatrix}\begin{bmatrix} 0 & -3 \\ -1 & 0 \end{bmatrix}\begin{bmatrix} 1 \\ 1 \end{bmatrix}} = \frac{3}{4} \quad \text{is not satisfactory;}$$

$$w_{12} = \frac{\begin{vmatrix} 2 & 3 \\ 2 & 0 \end{vmatrix}}{\begin{bmatrix} 1 & 1 \end{bmatrix}\begin{bmatrix} 0 & -3 \\ -2 & 2 \end{bmatrix}\begin{bmatrix} 1 \\ 1 \end{bmatrix}} = 2 \quad \text{is not satisfactory;}$$

$$w_{23} = \frac{\begin{vmatrix} -1 & 3 \\ 2 & 1 \end{vmatrix}}{\begin{bmatrix} 1 & 1 \end{bmatrix}\begin{bmatrix} 1 & -3 \\ -2 & -1 \end{bmatrix}\begin{bmatrix} 1 \\ 1 \end{bmatrix}} = \frac{7}{5} \quad \text{is not satisfactory;} \qquad (69.60)$$

$$w_{31} = \frac{\begin{vmatrix} 3 & -3 \\ 0 & 3 \end{vmatrix}}{\begin{bmatrix} 1 & 1 \end{bmatrix}\begin{bmatrix} 3 & 3 \\ 0 & 3 \end{bmatrix}\begin{bmatrix} 1 \\ 1 \end{bmatrix}} = 1 \quad \text{is provisionally satisfactory;}$$

$$w_{33} = \frac{\begin{vmatrix} -1 & 3 \\ 2 & 0 \end{vmatrix}}{\begin{bmatrix} 1 & 1 \end{bmatrix}\begin{bmatrix} 0 & -3 \\ -2 & -1 \end{bmatrix}\begin{bmatrix} 1 \\ 1 \end{bmatrix}} = 1 \quad \text{is provisionally satisfactory.}$$

We must now find the vectors which satisfy the relations (69.26) and (69.27).

Let us begin with the vectors with indices (3,1):

$$\left[\tfrac{1}{3} \ \tfrac{2}{3} \ 0\right] \begin{bmatrix} -1 & 3 & -3 \\ 2 & 0 & 3 \\ 2 & 1 & 0 \end{bmatrix} = [1 \ \ 1 \ \ 1],$$

$$\begin{bmatrix} -1 & 3 & -3 \\ 2 & 0 & 3 \\ 2 & 1 & 0 \end{bmatrix} \begin{Bmatrix} 0 \\ 2 \\ 3 \\ \tfrac{1}{3} \end{Bmatrix} = \begin{Bmatrix} 1 \\ 1 \\ \tfrac{2}{3} \end{Bmatrix} \leqslant \begin{Bmatrix} 1 \\ 1 \\ 1 \end{Bmatrix};$$

hence, these vectors form a new solution.

Let us pass on to the vectors with indices (3,3):

$$\left[\tfrac{1}{3} \ \tfrac{2}{3} \ 0\right] \begin{bmatrix} -1 & 3 & -3 \\ 2 & 0 & 3 \\ 2 & 1 & 0 \end{bmatrix} = [1 \ \ 1 \ \ 1],$$

$$\begin{bmatrix} -1 & 3 & -3 \\ 2 & 0 & 3 \\ 2 & 0 & 3 \end{bmatrix} \begin{Bmatrix} \tfrac{1}{2} \\ \tfrac{1}{2} \\ 0 \end{Bmatrix} = \begin{Bmatrix} 1 \\ 1 \\ \tfrac{3}{2} \end{Bmatrix}; \quad \text{these vectors are not suitable.}$$

In the end, the most generalized solution is

$$[X^*] = \lambda_1\left[\tfrac{1}{3} \ \tfrac{2}{3} \ 0\right] + \lambda_2\left[\tfrac{1}{3} \ \tfrac{2}{3} \ 0\right] = \left[\tfrac{1}{3} \ \tfrac{2}{3} \ 0\right],$$

$$[Y^*] = \mu_1\left[\tfrac{1}{5} \ \tfrac{3}{5} \ \tfrac{1}{5}\right] + \mu_2\left[0 \ \tfrac{2}{3} \ \tfrac{1}{3}\right],$$

(69.61)

with $\lambda_1 + \lambda_2 = 1$ and $\mu_1 + \mu_2 = 1$.

Example 3. Let us take the example of Table (69.2) reduced to that of Table 69.3.

$$[a] = \begin{bmatrix} 6 & 0 & 3 \\ 8 & -2 & 3 \\ 4 & 6 & 5 \end{bmatrix},$$

$$[a]^\Gamma = \begin{bmatrix} -28 & 18 & 6 \\ -28 & 18 & 6 \\ 56 & -36 & -12 \end{bmatrix}$$

(69.62)

It is useless to proceed further with this matrix, since it is single $(\mid a^r \mid = 0)$.

Let us successively consider the 9 submatrices of $[a]$, taking $[\alpha_{ij}]$ for the matrix corresponding to the minor of the element a_{ij} of $[a]$.

$$[\alpha_{11}] = \begin{bmatrix} -2 & 3 \\ 6 & 5 \end{bmatrix}, \qquad [\alpha_{11}]^r = \begin{bmatrix} 5 & -3 \\ -6 & -2 \end{bmatrix} \quad \text{is not satisfactory;} \quad (69.63)$$

$$[\alpha_{12}] = \begin{bmatrix} 8 & 3 \\ 4 & 5 \end{bmatrix}, \qquad [\alpha_{12}]^r = \begin{bmatrix} 5 & -3 \\ -4 & 8 \end{bmatrix} \quad \text{is not satisfactory;} \quad (69.64)$$

$$[\alpha_{13}] = \begin{bmatrix} 8 & -2 \\ 4 & 6 \end{bmatrix}, \qquad [\alpha_{13}]^r = \begin{bmatrix} 6 & 2 \\ -4 & 8 \end{bmatrix} \qquad\qquad (69.65)$$

is provisionally satisfactory and gives

$$[x]_{13} = \begin{bmatrix} 0 & \frac{1}{6} & \frac{5}{6} \end{bmatrix}, \qquad [y]_{13} = \begin{bmatrix} \frac{2}{3} & \frac{1}{3} & 0 \end{bmatrix};$$

$$[\alpha_{21}] = \begin{bmatrix} 0 & 3 \\ 6 & 5 \end{bmatrix}, \qquad [\alpha_{21}]^r = \begin{bmatrix} 5 & -3 \\ -6 & 0 \end{bmatrix} \quad \text{is not satisfactory;} \quad (69.66)$$

$$[\alpha_{22}] = \begin{bmatrix} 6 & 3 \\ 4 & 5 \end{bmatrix}, \qquad [\alpha_{22}]^r = \begin{bmatrix} 5 & -3 \\ -4 & 6 \end{bmatrix} \qquad\qquad (69.67)$$

is provisionally satisfactory and gives

$$[x]_{22} = \begin{bmatrix} \frac{1}{4} & 0 & \frac{3}{4} \end{bmatrix}, \qquad [y]_{22} = \begin{bmatrix} \frac{1}{2} & 0 & \frac{1}{2} \end{bmatrix};$$

$$[\alpha_{23}] = \begin{bmatrix} 6 & 0 \\ 4 & 6 \end{bmatrix}, \qquad [\alpha_{23}]^r = \begin{bmatrix} 6 & 0 \\ -4 & 6 \end{bmatrix} \qquad\qquad (69.68)$$

is provisionally satisfactory and gives

$$[x]_{23} = \begin{bmatrix} \frac{1}{4} & 0 & \frac{3}{4} \end{bmatrix}, \qquad [y]_{23} = \begin{bmatrix} \frac{3}{4} & \frac{1}{4} & 0 \end{bmatrix};$$

$$[\alpha_{31}] = \begin{bmatrix} 0 & 3 \\ -2 & 3 \end{bmatrix}, \qquad [\alpha_{31}]^r = \begin{bmatrix} 3 & -3 \\ 2 & 0 \end{bmatrix} \quad \text{is not satisfactory;} \quad (69.69)$$

$$[\alpha_{32}] = \begin{bmatrix} 6 & 3 \\ 8 & 3 \end{bmatrix}, \qquad [\alpha_{32}]^r = \begin{bmatrix} 3 & -3 \\ -8 & 6 \end{bmatrix} \quad \text{is not satisfactory;} \quad (69.70)$$

$$[\alpha_{33}] = \begin{bmatrix} 6 & 0 \\ 8 & -2 \end{bmatrix}, \qquad [\alpha_{33}]^r = \begin{bmatrix} -2 & 0 \\ -8 & 6 \end{bmatrix} \quad \text{is not satisfactory;} \quad (69.71)$$

For $[\alpha_{13}]$ and $[\alpha_{22}]$, let us calculate w:

$$w_{13} = \frac{\begin{vmatrix} 8 & -2 \\ 4 & 6 \end{vmatrix}}{[1 \quad 1]\begin{bmatrix} 6 & 2 \\ -4 & 8 \end{bmatrix}\begin{bmatrix} 1 \\ 1 \end{bmatrix}} = \frac{56}{12} = \frac{14}{3};$$

$$w_{22} = \frac{\begin{vmatrix} 6 & 3 \\ 4 & 5 \end{vmatrix}}{[1 \quad 1]\begin{bmatrix} 5 & -3 \\ -4 & 6 \end{bmatrix}\begin{bmatrix} 1 \\ 1 \end{bmatrix}} = \frac{18}{4} = \frac{9}{2}.$$

(69.72)

Let us check the relations (69.26) and (69.27):

$$\begin{bmatrix} 0 & \dfrac{1}{6} & \dfrac{5}{6} \end{bmatrix}\begin{bmatrix} 6 & 0 & 3 \\ 8 & -2 & 3 \\ 4 & 6 & 5 \end{bmatrix} = \begin{bmatrix} \dfrac{14}{3} & \dfrac{14}{3} & \dfrac{14}{3} \end{bmatrix},$$

(69.73)

$$\begin{bmatrix} 6 & 0 & 3 \\ 8 & -2 & 3 \\ 4 & 6 & 5 \end{bmatrix}\begin{Bmatrix} \dfrac{2}{3} \\ \dfrac{1}{3} \\ 0 \end{Bmatrix} = \begin{Bmatrix} 4 \\ \dfrac{14}{3} \\ \dfrac{14}{3} \end{Bmatrix} \leqslant \begin{Bmatrix} \dfrac{14}{3} \\ \dfrac{14}{3} \\ \dfrac{14}{3} \end{Bmatrix}.$$

(69.74)

Hence $[\alpha_{13}]$ gives the solution which is

$$[x^*] = \begin{bmatrix} 0 & \dfrac{1}{6} & \dfrac{5}{6} \end{bmatrix}, \quad [y^*] = \begin{bmatrix} \dfrac{2}{3} & \dfrac{1}{3} & 0 \end{bmatrix} \quad \text{and} \quad v = w = \frac{14}{3}. \quad (69.75)$$

Since $v = \frac{14}{3}$, then $[\alpha_{22}]$ cannot be satisfactory, as we shall check:

$$\begin{bmatrix} \dfrac{1}{4} & 0 & \dfrac{3}{4} \end{bmatrix}\begin{bmatrix} 6 & 0 & 3 \\ 8 & -2 & 3 \\ 4 & 6 & 5 \end{bmatrix} = \begin{bmatrix} \dfrac{9}{2} & \dfrac{9}{2} & \dfrac{9}{2} \end{bmatrix}$$

(69.76)

$$\begin{bmatrix} 6 & 0 & 3 \\ 8 & -2 & 3 \\ 4 & 6 & 5 \end{bmatrix}\begin{Bmatrix} \dfrac{1}{2} \\ 0 \\ \dfrac{1}{2} \end{Bmatrix} = \begin{Bmatrix} \dfrac{9}{12} \\ \dfrac{11}{2} \\ \dfrac{9}{2} \end{Bmatrix}$$

(69.77)

is not less than or equal to $\begin{Bmatrix} \dfrac{9}{2} \\ \dfrac{9}{2} \\ \dfrac{9}{2} \end{Bmatrix}.$

We should find equally that $[\alpha_{23}]$ is not suitable.

Remark Concerning this Method. It would be advisable to consider the results for all the minors of the order $2 \leqslant r \leqslant \min(m, n)$, but their number quickly becomes unmanageable, and even with a very powerful computer the calculations would be much too lengthy.

70. Reducing a Rectangular Game to a Linear Program

A. First Method

Let us consider the relations in (67.2).

$$
\begin{array}{lll}
(0) & y_1 + y_2 + \cdots + y_n & = 1, \\
(1) & a_{11} y_1 + a_{12} y_2 + \cdots + a_{1n} y_n \leqslant v \\
(2) & a_{21} y_1 + a_{22} y_2 + \cdots + a_{2n} y_n \leqslant v \\
& \qquad \qquad \vdots \\
(m) & a_{m1} y_1 + a_{m2} y_2 + \cdots + a_{mn} y_n \leqslant v & \quad y_j \geqslant 0, \quad j = 1, 2, \ldots, n.
\end{array}
\tag{70.1}
$$

Dantzig has proved[1] that the search for an optimal policy for player B by means of the above relations is equivalent to finding the solution of a linear program by the following method:

(1) Replace the inequalities which are not strict by equations, adding variables of deviation. It follows that

$$
\begin{array}{lll}
(0) & y_1 + y_2 + \cdots + y_n & = 1 \quad y_j \geqslant 0 \\
(1) & a_{11} y_1 + a_{12} y_2 + \cdots + a_{1n} y_n + z_1 = v & \quad j = 1, 2, \ldots, n \\
(2) & a_{21} y_1 + a_{22} y_2 + \cdots + a_{2n} y_n + z_2 = v & \quad z_i \geqslant 0, \\
& \qquad \qquad \vdots & \quad i = 1, 2, \ldots, m \\
(m) & a_{m1} y_1 + a_{m2} y_2 + \cdots + a_{mn} y_n + z_m = v
\end{array}
\tag{70.2}
$$

(2) Subtract line (1) from lines (2), (3),..., (m). This is an arbitrary choice, and we could equally subtract a line other than (1) from the remainder. It follows that

$$
\begin{array}{l}
(1) \quad v = a_{11} y_1 + a_{12} y_2 + \cdots + a_{1n} y_n + z_1, \\
(0) \quad y_1 + y_2 + \cdots + y_n = 1, \\
(2) \quad (a_{21} - a_{11}) y_1 + (a_{22} - a_{12}) y_2 + \cdots + (a_{2n} - a_{1n}) y_n + z_2 - z_1 = 0, \\
\qquad \qquad \vdots \\
(m) \quad (a_{m1} - a_{11}) y_1 + (a_{m2} - a_{12}) y_2 + \cdots + (a_{mn} - a_{1n}) y_n + z_m - z_1 = 0.
\end{array}
\tag{70.3}
$$

---

[1] See "A proof of the equivalence of the programming problem and the game problem," G. B. Dantzig, *Activity Analysis of Production and Allocation*, Wiley, New York, 1951.

(3)   The linear program:

(1) $[\text{MIN}]\, F = a_{11}y_1 + a_{12}y_2 + \cdots + a_{1n}y_n + z_1 \,,$

(0) $y_1 + y_2 + \cdots + y_n = 1,$

(2) $(a_{21} - a_{11})y_1 + (a_{22} - a_{12})y_2 + \cdots + (a_{2n} - a_{1n})y_n + z_2 - z_1 = 0,$
$$\vdots$$

(m) $(a_{m1} - a_{11})y_1 + (a_{m2} - a_{12})y_2 + \cdots + (a_{mn} - a_{1n})y_n + z_m - z_1 = 0,$

$y_j \geqslant 0, \quad j = 1, 2,\ldots, n; \qquad z_i \geqslant 0, \quad i = 1, 2,\ldots, m, \qquad (70.4)$

gives B's optimal strategy and the value of the game. This fact is clear, since, as (70.1) always has solution, the program (70.4) must always have one. In addition, v is, by definition, the smallest quantity for which (70.1) has a solution, and this definition is also valid for (70.4). Hence, these two linear systems are equivalent.

Similarly, an optimal strategy for A can be obtained by taking (70.1) and forming the linear program:

(1) $[\text{MAX}]\, G = a_{11}x_1 + a_{21}x_2 + \cdots + a_{m1}x_m - u_1 \,,$

(0) $x_1 + x_2 + \cdots + x_m = 1,$

(2) $(a_{12} - a_{11})x_1 + (a_{22} - a_{21})x_2 + \cdots + (a_{m2} - a_{m1})x_m - u_2 + u_1 = 0,$
$$\vdots$$

(n) $(a_{1n} - a_{11})x_1 + (a_{2n} - a_{21})x_2 + \cdots + (a_{mn} - a_{m1})x_m - u_n + u_1 = 0,$

$x_i \geqslant 0, \quad i = 1, 2,\ldots, m; \qquad u_j \geqslant 0, \quad j = 1, 2,\ldots, n. \qquad (70.5)$

Lastly we have

$$[\text{MIN}]\, F = [\text{MAX}]\, G = v \qquad (70.6)$$

Example.

TABLE 70.1

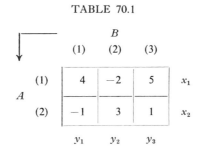

Let us consider the rectangular game of Table 70.1 in which the relations giving an optimal strategy for A are

$$x_1 + x_2 = 1,$$
$$4x_1 - x_2 \geqslant v,$$
$$-2x_1 + 3x_2 \geqslant v,$$
$$5x_1 + x_2 \geqslant v,$$

(70.7)

from which we obtain the linear program:

$$[\text{MAX}]\ G = 4x_1 - x_2 - u_1, \qquad x_1, x_2, u_1, u_2, u_3 \geqslant 0,$$
$$-6x_1 + 4x_2 + u_1 - u_2 = 0,$$
$$x_1 + 2x_2 + u_1 - u_3 = 0,$$
$$x_1 + x_2 = 1.$$

(70.8)

Taking $x_3 = u_1$, $x_4 = u_2$, $x_5 = u_3$, it follows:

$$[\text{MAX}]\ G = 4x_1 - x_2 - x_3,$$
$$-6x_1 + 4x_2 + x_3 - x_4 = 0,$$
$$x_1 + 2x_2 + x_3 - x_5 = 0,$$
$$x_1 + x_2 = 1.$$

(70.9)

The solution of this program obtained by the use of the simplex[1] gives

$$x_1 = \tfrac{2}{5}, \qquad x_2 = \tfrac{3}{5}, \qquad x_3 = 0, \qquad x_4 = 0, \qquad x_5 = \tfrac{8}{5}; \qquad \max G = 1. \quad (70.10)$$

Similarly, for B's strategy we have

$$y_1 + y_2 + y_3 = 1,$$
$$4y_1 - 2y_2 + 5y_3 \leqslant v,$$
$$-y_1 + 3y_2 + y_3 \leqslant v;$$

(70.11)

from which we obtain the linear program:

$$[\text{MIN}]\ F = 4y_1 - 2y_2 + 5y_3 + z_1, \qquad y_1, y_2, z_1, z_2, z_3 \geqslant 0.$$
$$-5y_1 + y_2 - 4y_3 - z_1 + z_2 = 0$$
$$y_1 + y_2 + y_3 = 1$$

(70.12)

---

[1] See A. Kaufmann, *Methods and Models of Operations Research,* Chaps. 2 and 7. Prentice-Hall, Englewood Cliffs, New Jersey, 1963.

Substituting $y_4 = z_1$, $y_5 = z_2$; it follows that

$$[\text{MIN}]\, F = 4y_1 - 2y_2 + 5y_3 + y_4 ,$$
$$-5y_1 + y_2 - 4y_3 - y_4 + y_5 = 0, \qquad (70.13)$$
$$y_1 + y_2 + y_3 = 1.$$

The solution of this program is

$$y_1 = \tfrac{1}{2}, \qquad y_2 = \tfrac{1}{2}, \qquad y_3 = 0, \qquad y_4 = 0, \qquad y_5 = 2;\ \min F = 1. \quad (70.14)$$

B. ANOTHER METHOD

The following method is valid if we know beforehand that the value of the game is positive, a fact which can always be ascertained by adding a number of sufficient magnitude to all the terms, which does not alter the optimal strategies, but merely increases the value of the game.
Let

$$\sum_{i=1}^{m} x_i = 1,$$
$$\sum_{i=1}^{m} a_{ij} x_i \geqslant v, \qquad j = 1, 2,..., n, \qquad (70.15)$$

the relations enable us to calculate an optimal strategy for A. By assumption $v > 0$, and we assume

$$\bar{x}_i = x_i/v, \qquad i = 1, 2,..., m; \qquad (70.16)$$

the relations (70.15) become

$$\sum_{i=1}^{m} \bar{x}_i = 1/v,$$
$$\sum_{i=1}^{m} a_{ij}\bar{x}_i \geqslant 1, \qquad j = 1, 2,..., n. \qquad (70.17)$$

But finding the maximum for v is equivalent to finding the minimum for $1/v$ in (70.17), whereby we obtain the following linear program:

$$[\text{MIN}]\, F = \sum_{i=1}^{m} \bar{x}_i ,$$
$$\sum_{i=1}^{m} a_{ij}\bar{x}_i \geqslant 1, \qquad j = 1, 2,..., n. \qquad (70.18)$$

Using similar notation for B, we obtain

$$[\text{MAX}]\ G = \sum_{j=1}^{n} \bar{y}_j \,,$$

(70.19)

$$\sum_{j=1}^{n} a_{ij}\bar{y}_j \leqslant 1.$$

Hence, we are finally led to the calculation of a linear program or its dual. Our decided preference is for this second method, which does not require artificial variables, and introduces a second positive member.

71. Reducing a Linear Program to a Rectangular Game

A. SYMMETRICAL EXTENSION OF A RECTANGULAR GAME

We apply the term *symmetrical extension* of a rectangular game

$$a_{ij}\,, \qquad i = 1, 2,..., m; \quad j = 1, 2,..., n,$$

to a symmetrical game[1] of $m + n + 1$ lines and columns formed in the order of its main diagonal by a zero matrix $m \times m$, a zero matrix $n \times n$, and a zero matrix 1×1, the other elements being formed as in Tables 71.1 and 71.2. The game of Table 71.2 is the symmetrical extension of the game in 71.1.

The solution of the symmetrical extension is formed by an optimal strategy for A followed by one for B.

Given $\tilde{x}_i$, $i = 1, 2,..., m + n + 1$, an optimal strategy for A (or B) in the symmetrical extension; then

(a)   An optimal strategy for A in the given game has the value:

$$x_1 = \tilde{x}_1 \,, \qquad x_2 = \tilde{x}_2 \,,..., \qquad x_m = \tilde{x}_m \,;$$

(71.1)

TABLE 71.1

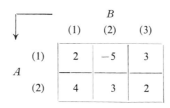

---

[1] In other words, a game on an *antisymmetrical matrix*.

TABLE 71.2

		(1)	(2)	(3)	(4)	(5)	(6)
					B		
	(1)	0	0	2	-5	3	-1
	(2)	0	0	4	3	2	-1
A	(3)	-2	-4	0	0	0	1
	(4)	5	-3	0	0	0	1
	(5)	-3	-2	0	0	0	1
	(6)	1	1	-1	-1	-1	0

(b)   An optimal strategy for B in the given game has the value:

$$y_1 = \tilde{x}_{m+1}, \qquad y_2 = \tilde{x}_{m+2}, ..., \qquad y_n = \tilde{x}_{m+n}; \qquad (71.2)$$

(c)   The value v of the given game is

$$v = \tilde{x}_{m+n+1}. \qquad (71.3)$$

B. REDUCING A LINEAR PROGRAM TO A GAME ON A RECTANGULAR MATRIX

Given a linear program[1]:

$$[\text{MIN}]\, f = c_1 x_1 + c_2 x_2 + \cdots + c_n x_n,$$
$$a_{11}x_1 + a_{12}x_2 + \cdots + a_{1n}x_n \geqslant b_1,$$
$$a_{21}x_1 + a_{22}x_2 + \cdots + a_{2n}x_n \geqslant b_2,$$
$$\vdots$$
$$a_{m1}x_1 + a_{m2}x_2 + \cdots + a_{mn}x_n \geqslant b_m, \qquad (71.4)$$
$$x_i \geqslant 0, \qquad i = 1, 2, ..., n.$$

The optimal solution of this program and that of its dual correspond to the strategy of A (or B) in the rectangular game of Table 71.3, with the obvious reservation that the former must have a solution.

Table 71.3 has an optimal strategy for A formed by

$$\tilde{x}_1, \tilde{x}_2, ..., \tilde{x}_n, \tilde{x}_{n+1}, ..., \tilde{x}_{n+m}, \tilde{x}_{n+m+1},$$

[1] We use the symbols x and y equally for the variables which enter into the mixed strategies of A and/or B as for those of a linear program and its dual.

TABLE 71.3

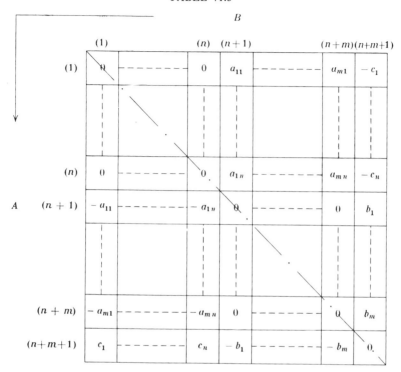

with $x_{n+m+1} \neq 0$, there is a corresponding optimal solution of the linear program (71.4) which is formed by

$$x_i = \frac{\tilde{x}_i}{\tilde{x}_{n+m+1}}, \qquad i = 1, 2, ..., n, \tag{71.5}$$

while the solution of the dual is given by

$$y_j = \frac{\tilde{x}_{n+j}}{\tilde{x}_{n+m+1}}, \qquad j = 1, 2, ..., m. \tag{71.6}$$

Example. Given the linear program:

$$[\text{MIN}]f = 10x_1 + 4x_2, \qquad x_1, x_2 \geqslant 0,$$
$$x_1 \geqslant 4,$$
$$x_2 \geqslant 6, \tag{71.7}$$
$$x_1 + 2x_2 \geqslant 20,$$
$$2x_1 + x_2 \geqslant 17;$$

the corresponding rectangular game is given by Table (71.4). By developing the inequalities in this game, which are not strict, we find:

$$\tilde{x}_1 + \tilde{x}_2 + \tilde{x}_3 + \tilde{x}_4 + \tilde{x}_5 + \tilde{x}_6 + \tilde{x}_7 = 1$$
$$-\tilde{x}_3 - \tilde{x}_5 - 2\tilde{x}_6 + 10\tilde{x}_7 \geqslant 0,$$
$$-\tilde{x}_4 - 2\tilde{x}_5 - \tilde{x}_6 + 4\tilde{x}_7 \geqslant 0,$$
$$\tilde{x}_1 - 4\tilde{x}_7 \geqslant 0,$$
$$\tilde{x}_2 - 6\tilde{x}_7 \geqslant 0,$$ (71.8)
$$\tilde{x}_1 + 2\tilde{x}_2 - 20\tilde{x}_7 \geqslant 0,$$
$$2\tilde{x}_1 + \tilde{x}_2 - 17\tilde{x}_7 \geqslant 0,$$
$$-10\tilde{x}_1 - 4\tilde{x}_2 + 4\tilde{x}_3 + 6\tilde{x}_4 + 20\tilde{x}_5 + 17\tilde{x}_6 \geqslant 0.$$

By changing the variables in (71.5) and (71.6), all the inequalities, except the last one, in (71.8) become

$$y_1 + y_3 + 2y_4 \leqslant 10,$$
$$y_2 + 2y_3 + y \leqslant 4,$$
$$x_1 \geqslant 4,$$
$$x_2 \geqslant 6,$$ (71.9)
$$x_1 + 2x_2 \geqslant 20,$$
$$2x_1 + x_2 \geqslant 17.$$

TABLE 71.4

		(1)	(2)	(3)	(4)	(5)	(6)	(7)
	(1)	0	0	1	0	0	2	−10
	(2)	0	0	0	1	2	1	−4
	(3)	−1	0	0	0	0	0	4
A	(4)	0	−1	0	0	0	0	6
	(5)	−1	−2	0	0	0	0	20
	(6)	−2	−1	0	0	0	0	17
	(7)	10	4	−4	−6	−20	−17	0

(Column header group: B)

The inequalities of the x variables in (71.9) correspond to those of program (71.7) and those of the y variables correspond to the dual. Hence, the solution of the game provides a solution for the problem and its dual. If we now consider the last inequality of (71.8), we have

$$10x_1 + 4x_2 \leqslant 4y_1 + 6y_2 + 20y_3 + 17y_4 , \qquad (71.10)$$

which (as can be proved),[1] must be an inequality in which the two members are equal to the optimum of the program and its dual. This final solution is the optimal one for the program and its dual.

B. ANOTHER METHOD

Let us assume that in the following linear program all the b_i terms $(i = 1, 2,..., m)$ and c_j terms $(j = 1, 2,..., n)$ are positive,[2] a condition which considerably restricts the use of this method, though it leads to a much simpler rectangular game than the previous one. Let

$$[\text{MAX}]\ G = \sum_{j=1}^{n} c_j x_j , \qquad x_j \geqslant 0, \qquad j = 1, 2,..., n,$$

$$\sum_{j=1}^{n} a_{ij} x_j \leqslant b_i , \qquad i = 1, 2,..., m. \qquad (71.11)$$

Let us assume that

$$\bar{y}_j = c_j x_j / G, \qquad j = 1, 2,..., n. \qquad (71.12)$$

The relations (71.11) become

$$\sum_{j=1}^{n} \bar{y}_j = 1, \qquad (71.13)$$

$$\sum_{j=1}^{n} \frac{a_{ij}}{b_i c_j} \bar{y}_j \leqslant 1/G, \qquad i = 1, 2,..., m. \qquad (71.14)$$

Let us also assume

$$A_{ij} = \frac{a_{ij}}{b_i c_j}, \qquad w = 1/G, \qquad (\min w = 1/\max G); \qquad (71.15)$$

---

[1] See footnote reference to Dantzig, p. 441.

[2] This is a sufficient condition if the program is to have a solution, but it is not, fortunately, a necessary one.

it follows that

$$\sum_{j=1}^{n} \bar{y}_j = 1, \qquad\qquad (71.16)$$

$$\sum_{j=1}^{n} A_{ij}\bar{y}_j \leqslant w. \qquad\qquad (71.17)$$

We are led back to the search for B's optimal mixed strategy in the rectangular game

$$\overset{\displaystyle\downarrow^{\!-B}}{A[A_{ij}]}. \qquad\qquad (71.18)$$

Dealing with the dual, let:

$$[\text{MIN}]\, F = \sum_{i=1}^{m} b_i y_i\,, \qquad y_i \geqslant 0, \qquad i = 1, 2,..., m,$$

$$\sum_{i=1}^{m} a_{ij} y_j \geqslant c_j\,, \qquad j = 1, 2,..., n. \qquad (71.19)$$

Let us state

$$\bar{x}_i = b_i y_i / F, \qquad i = 1, 2,..., m. \qquad (71.20)$$

The relations (71.19) become

$$\sum_{i=1}^{m} \bar{x}_i = 1 \qquad\qquad (71.21)$$

$$\sum_{i=1}^{m} \frac{a_{ij}}{b_i c_j}\, \bar{x}_i \geqslant 1/F, \qquad j = 1, 2,..., n. \qquad (71.22)$$

Let us further assume that

$$A_{ij} = \frac{a_{ij}}{b_i c_j}\,, \qquad u = 1/F, \qquad (\max u = 1/\min F). \qquad (71.23)$$

It follows that

$$\sum_{i=1}^{m} \bar{x}_i = 1, \qquad\qquad (71.24)$$

$$\sum_{i=1}^{m} A_{ij} x_i \geqslant u \qquad\qquad (71.25)$$

We are led back to the search for A's mixed optimal strategy in the rectangular game (71.18).

Example. Given the linear program:

$$[\text{MAX}]\ G = 4x_1 + 3x_2,$$
$$x_1 \leqslant 4000,$$
$$x_2 \leqslant 6000, \tag{71.26}$$
$$x_1 + \tfrac{2}{3}x_2 \leqslant 6000.$$

The changes of the variable (71.12) and (71.15) give

$$\bar{y}_1 = 4x_1/G, \qquad \bar{y}_2 = 3x_2/G. \tag{71.27}$$

It follows

$$\bar{y}_1 + \bar{y}_2 = 1,$$
$$\bar{y}_1/16,000 \leqslant w,$$
$$\bar{y}_2/18,000 \leqslant w, \tag{71.28}$$
$$\bar{y}_1/24,000 + \bar{y}_2/27,000 \leqslant w.$$

We are led back to the solution of the game in Table 71.5, and finally obtain

$$\bar{y}_1 = \tfrac{4}{13}, \qquad \bar{y}_2 = \tfrac{9}{13}, \qquad w = v = \tfrac{1}{26,000}, \tag{71.29}$$

whence

$$x_1 = 2000, \qquad x_2 = 6000, \qquad \max G = 26,000. \tag{71.30}$$

TABLE 71.5

		B	
		(1)	(2)
	(1)	$\dfrac{1}{16,000}$	0
A	(2)	0	$\dfrac{1}{18,000}$
	(3)	$\dfrac{1}{24,000}$	$\dfrac{1}{27,000}$

Observation. It is sometimes possible, and also convenient, by making an appropriate change in the variable, to convert a linear program in which some of the b_i and c_j terms are negative, into one in which they are all positive.

72. Evaluation by Multistage Adaptation

Let us explain with the help of an example a nonanalytic method of calculating the value of a game. This method depends on the following principle:

"The moves of each player are based on those previously made by his opponent"; or, expressed differently: "Each player makes a decision in conformity with the history of the game, which is known to him."[1]

Let us suppose that in Table 72.1 A chooses line (1). We copy this line at the foot of the matrix and mark the lowest number in it with an asterisk. Since we shall have to count the number of times each line or column has been chosen, we place an asterisk to the right of line (1), using this sign to assist our addition. On the right of the matrix we now reproduce the third column, which corresponds to the lowest value (-2) found in the first line, and this is assumed to be B's choice. We have now obtained line (5) and column (5) of Table 72.2.

TABLE 72.1

		(1)	(2)	(3)	(4)	
				B		
	(1)	-1	0	-2	1	$*-2$
	(2)	-2	-1	2	1	$2*$
A	(3)	-1	0	1	-2	1
	(4)	1	-1	-1	-1	-1
		-1	0	$-2*$	1	

Knowing B's choice, A will choose line (2) corresponding to the largest element in column (5), and the 2 in the latter is marked with an asterisk. Lines (2) and (5) are now added to form line (6) in which B will choose the smallest number; this is marked by an asterisk.

Next B chooses column (1) which will be added to column (5) to form column (6) in which A will select the highest number (0), which appears three times. As A makes an arbitrary choice of line (2), this is marked, and is then added to line (6) to form line (7); and so on.

---

[1] In the case of a struggle with an intelligent adversary, we are again dealing with *adaptive behavior* of which we gave an outline in Section 17.

TABLE 72.2. Choose the highest number

	B																			
A	(1)	(2)	(3)	(4)	(5)	(6)	(7)	(8)	(9)	(10)	(11)	(12)	(13)	(14)	(15)	(16)	(17)	(18)	(19)	(20)
(1)	−1	0	−2	1	−2*	−3	−4	−5	−5	−5	−5	−5	−5	5	−4*	−4*	−4*	−4*	−6	−7
(2)	−2	−1	2	1	2*	0*	−2	−4	−5	−6	−7	−8	−9	−10	−9	−10	−11	−12	−10	−12
(3)	−1	0	1	−2	1	0	−1	−2	−2	−2	−2	−2*	−2*	−2*	−4	−4	−4	−4	−3*	−4*
(4)	1	−1	−1	−1	−1	0	1*	2*	1*	0*	−1*	−2	−3	−4	−5	−6	−7	−8	−9	−8
(5)	−1	0	−2*	1																
(6)	−3*	−1	0	2																
(7)	−5*	−2	2	3																
(8)	−4*	−3	1	2																
(9)	−3	−4*	0	1																
(10)	−2	−5*	−1	0																
(11)	−1	−6*	−2	−1																
(12)	0	−7*	−3	−2																
(13)	−1	−7*	−2	−4																
(14)	−2	−7	−1	−6																
(15)	−3	−7*	0	−8*																
(16)	−4	−7*	−2	−7																
(17)	−5	−7*	−4	−6																
(18)	−6	−7*	−6	−5																
(19)	−7	−7	−8*	−4																
(20)	−8*	−7	−7	−6																

Choose the lowest number

453

If we stop at the 20th line and 19th column, which is enough to make our method clear, we find:

A has played:	5	times line	(1),
	2		(2),
	4		(3),
	5		(4);
B has played:	4	times column	(1),
	9		(2),
	2		(3),
	1		(4).

After 16 moves by each player the gains and losses are between $-\frac{8}{16}$ and $-\frac{4}{16}$, or -0.50 and -0.25. At this stage the strategies are

for A: $\frac{5}{16}$, $\frac{2}{16}$, $\frac{4}{16}$, $\frac{5}{16}$;

for B: $\frac{4}{16}$, $\frac{9}{16}$, $\frac{2}{16}$, $\frac{1}{16}$.

With a larger number of moves the results become increasingly accurate, and we find after 100 that

$$v = -0.44 \qquad (72.1)$$

Mixed optimal strategy for A: 0.29, 0.11, 0.25, 0.35;   (72.2)

Mixed optimal strategy for B: 0.26, 0.42, 0.16, 0.16.   (72.3)

The absolute solution is very close:

$$v = -\tfrac{4}{9} \qquad (72.4)$$

Mixed optimal strategy for A: $\frac{8}{27}$, $\frac{3}{27}$, $\frac{7}{27}$, $\frac{9}{27}$;   (72.5)

Mixed optimal strategy for B: $\frac{5}{18}$, $\frac{7}{18}$, $\frac{3}{18}$, $\frac{3}{18}$.   (72.6)

This method enables us to calculate the value of the game and the optimal strategies, but it sometimes requires a large number of moves, especially if the number of lines and/or columns exceeds 10. It was introduced by Brown,[1] and the convergence of the procedure by Robinson.[2]

[1] Brown, G. W., "Iterative solution of games by fictitious play," *Activity Analysis of Production and Allocation*, Wiley, New York, 1951.

Brown, G. W., and Koopmans, T. C., "Computational suggestions for maximizing a linear function subject to linear inequalities," *Activity Analysis of Production and Allocation*, Wiley, New York, 1951.

Brown, G. W., and von Neumann, J., Solutions of games by differential equations. Contribution to the theory of games (H. Kuhn and A. J. Tucker, eds.) *Annals of Mathematical Studies*, No. 24, Princeton University Press, New York, 1950.

[2] Robinson, J., An Iterative Method of Solving a Game, *Ann. Math.* 54, 296, 301 (1951).

73. Finding Bayes's Optimal Strategy in a "Game against Nature"

A. USE OF A RULE FOR SCORING

We shall now give the calculations for finding the set (complete class) of Bayes's strategies, to which we referred in Section 27.[1]

The relation of the *a posteriori* probabilities r_n, if we know X_1, X_2,..., X_n, is expressed:

$$r_n = r \cdot Y_1 \cdot Y_2 \cdot \cdots \cdot Y_n. \tag{73.1}$$

The double inequality which leads to a new draw can therefore be expressed:

$$r'/r < Y_1 \cdot Y_2 \cdot \cdots \cdot Y_n < r''/r. \tag{73.2}$$

Taking the logarithm of the two members, and assuming:

$$-b = \ln \frac{r'}{r''}, \qquad a = \ln \frac{r''}{r'} \quad \text{and} \quad Z_i = \ln Y_i, \tag{73.3} \tag{73.4} \tag{73.5}$$

we obtain the equivalence rule:

$$-b < \sum_{i=1}^{n} Z_i < a. \tag{73.6}$$

It is a question of *a rule for scoring*, well known in statistics. We start with a score of 0 and attribute a value z to the result of each draw, which is added algebraically to the aggregate of the quantities already obtained. As long as the score remains within the prescribed limits, we continue to make draws; if not, we make the decision indicated by the bound which we have passed.

In the case of the example given in Section 27, we have initially:

$$\text{if } X = \text{red} \qquad Z = \ln \frac{\frac{1}{3}}{\frac{2}{3}} = -\ln 2,$$
$$\text{if } X = \text{yellow} \quad Z = \ln \frac{\frac{2}{3}}{\frac{1}{3}} = \ln 2; \tag{73.7}$$

so that Bayes's strategies after division by $\ln 2$ take the form:

(a) Leave zero,

(b) Count (-1) for a red ball,

[1] Still following d'Herbemont's text.

(c)   Count $(+1)$ for a yellow ball,

(d)   Decide P_1 if the score becomes less than or equal to $-b/\ln 2$,

(e)   Decide P_2 if the score becomes greater than or equal to $a/\ln 2$,

(f)   Otherwise, continue to make draws.

Given an uncertain variable Z for which the law of probability is $g(z)$, we call:

$\bar{n}(-b, a)$ the expected number of values to be drawn before reaching or exceeding one of the limits;

$\varpi(-b, a)$ the probability of reaching or passing the higher bound (hence $1 - \varpi$ is the similar probability for the lower bound).

The exact relation which determines $\bar{n}$ and ϖ as a function of $g(z)$ and of the bounds is not known, but approximate values can be obtained in the following manner:

We seek the root which was not 0 in the equation in t:

$$\sum_z e^{tz} g(z) = 1, \tag{73.8}$$

and we assume

$$E = \sum_z z g(z). \tag{73.9}$$

We then have, if $-b < 0 < a$,

$$\varpi(-b, a) \approx \frac{1 - e^{-bt}}{1 - e^{-(a+b)t}}, \tag{73.10}$$

$$\bar{n}(-b, a) \approx \frac{a\varpi(-b, a) + b[1 - \varpi(-b, a)]}{E}. \tag{73.11}$$

These formulas are strict when the score can only leave the bounds by passing exactly through them. This is true for Bayes's strategies in this example, since by taking

$$Z = \frac{1}{\ln 2} \ln \frac{\pi_{E_1}(X)}{\pi_{E_2}(X)},$$

the score, which is then a sum of values equal to ± 1, can only be an integer, and the scoring rule is not altered if we take the smallest integer which is equal to or larger than a and b.

If crossing bounds a and b costs W_a and W_b, respectively, with 1 still the cost of the cost of drawing a value Z, the scoring rule counts, as an average:

$$\rho(-b, a) = \bar{n}(-b, a) + \varpi(-b, a)W_a + [1 - \varpi(-b, a)]W_b ; \tag{73.12}$$

this loss depends on $g(z)$.

If Z may follow one of the two laws of probability $g_1(z)$ or $g_2(z)$, depending on the state of nature, we have two average corresponding losses:

$$\rho_1(-b, a) \qquad \text{and} \qquad \rho_2(-b, a).$$

In our example

$$g_1(-1) = \tfrac{2}{3}, \qquad g_1(+1) = \tfrac{1}{3}, \tag{73.13}$$

$$g_2(-1) = \tfrac{1}{3}, \qquad g_2(+1) = \tfrac{2}{3}. \tag{73.14}$$

So that t_1 and t_2 are given by

$$e^{-t_1} g_1(-1) + e^{t_1} g_1(+1) = 1, \tag{73.15}$$

$$e^{-t_2} g_2(-1) + e^{t_2} g_2(+1) = 1, \tag{73.16}$$

which have the respective roots:

$$t_1 = \ln 2 \tag{73.17}$$

and

$$t_2 = -\ln 2. \tag{73.18}$$

Hence we obtain ρ_1 and ρ_2 in an explicit form. We can than obtain r' and r'' by saying that, for $r = r'$ and $r = r''$, the loss is the same for a decision whether or not there is a draw. In fact, the limits of the score which correspond to these two cases are

$$\text{against } r = r'': \quad -b \ln 2 = \ln \frac{r'}{r''} = -b_0, \quad a \ln 2 = \ln \frac{r''}{r''} = 0. \tag{73.19}$$

$$\text{against } r = r': \quad -b \ln 2 = \ln \frac{r'}{r'} = -0, \quad a \ln 2 = \ln \frac{r''}{r'} = b_0. \tag{73.20}$$

In this example we find

$b_0 = 7$ (corrected to the integer above, since the score only includes whole numbers.)

$r' = 0.075$,

$r'' = 4.866$.

Hence, the set of Bayes's strategies (which depend on the parameter r) against r are

if $r \leqslant 0.075$	decide P_1 without a draw,	
if $r \geqslant 4.866$	decide P_2 without a draw,	(73.21)
if $0.075 < r < 4.866$	use the scoring rule enunciated above,	

using the bounds:

$$-\left[\frac{\ln r/0.075}{\ln 2}\right]$$ (73.22)

and

$$\left[\frac{\ln 4.866/r}{\ln 2}\right],$$ (73.23)

where the brackets show that we must take the smallest integer which is equal or greater.

These pure strategies are shown in Fig. 73.1, where the bounds corresponding to the score are given beside each point.

Expressed as a generalization, if there are two states of nature and two final decisions, we take the cost of a draw, which is constant, as the

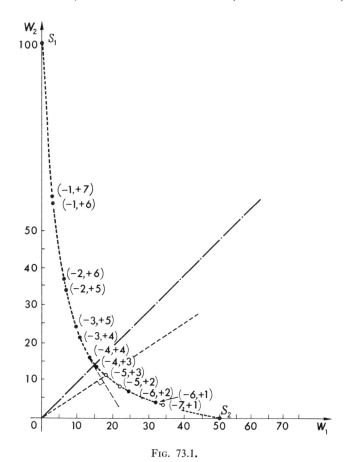

FIG. 73.1.

unit of money. In addition, since r' and r'' are not altered by a translation of the set of points representing strategies, we can always express the loss matrix in the form given in Table 73.1 with $w_{12} > 0$ and $w_{21} > 0$ by adding or removing a suitable number from each line (we obviously assume that one of the strategies does not dominate the other, for the problem would then become trivial).

TABLE 73.1

Nature

	E_1	E_2
(1)	0	$-w_{12}$
(2)	$-w_{21}$	0

The knowledge of the two laws of probability $\pi_{E_1}(X)$ and $\pi_{E_2}(X)$ which correspond to the states E_1 and E_2 of nature enable us to find the laws $g_1(z)$ and $g_2(z)$ of

$$z = \frac{\pi_{E_2}(X)}{\pi_{E_1}(X)}.$$

If we can determine $\rho_1(-b, a)$ and $\rho_2(-b, a)$, then r' and r'' are respective solutions of the two equations:

$$w_{12} = 1 + r' + w_{12} \sum_{z \geqslant 0} g_1(z) + r'w_{21} \sum_{z \leqslant -b_0} g_2(z)$$

$$+ \sum_{-b_0 < z < 0} [\rho_1(-b_0 + z, -z) g_1(z) + r'\rho_2(-b_0 + z, -z) g_2(z)] \quad (73.24)$$

$$w_{21} = 1 + r'' + w_{12} \sum_{z \geqslant b_0} g_1(z) + r''w_{21} \sum_{z \leqslant 0} g_2(z)$$

$$+ \sum_{0 < z < b_0} [\rho_1(-z, b_0 - z) g_1(z) + r''\rho_2(-z, b_0 - z) g_2(z)], \quad (73.25)$$

where

$$b_0 = \ln r'' - \ln r'. \quad (73.26)$$

Conversely, if we obtain r' and r'', we can construct the corresponding table of loss; the two equations are linear in w_{12} and w_{21}.

We can also find r' and r'' by successive approximations. To do so, we take p for the *a priori* probability of E_1 and $\rho(p)$ for the loss of

Bayes's strategy against p. The method consists of finding an increasing sequence of lower bounds $L_n''(p)$ and a decreasing sequence of upper bounds $L_n'(p)$:

$$L_n'' < \rho(p) < L_n' \tag{73.27}$$

which approach $\rho(p)$.

To do so, we use the formulas:

$$L_n(p) = \min\left[\psi(p), 1 + \sum_X \{p\pi_{E_1}(X)\right.$$

$$\left. + (1-p)\pi_{E_2}(X)\}L_{n-1}\left(\frac{p\pi_{E_1}(X)}{p\pi_{E_1}(X) + (1-p)\pi_{E_2}(X)}\right)\right] \tag{73.28}$$

with

$$\psi(p) = \min[pw_{12}, (1-p)w_{21}]. \tag{73.29}$$

L_n' is obtained for the initial value $L_0'(p) = \Psi(p)$; and L_n'' by taking $L_0''(p) = 0$.

Each function $L_n(p)$ has two corresponding values of p: p_1 and $p_2 < p_1$, which are the two solutions of the equation:

$$\psi(p) = 1 + \sum_X [p\pi_{E_1}(X) + (1-p)\pi_{E_2}(X)]L_n\left[\frac{p\pi_{E_1}(X)}{p\pi_{E_1}(X) + (1-p)\pi_{E_2}(X)}\right],$$

$$\tag{73.30}$$

then $(1-p_1)/p_1$ approaches r' and $(1-p_2)/p_2$ approaches r''.

Lastly, Fig. 73.1 enables us to find the optimal strategy. The intersection with the bisector shows that we have a weighting of pure strategies with bounds $(-4, 3)$ and $(-4, 4)$, or 0.586 and 0.413. The perpendicular from 0 to the segment joining the bounds gives the slope of r_{mM} of the minimax strategy for nature:

$$r_{mM} = \frac{0.378}{0.621};$$

hence the value of the game (the minimax loss for the statistician) is 14.64.

In [J4], [J7], and [J21] the reader will find important material about the statistical theory of decisions, which is one of the fundamental developments of the theory of games of strategy.

74. Game with a Continuous Function[1]

A. ANALYTIC ASPECT

(1)   If the choice of one or both the players is no longer made from a discrete set, but from a continuous set, the game is said to be *infinite*. It is natural to consider the choices x and y, $x \in$ **A**, and $y \in$ **B** as a function of payment $M(x, y)$, and for convenience, we often bring the sets **A** and **B** back to the interval $[0, 1]$.

Games of this type have been studied by Ville, who has shown that we may be faced by different eventualities.

(a)   At the same time we have

$$\underset{0 \leqslant x \leqslant 1}{\text{MAX}} \left[\underset{0 \leqslant y \leqslant 1}{\text{MIN}} M(x, y) \right] = v_1 \tag{74.1}$$

and

$$\underset{0 \leqslant y \leqslant 1}{\text{MIN}} \left[\underset{0 \leqslant x \leqslant 1}{\text{MAX}} M(x, y) \right] = v_2 . \tag{74.2}$$

Moreover, if $M(x_0, y_0) = v_1 = v_2$, (x_0, y_0) is a saddle point. The case where $v_1 \neq v_2$ will be considered in (2).

(b) v_1 and v_2 do not exist simultaneously.

(c) Neither v_1 nor v_2 exist:

$$\underset{0 \leqslant x \leqslant 1}{\text{SUP}} \left[\underset{0 \leqslant y \leqslant 1}{\text{INF}} M(x, y) \right] \quad \text{and} \quad \underset{0 \leqslant y \leqslant 1}{\text{INF}} \left[\underset{0 \leqslant x \leqslant 1}{\text{SUP}} M(x, y) \right] \tag{74.3}$$

can exist and eventually be equal. If $M(x, y)$ is a function bounded on the unit square and has a saddle point, the equality is verified.

(2)   If $M(x, y)$ is the function of payment of a game, x and $y \in [0, 1]$, and the first player A chooses in accordance with the distributive function F, and the second player B in accordance with G, then the expected value for A is

$$\int_0^1 M(x, y) \, dF(x)$$

for each y, and his total expected value

$$\int_0^1 \int_0^1 M(x, y) \, dF(x) \, dG(y) = E(F, G).$$

[1] This section and the following have been prepared in collaboration with Robert Faure, Scientific Adviser to the *Compagnie des Machines Bull*.

If M is continuous for $x, y \in [0, 1]$, then

$$\int_0^1 \int_0^1 M(x, y) \, dG(y) \, dF(x) = \int_0^1 \int_0^1 M(x, y) \, dF(x) \, dG(y). \qquad (74.4)$$

In these conditions, since $\mathbf{D}$ is the set of the functions of distribution,

$$v_1 = \operatorname*{MAX}_{F \in \mathbf{D}} [\operatorname*{MIN}_{G \in \mathbf{D}} E(F, G)], \qquad (47.5)$$

$$v_2 = \operatorname*{MIN}_{G \in \mathbf{D}} [\operatorname*{MAX}_{F \in \mathbf{D}} E(F, G)] \qquad (74.6)$$

are equal, and there is a saddle point (F_0, G_0) such that

$$E(F, G_0) \leqslant E(F_0, G_0) \leqslant E(F_0, C). \qquad (74.7)$$

F_0 is an optimal mixed strategy for A, and G_0 is the same for B; (F_0, G_0) is the solution of the game and $v = v_1 = v_2$ its value.

B. VILLE'S THEOREM

If $M(x, y)$ is a continuous function with two variables in the closed unit square, we have

$$\operatorname*{MAX}_{F \in \mathbf{D}} [\operatorname*{MIN}_{G \in \mathbf{D}} \int_0^1 \int_0^1 M(x, y) \, dF(x) \, dG(y)] = \operatorname*{MIN}_{G \in \mathbf{D}} [\operatorname*{MAX}_{F \in \mathbf{D}} \int_0^1 \int_0^1 M(x, y) \, dG(y) \, dF(x)]; \qquad (74.8)$$

consequently, the following expressions are equivalent:

(1) F_0 is an optimal strategy for A and G_0 for B;

(2) $$\int_0^1 \int_0^1 M(x, y) \, dF(x) \, dG_0(y) \leqslant \int_0^1 \int_0^1 M(x, y) \, dF_0(x) \, dG_0(y) \qquad (74.9)$$
$$\leqslant \int_0^1 \int_0^1 M(x, y) \, dF_0(x) \, dG(y).$$

Since the value of the game is v, the distributive function is optimal for A if and only if $\forall y \in [0, 1]$:

$$v \leqslant \int_0^1 M(x, y) \, dF_0(x); \qquad (74.10)$$

correspondingly, G_0 is optimal for B if and only if $\forall x \in [0, 1]$:

$$\int_0^1 M(x, y) \, dG_0(y) \leqslant v. \qquad (74.11)$$

Reciprocally, if it is possible to find a value for v which is real, unique, and which satisfies these two inequalities, then v is the value of the game, and (F_0, G_0) is the solution of the game.

If we state

$$H(x) = \int_0^1 M(x, y)\, dG_0(y), \qquad (74.12)$$

and

$$K(y) = \int_0^1 M(x, y)\, dF_0(x); \qquad (74.13)$$

then

$$v = \underset{x}{\text{MAX}}\, H(x) = \underset{y}{\text{MIN}}\, K(y). \qquad (74.14)$$

These theorems become complicated if F_0 and G_0 are only continuous in parts.

Finally, if

$$M(x, y) = -M(y, x), \qquad (74.15)$$

the game is said to be "symmetrical," any optimal strategy for A is also one for B, and the value of the game is zero (hence it is equitable).

C. EXAMPLE OF A GAME WITH A CONTINUOUS FUNCTION

The search for the solution of such a game is usually very difficult, and it can only be found in some special cases. Let us give an example without showing the calculations; let

$$M(x, y) = 16(x - y)^2. \qquad (74.16)$$

This surface is shown in Fig. 74.1, and is a case of a function which is convex in relation to x and to y, as will be realized by considering the curves corresponding to $x = 0, \frac{1}{4}, \frac{1}{2}, \frac{3}{4}, 1$ or $y = 0, \frac{1}{4}, \frac{1}{2}, \frac{3}{4}, 1$.

A rather lengthy calculation enables us to show that the optimal strategy for A is

$$F_0(x) = \tfrac{1}{2}, \qquad 0 \leqslant x < 1$$
$$= 1, \qquad x = 1 \qquad (74.17)$$

and that of B is

$$G_0(y) = 0, \qquad 0 \leqslant y < \tfrac{1}{2}$$
$$= 1, \qquad \tfrac{1}{2} \leqslant y \leqslant 1. \qquad (74.18)$$

The value of the game is 4.

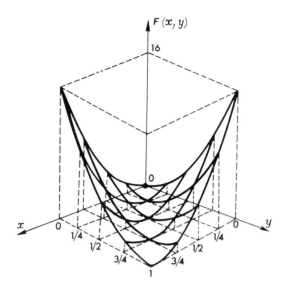

FIG. 74.1.

In Table 74.1 we have shown the game by a rectangular matrix which corresponds to the values for x and y given above.

TABLE 74.1

		B				
		$y = 0$	$y = \frac{1}{4}$	$y = \frac{1}{2}$	$y = \frac{3}{4}$	$y = 1$	
		(1)	(2)	(3)	(4)	(5)	
$x = 0$	(1)	0	1	4	9	16	$x_1 = \frac{1}{2}$
$x = \frac{1}{4}$	(2)	1	0	1	4	9	$x_2 = 0$
A $x = \frac{1}{2}$	(3)	4	1	0	1	4	$x_3 = 0$
$x = \frac{3}{4}$	(4)	9	4	1	0	1	$x_4 = 0$
$x = 1$	(5)	16	9	4	1	0	$x_5 = \frac{1}{2}$

$y_1 = 0 \quad y_2 = 0 \quad y_3 = 1 \quad y_4 = 0 \quad y_5 = 0$

75. Games with n Players and a Zero Sum ($n > 2$)

A. STRATEGY WHEN THERE ARE PARTNERS

In this rectangular game with n players and a zero value, there are n moves. At the ith move, player J_i, who has no information of the previous moves, chooses a number x_i from a set $\mathbf{C}_i$. When the n plays have taken place, he receives the total payment.

$$M_i(x_i, x_2, ..., x_n) \tag{75.1}$$

If the function of payment M_i is *objective and transferable*, the study of such games becomes that of the partnerships or coalitions which the players can form among themselves.

The set of players $\mathbf{N} = \{1, 2, ..., n\}$ can be divided into two complementary subsets $\mathbf{T}$ and $\mathbf{\bar{T}}$, and the situation of these subsets can be examined in relation to each other, this situation being expressed as a choice in the cartesian product

$$\mathbf{C}_{i_1} \times \mathbf{C}_{i_2} \times \cdots \times \mathbf{C}_{i_n}$$

for $\mathbf{T}$ and by a choice in the cartesian product $\mathbf{C}_{j_1} \times \mathbf{C}_{j_2} \times \cdots \times \mathbf{C}_{j_n}$ for $\mathbf{\bar{T}}$.

Let us assume

$$A_j(x_{j_1}, ..., x_{j_r}), \tag{75.2}$$

$$B_k(x_{k_1}, ..., x_{k_s}) \tag{75.3}$$

and

$$M_i(x_i, ..., x_n) = M_i(A_j, B_k). \tag{75.4}$$

The function for the total payment, when the players in subset $\mathbf{T}$ use the strategies A_j, is

$$\sum_{i \in \mathbf{T}} M_i(A_j, B_k) = M_{\mathbf{T}}(A_j, B_k). \tag{75.5}$$

We have

$$M_{\mathbf{\bar{T}}} = -M_{\mathbf{T}}. \tag{75.6}$$

If $\mathbf{C}$ contains u elements, a mixed strategy for $\mathbf{T}$ is an element of $\mathbf{S}_u$ and for $\mathbf{\bar{T}}$ is an element of $\mathbf{S}_v$.

If $\mathbf{T}$ uses a mixed strategy: $\alpha = (\alpha_1, \alpha_2, ..., \alpha_u)$ and $\mathbf{\bar{T}}$ a mixed strategy $\beta = (\beta_1, \beta_2, ..., \beta_v)$, the expected value for $\mathbf{T}$ is

$$\sum_{k=1}^{v} \sum_{j=1}^{u} M_{\mathbf{T}}(A_j, B_k)\alpha_j\beta_k = E_{\mathbf{T}}(\alpha, \beta) \tag{75.7}$$

and that of $\overline{\mathbf{T}}$:

$$\sum_{k=1}^{v}\sum_{j=1}^{u} M_{\overline{\mathbf{T}}}(A_j , B_k)\alpha_j\beta_k = E_{\mathbf{T}}(\alpha, \beta). \tag{75.8}$$

For every j and k:

$$M_{\overline{\mathbf{T}}} = M_{\mathbf{T}} , \tag{75.9}$$

and hence

$$E_{\overline{\mathbf{T}}}(\alpha, \beta) = -E(\alpha, \beta). \tag{75.10}$$

B. CHARACTERISTIC FUNCTION

In accordance with the results obtained for a game with two players:

$$\underset{\alpha\in\mathbf{S}_u}{\text{MAX}}[\underset{\beta\in\mathbf{S}_v}{\text{MIN}}\, E_{\mathbf{T}}(\alpha, \beta)] = \underset{\beta\in\mathbf{S}_v}{\text{MIN}}[\underset{\alpha\in\mathbf{S}_u}{\text{MAX}}\, E_{\mathbf{T}}(\alpha, \beta)] = v(\mathbf{T}) \tag{75.11}$$

and we take this common value $v(\mathbf{T})$, defined for any subset $\mathbf{T}$ of $\mathbf{N}$, as the *characteristic function* of the game which has the following properties:

(1) $v(\mathbf{N}) = 0;$ $\tag{75.12}$

(2) $v(\overline{\mathbf{T}}) = -v(\mathbf{T});$

$\tag{75.13}$

(3)   if $\mathbf{R}$ and $\mathbf{T}$ are disjoint in $\mathbf{N}$:

$$v(\mathbf{R} \cup \mathbf{T}) \geqslant v(\mathbf{R}) + v(\mathbf{T}). \tag{75.14}$$

Conversely, if $\varnothing = \mathbf{N} - \mathbf{N}$ and if $\mathbf{T}_1 , \mathbf{T}_2 ,..., \mathbf{T}_r$ are disjoint subsets of $\mathbf{N}$:

(a) $v(\varnothing) = 0;$ $\tag{75.15}$

(b) $v(\mathbf{T}_1 \cup \mathbf{T}_2 \cup \cdots \cup \mathbf{T}_r) \geqslant v(\mathbf{T}_1) + v(\mathbf{T}_2) + \cdots + v(\mathbf{T}_r),$ $\tag{75.16}$

and, if the reunion of the disjoint $\mathbf{T}_i$ terms covers $\mathbf{N}$:

$$v(\mathbf{T}_1) + \cdots + v(\mathbf{T}_r) \leqslant 0. \tag{75.17}$$

C. STRATEGIC EQUIVALENCE

The concept of *strategic equivalence* is essential for the treatment of games with n players $(n > 2)$. We usually introduce equivalence $\mathbf{S}$ defined as follows:

Given Γ and Γ' two games with n players and a zero sum, where the set of choices and functions of payment are, respectively,

$\mathbf{C}_1 , \mathbf{C}_2 ,..., \mathbf{C}_n$ and $\mathbf{C}_1', \mathbf{C}_2',..., \mathbf{C}_n';$ $M_1 , M_2 ,..., M_n$ and $M_1', M_2',..., M_n'$

we say that Γ and Γ'' have **S**-equivalence if there are functions f_1, $f_2,...,f_n$, real numbers a_1, a_2 ,..., a_n , and a positive real number k such that

(1) $\sum_{i=1}^{n} a_i = 0$;

(2) for $i = 1, 2,..., n$, the f_i terms establish a biunivocal relation (75.18)
between the $\mathbf{C}_i$ and the $\mathbf{C}_i'$ terms;

(3) for each $(x_1 , x_2 ,..., x_n)$ of the cartesian product $\mathbf{C}_1 \times \mathbf{C}_2 \times \cdots \times \mathbf{C}_n$:

$$M_i(x_1 , x_2 ,..., x_n) = kM_i'[f_1(x_1), f_2(x_2),..., f_n(x_n)] + a_i . \quad (75.19)$$

If Γ and Γ'' are two games with n players, a zero sum, and **S**-equivalence with a_i and k as defined above, and v and v' are their respective characteristic functions, we have $\forall \mathbf{T} \subset \mathbf{N}$:

$$v(\mathbf{T}) = kv'(\mathbf{T}) + \sum_{i \in \mathbf{T}} a_i . \quad (75.20)$$

Any game with n players can be divided into equivalence classes in accordance with the **S** relation.

If we put a game with n players, with characteristic function v, in the form:

$$v(\{1\}) = v(\{2\}) = \cdots = v(\{n\}) = \gamma, \quad (75.21)$$

where γ can take the value 0 or the value (-1), it is called *subform reduced modulus* γ. If v is the characteristic function in the reduced modulus form γ and $\mathbf{T}$ is a subset of $\mathbf{N}$, and contains p elements:

$$p\gamma \leqslant v(\mathbf{T}) \leqslant (p - n)\gamma. \quad (75.22)$$

In particular, if $\gamma = 0$, $\forall \mathbf{T} \subset \mathbf{N}$:

$$v(\mathbf{T}) = 0. \quad (75.23)$$

If v and v' are characteristic functions with **S**-equivalence, in the reduced form $\forall \mathbf{T} \subset \mathbf{N}$:

$$v'(\mathbf{T}) = v(\mathbf{T}). \quad (75.24)$$

Every characteristic function has equivalence with one, and only one, characteristic function in the reduced form.

A game is said to be *essential* if $\gamma = -1$ and *inessential* if $\gamma = 0$. A game is *inessential* if and only if its characteristic function v is such that

$$\sum_{i=1}^{n} v(\{i\}) = 0; \tag{75.25}$$

or also if and only if, with **R** and **T** disjoint:

$$v(\mathbf{R} \cup \mathbf{T}) = v(\mathbf{R}) + v(\mathbf{T}). \tag{75.26}$$

D. SOLUTION OF GAMES WITH n PLAYERS

The payments in a partnership with given fees of adherence, where x_i is the total handed over to the ith player, can be represented by a vector:

$$(x_1, x_2, ..., x_n). \tag{75.27}$$

If the game has a zero sum:

$$\sum_{i=1}^{n} x_i = 0, \tag{75.28}$$

and we find

$$x_i \geqslant v(\{x_i\}), \tag{75.29}$$

since, in the contrary case, the solution would be lacking in interest.

A vector such as (75.27) with the conditions (75.28) and (75.29) is called a *charge*.

If there are two such vectors $(y_1, y_2, ..., y_n)$ and $(x_1, x_2, ..., x_n)$ in a game with characteristic function v and $\mathbf{T} \subset \mathbf{N}$, we say that the first vector *dominates* the second, expressing this as

$$(y) \underset{\mathbf{T}}{>} (x),$$

if **T** satisfies:

(1) $\mathbf{T} \neq \varnothing,$ \hfill (75.30)

(2) $v(\mathbf{T}) \geqslant \sum_{i \in \mathbf{T}} y_i,$ \hfill (75.31)

(3) $y_i > x_i, \qquad \forall i \in \mathbf{T}.$ \hfill (75.32)

A set **A** of charges for a game with n players is called *the solution* if:

(1)   There is no *relation of domination* between them when they are taken in pairs;

(2)   Each charge which is not included in **A** is dominated by one found in **A**.

Two games with n players with characteristic functions v and v' are called *isomorphic* when there is a biunivocal correspondence between v and v' so that, if α, β are charges of v, and α', β' of v' such that $\alpha \Leftrightarrow \alpha'$ and $\beta \Leftrightarrow \beta'$, then

$$\alpha \underset{\mathsf{T}}{\gtrless} \beta \qquad \text{in} \quad v, \tag{75.33}$$

if and only if

$$\alpha' \underset{\mathsf{T}}{\gtrless} \beta' \qquad \text{in} \quad v'. \tag{75.34}$$

The isomorphic relation is one of equivalence.

If two games v and v' have **S**-equivalence with the constant k and values a_1, a_2,..., a_n, and if (x) is a deduction of one of them, a biunivocal correspondence:

$$(x_1, x_2,..., x_n) \Leftrightarrow (kx_1 + a_1, kx_2 + a_2,..., kx_n + a_n) \tag{75.35}$$

exists, and the relation $\Leftrightarrow$ establishes an *isomorphism* between v and v'.

Hence, if the set **A** is the solution of v, set **A**$'$ formed from the deductions

$$(kx_1 + a_1, kx_2 + a_2,..., kx_n + a_n),$$

such that

$$(x_1, x_2,..., x_n) \in \mathbf{A},$$

is the solution of v'.

Example. Let us consider a game of *black and white* with three players, such that if two players choose the same color, the third player is the loser for that move. For player 1 we find

$$M_1(w, w, w) = M_1(b, b, b) = 0$$

$$M_1(w, w, b) = M_1(b, b, w)$$

$$= M_1(w, b, w) = M_1(b, w, b) = 1 \tag{75.36}$$

$$M_1(w, b, b) = M_1(b, w, w) = -2.$$

Let us now suppose that there is a partnership $\{2, 3\}$:

	(1)	(2)	(3)	(4)
	ww	wb	bw	bb
$w \cdots$	0	1	1	-2
$b \cdots$	-2	1	1	0

(75.37)

Only columns (1) and (4) are of interest to the partnership, and we obtain

$$v(\{1\}) = -1, \qquad v(\{2, 3\}) = 1, \tag{75.38}$$

and, by analogy:

$$v(\{2\}) = v(\{3\}) = -1, \tag{75.39}$$

$$v(\{1, 3\}) = v(\{1, 2\}) = 1. \tag{75.40}$$

Obviously,

$$v(\varnothing) = v(\{1, 2, 3\}) = 0. \tag{75.41}$$

76. Games with n Players ($n \geqslant 2$) with a Nonzero Sum

A. CONCEPT OF CHARGE

In the majority of business games, the total gains are not equal to the total of the losses. A game of the kind we are discussing can be turned into one with $n + 1$ players with a zero value, in which the fictitious player takes no part in the game (that is to say, he cannot adhere to any partnership or transfer his profit).

We use the term *charge* in a game with n players, a nonzero sum, with characteristic function v, for a vector:

$$(x_1, x_2, ..., x_n) \tag{76.1}$$

such that

$$\sum_i x_i = v(\{1, 2, ..., n\}) \tag{76.2}$$

and

$$x_i \geqslant v(\{i\}), \qquad i = 1, 2, ..., n. \tag{76.3}$$

Similar considerations to those outlined in the previous chapter therefore apply to games with a nonzero sum.

B. EXAMPLE

Given a game with two players A and B with a nonzero sum, defined by the following matrices of gain,

	(1)	(2)
(1)	1	−2
(2)	−1	1

$$(76.4)$$

Gains of A

$$\begin{array}{cc} & (1) \quad (2) \\ (1) & \begin{array}{|c|c|} \hline 1 & 3 \\ \hline 4 & -1 \\ \hline \end{array} \\ (2) & \end{array},$$

(76.5)

Gains of B

we have

$v(\{1\}) = -\frac{1}{5}$,   with the mixed strategy: $x_1 = \frac{2}{5}$,

$x_2 = \frac{3}{5}$;   (76.6)

$v(\{2\}) = \frac{13}{7}$,   with the mixed strategy: $y_1 = \frac{4}{7}$,

$y_2 = \frac{3}{7}$,   (76.7)

where $v(\{1\})$ and $v(\{2\})$ are related to the fictitious games with a zero sum (76.4) and (76.5).

Since:

$$\begin{bmatrix} 1 & -2 \\ -1 & 1 \end{bmatrix} + \begin{bmatrix} 1 & 3 \\ 4 & -1 \end{bmatrix} = \begin{bmatrix} 2 & 1 \\ 3 & 0 \end{bmatrix},$$

(76.8)

the solution (x, y) of the game is such that

$$x + y = 3, \qquad x \geqslant -\tfrac{1}{5}, \qquad y \geqslant \tfrac{13}{7};$$

(76.9)

that is to say,

$$x = -\tfrac{1}{5} + \tfrac{47}{35}\theta,$$ (76.10)

$$y = \tfrac{13}{7} + \tfrac{47}{35}(1 - \theta),$$ (76.11)

with $0 < \theta < 1$, where θ represents an understanding or arbitration between the players.

Conclusion

For the reader who began this book with the wish to obtain an insight into operations research, a number of milestones have now been passed, but he will still need to perfect his knowledge and should not fail to consult those works given in the bibliography which will be of most value in his particular profession. With the understanding he has acquired of the models and methods employed, he will realize that the most delicate part of economic research and analysis is the collection of suitable data, and that unless there is a satisfactory collaboration between the statistician and the economist the data will not lead to a correct analysis. In any event, he now knows the type of reasoning to use and the procedures at his disposal; and in the desire to improve his knowledge will wish to "recycle" (to use a term which is at present popular) as soon as he is free to do so.

In another volume which is now in preparation, new models and methods introduced in recent works will be explained. Thanks to the encouragement which we have received, we believe we can remove the difficulties which confront anyone who is anxious to take a modest but active part in applying theoretic knowledge to practical purposes. It is true that the applied mathematician is often regarded with disfavor, both by the pure mathematician and by the technician, because he is thought insufficiently precise by the former and too "mathematical" by the latter. For ourselves, mathematics are only a means, though a marvelous one, of obtaining a better understanding of nature and of man.

The material in the first and second volumes of this work has been taught or has formed the subject of lectures and conferences in many countries: Belgium, Switzerland, Spain, Portugal, Lebanon, India, Japan, Canada, U.S.A., Mexico, Brazil, Uruguay, and Argentina, from 1959 to 1963. Both students and others in the audiences have helped the author with the fruits of their imagination and experience, and their friendly collaboration proved a sufficient recompense for the many tiring journeys.

By showing the similarity of human problems in different countries, operations research can effectively contribute to a better understanding between nations: more than ever before, mathematics has become a universal language.

This bibliography, which is a continuation of that in Volume I,[1] is arranged under three headings:

IX. Theory of Graphs and Their Applications
X.   Dynamic Programming
XI. Theory of Games of Strategy

The following abbreviations are used:

J.O.R.S.A., Journal of the Operations Research Society of America (U.S.A.).

N.R.L.Q., Naval Research Logistics Quarterly. Office of Naval Research (U.S.A.).

Rand Report, Rand Corporation, Santa-Monica (U.S.A.).

O.R.Q., Operations Research Quarterly (Great Britain).

R.R.O., Revue de Recherche Opérationelle (France).

B.C.R.O., Bolletino del Centro per la Ricerca Operativa (Italy).

C.C.E.R.O., Cahiers du Centre d'Études de Recherche Opérationelle (Belgium).

J.S.I.A.M., Journal of Society for Industrial and Applied Mathematics (U.S.A.).

Manag. Sc., Management Science (U.S.A.).

IX. Theory of Graphs and Their Applications

1. WORKS

H1.   G. Avondo-Bodino, *Economic Applications of the Theory of Graphs*, Gordon and Breach, New York, 1962.

H2.   C. Berge, *Théorie des graphes et ses applications*, Dunod, Paris, 1958.

H3.   C. Berge and A. Ghouila-Houri, *Programmes, jeux et réseaux de transport*, Dunod, Paris, 1962.

H4.   M. Denis-Papin, R. Faure, and A. Kaufmann, *Cours de calcul booléien appliqué*, Albin Michel, Paris, 1963.

H5.   F. Harary and R. Z. Norman, *Graph theory as a mathematical model in social science*, Univ. of Michigan, 1953.

H6.   D. König, *Theorie der Endlichen und Unendlichen Graphen*, Akad. Verl. M.B.H., Leipzig, 1936; or Chelsea, New York, 1950.

H7.   K. Lewin, *Principle of topological psychology*, McGraw-Hill, New York, 1936.

H8.   A. Sainte-Lague, *Avec des nombres et des lignes*, Vuibert, Paris, 1943.

H9.   *Proceedings of Symposia in Applied Mathematics*, Vol. X, "Combinatorial analysis," *Am. Math. Soc.* (1960).

---

[1] English language edition: A. Kaufmann, *Methods and Models of Operations Research*. Prentice-Hall, Englewood Cliffs, New Jersey, 1963.

2. ARTICLES

H10. P. Appel, "Le problème géométrique des déblais et remblais," *Mem. sci. math.* **27** (1928).

H11. R. Bellman, "Combinatorial Processes and Dynamic Programming," *Proc. Symp. Appl. Math.–Combinatorial Analysis*, p. 217, 1960.

H12. C. Berge, "Two theorems in graph theory," *Proc. Natl. Acad. Sci.* **43**, 842 (1957).

H13. C. Berge, Sur le couplage maximum d'un graphe," *C. R. Acad. Sci.* **247**, 258 (1958).

H14. C. Berge, "Sur l'équivalence du problème de transport généralisé et du problème des réseaux," *C. R. Acad. Sci.* **251**, 324 (1960).

H15. C. Berge, "Les problèmes de flot et de tension," *C.C.E.R.O.* **3**, 69 (1961).

H16. R. Bott and J. P. Mayberry, "Matrices and trees," *Economic Activity Analysis*, p. 391. Wiley, New York, 1954.

H17. P. Camion, "Quelques propriétés des chemins et des circuits hamiltoniens dans la théorie des graphes," *C.C.E.R.O.* **2**, 10 (1960).

H18. A. Charnes and W. W. Cooper, "Nonlinear network flows and convex programming over incidence matrices," *N.R.L.Q.* **5**, 231 (1958).

H19. G. B. Dantzig, "On the shortest route through a network," *Manag. Sci.* **6**, 187 (1960).

H20. J. Ergévary, "Matrixok kombinatorius tulajdonsagairol," *Mat. Fiz. Lapok* **16** (1931).

H21. J. Errera, "Du coloriage des cartes," Thesis, Brussels, 1921; *Mathesis* **36**, 56 (1922).

H22. C. Flament, "Nombre de cycles complets dans un réseau de communications," *Bull. centre études et recherches psychotechniques* **3** (1959).

H23. C. Flament, "Analyse des structures préférentielles intransitives," *Proc. of the Second Inter. Conf. of O.R.*, London, p. 150, 1960.

H24. M. M. Flood, "The travelling salesman problem," *J.O.R.S.A.* **4**, 61 (1956).

H25. M. M. Flood, "An alternating proof of a theorem of König as an algorithm for Hitchcock distribution problem-10th on Applied Math.," *Am. Math. Soc.* **1960**.

H26. L. R. Ford and D. R. Fulkerson, "Maximal flow through a network," *Canadian J. Math.* **8**, 399 (1956).

H27. L. R. Ford and D. R. Fulkerson, "Dynamic network flow," *Rand Report*, p. 967, 1956.

H28. L. R. Ford, "Network Flow Theory," *Rand Report*, p. 923, 1956.

H29. L. R. Ford and D. R. Fulkerson, "A primal-dual algorithm for the capacited Hitchcock problem," *Rand Report*, p. 827, 1956.

H29a. D. Gale, "A Theorem on Flows in Network," *Pacific J. Math.* **7**, 1073 (1957).

H30. V. D. Foulkes, "Directed graphes and assembly schedules," *Proc. Symp. Appl. Math.–Combinatorial Analysis*, p. 281, 1960.

H30a. G. de Ghellinck, "Aspects de la notion de dualité en théorie des graphes," *C.C.E.R.O.* **3**, 94 (1961).

H31. A. Ghouila-Houri, "Recherche du flot maximum dans certains réseaux lorsqu'on impose une condition de bouclage," *Proc. Second Intl. Conf. O.R.*, London, p. 156, 1960.

H32. A. Ghouila-Houri, "Sur l'existence d'un flot ou d'une tension prenant ses valeurs sur un groupe abélien," *C. R. Acad. Sci.* **250**, 3931 (1960).

H33. A. Ghouila-Houri, "Une généralisation de l'algorithme de Ford-Fulkerson," *C. R. Acad. Sci.* **250**, 457 (1960).

H34. R. E. Gomory and T. C. Hu, "Multi-terminal networks," *I.B.M. Report R. C.* **318**, 1960.

H35. F. Harary and G. E. Uhlenbeck, "On some generalization of rooted trees," Abstract 131, *Bull. Am. Math. Soc.* **58**, 168 (1952).

H36. R. Kalaba, "On some communication network problems," *Proc. Symp. Appl. Math.–Combinatorial analysis*, p. 261, 1960.

H37. M. Kreweras, "Peut-on former un réseau donné avec des parties finies d'un ensemble dénombrable ?," *C. R. Acad. Sci.* **222**, 1025 (1946).

H38. M. Kreweras, "Extension d'un théorème sur les répartitions en classes," *C. R. Acad. Sci.* **222**, 431 (1946).

H39. J. B. Kruskal, "On the shortest spanning subtree of a graph," *Proc. Am. Math. Soc.* **7**, 48 (1956).

H40. H. J. Leavitt, "Some effect of certain communication patterns on groupe performance," *J. Abnormal Social Psychol.* **46**, 28–50 (1951).

H41. S. MacLane, "A structural characterisation of planar combinatorial graphs," *Duke Math. J.* **3**, 466 (1937).

H42. D. G. Malcolm, J. H. Roseboom, C. E. Clark, and W. Fazar, "Application of a technique for research and development program evaluation," *J.O.R.S.A.* **7**, 646–669 (1959).

H43. Y. Malgrange, "Présentation d'un algorithme. Application à certains problèmes de la théorie des graphes," *Publication BULL. Utilisation des calculateurs électroniques en R.O.*, 1961.

H44. G. Matthys and M. Ricard, "Étude du débit maximal entre deux triages," *R.R.O.*, 2nd ed., No. 15, 1960.

H45. G. Matthys, "Flow optimum dans un réseau à capacités de faisceaux," *Proc. Second Intl. Conf. O.R.*, London, p. 164, 1960.

H46. Monge, "Déblai et remblai," *Mem. Acad. Sci.* (1781).

H47. R. Z. Norman and M. O. Rabin, "An algorithm for a minimum cover of a graph," *Notices Am. Math. Soc.* **15**, 193 (1891).

H48. S. Okada, "Algebraïc and topological foundations of network synthesis," *Proc. Symp. Modern Network Synthesis*, p. 283, 1955.

H49. A. Orden, "The transshipment problem," *Manag. Sci.* **2** (1956).

H50. J. Pétersen, "Die Theorie der regulären Graphs," *Acta Math.* **15**, 193 (1891).

H51. C. de Polignac, "Théorie des ramifications," *Bull. Soc. Math. France* **8**, 120 (1880).

H52. M. Pollack and W. Wiebenson, "Solution of the shortest route problem (a review)," *J.O.R.S.A.* **8**, 224 (1960).

H53. G. Polya, "Sur le nombre des isomères de certains composés chimiques," *C. R. Acad. Sci.* **202**, 1554 (1936).

H54. R. Radner and A. Tritter, "Communication in networks," *Cowles Comm. Paper*, 2098, 1954.

H55. B. Roy, "Contribution de la théorie des graphes à l'étude de certains problèmes linéaires," *C. R. Acad. Sci.* **248**, 2437 (1959).

H56. B. Roy, "Cheminements et connexité dans les graphes, application aux problèmes d'ordonnancement," *Thèse*, Paris, 1962.

H57. W. T. Tutte, "On Hamiltonian circuits," *J. London Math. Soc.* **21**, 99 (1946).

H58. H. Whitney, "Nonseparable and planar graphs," *Trans. Am. Math. Soc.* **34**, 339 (1932).

H59. H. Whitney, "Congruent graphs and the connectivity of graphs," *Am. Math. Soc.* (1932).

H60. H. Whitney, "Planar graphs," *Fund. Math.* **21**, 73 (1933).

X. Dynamic Programming

1. WORKS*

I1.　R. Bellman, *Dynamic Programming.* Princeton Univ. Press, Princeton, New Jersey, 1957.

I2.　R. Bellman, *Adaptive Control Processes, a Guided Tour.* Princeton Univ. Press, Princeton, New Jersey, 1961.

I3.　R. Bellman and S. Dreyfus, *Applied Dynamic Programming.* Princeton Univ. Press, Princeton, New Jersey, 1962.

I4.　A. T. Bharuche-Reid, *Elements of the Theory of Markov Processes and their Applications.* McGraw-Hill, New York, 1960.

I5.　R. Companys, unpublished work.

I6.　R. Howard, *Dynamic Programming and Markov Processes.* Technology Press, MIT, Wiley, New York, 1960.

I7.　Kai Lai Chung, *Markov Chains with Stationary Transition Probabilities.* Springer-Verlag, Berlin, 1960.

I8.　A. Kaufmann and R. Cruon, *La programmation dynamique et ses applications.* Dunod, to appear.

I9.　P. Rosenstiehl and A. Ghouila-Houri, *Les choix économiques. Décisions séquentielles et simulation.* Dunod, Paris, 1960.

I10.　*Proceedings of Symposia in Applied Mathematics*, Vol. X, Combinatorial analysis, *Am. Math. Soc.* (1960).

I11.　J. C. Kemeny and J. L. Snell, *Finite Markov Chains.* Van Nostrand, Princeton, New Jersey, 1960.

2. ARTICLES

I11a.　R. Bellman, "The theory of dynamic programming," *Bull. Am. Math. Soc.* **60**, 503–515 (1954).

I12.　R. Bellman, "Equipment replacement policy," *J.S.I.A.M.* **3**, 133–136 (1955).

I13.　R. Bellman, "Dynamic programming and Lagrange multipliers," *Proc. Natl. Acad. Sci.* **42**, 767–769 (1956).

I14.　R. Bellman, "On the application of the theory of dynamic programming to the study of control processes, *Proc. Symp. Control Processes*, Polytechnic Institute of Brooklyn, New York, p. 199–213 (1956).

I15.　R. Bellman, "Mathematical aspects of scheduling theory," *J.S.I.A.M.* **4**, 168–205 (1956).

I16.　R. Bellman, "On the theory of dynamic programming. A warehousing problem," *Manag. Sci.* **2**, 272–276 (1956).

I17.　R. Bellman, "Dynamic programming and the smoothing problem," *Manag. Sci.* **3**, 111–113 (1956).

I18.　R. Bellman, "On the computational solution of linear programming problems involving almost block diagonal matrices," *Manag. Sci.* **3**, 403–406 (1957).

I19.　R. Bellman, "On the computational solution of dynamic programming processes. On a cargo loading problem," Rand Report R. M. 1746, 1956.

* I4 and I7 deal only with Markovian decision chains and processes. I10 contains several sections on dynamic programming, and there are numerous references to the subject in I1, 2, 3, 8, and 9.

I20.   R. Bellman, "On the computational solution of dynamic programming processes. A smoothing problem, Rand Report R. M. 1749, 1957.

I21.   R. Bellman, "On the computational solution of dynamic programming processes. The flyaway kit problem," Rand Report R. M. 1889, 1957.

I22.   R. Bellman and R. Kalaba, "On the role of dynamic programming in statistical communication theory," *Trans. I.R.E.* **IT-3**, 197–203 (1957).

I23.   R. Bellman, "Dynamic programming and the reliability of multicomponent devices," *J.O.R.S.A.* **6**, 200–206 (1958).

I24.   R. Bellman, "Communication processes involving learning and random duration," *Convention I.R.E. Information Theory*, Part 4, pp. 16–20, 1958.

I25.   R. Bellman, "Dynamic programming and stochastic control processes," *Information and control* **1**, 228–239 (1958).

I26.   R. Bellman, "Some new techniques in the dynamic programming solutions of variational problems," *Quart. Appl. Math.* **16**, 295–305 (1958).

I27.   R. Bellman and Dreyfus, "An application of dynamic programming to the determination of optimal satellite trajectory problems," *J. British Interplanetary Soc.* **17**, 78–83 (1959).

I28.   R. Bellman and S. Dreyfus, "Functional approximation and dynamic programming," *Math. tables* **XIII**, 247–251 (1959).

I29.   R. Bellman and R. Kalaba, "Reduction of dimensionality, Dynamic programming and control processes," *Rand Report*, p. 1694, 1960.

I30.   R. Beckwith, "Analytic and computational aspects of dynamic programming processes of high dimension," *Thèse*, Purdere Univ., 1959.

I31.   L. Bosset, "Résolution de certaines équations en programmation dynamique." *Séminaire d'Analyse numérique de l'A.F.C.A.L.T.I.* (1960).

I32.   H. Cartaino and S. Dreyfus, "Applications of dynamic programming to the minimum time-to-climb problem," *Aeronautical Engr. Rev.* (1957).

I33.   M. Cuenod, "Méthode de calcul à l'aide de suites," *Sciences technique*, P. Feissly, Lausanne, 1955.

I34.   G. B. Dantzig, "On the status of multistage linear programming problems," *Management Science* **6**, No. 1 (1959).

I35.   S. Dreyfus, "A generalized equipment replacement study," *J.S.I.A.M.* **8**, 425–435 (1960).

I36.   F. d'Epenoux, "Sur un problème de production et de stockage dans l'aléatoire," *R.R.O.* No. 14, 1960.

I37.   R. Fortet, "Propriétés des applications de transition des programmations dynamiques," *Revue METRA*, **II**, No. 1, 79–97 (1963).

I38.   S. Johnson and S. Karlin, "A Bayes model in sequential design," *Ann. Math. Statut.* (1956).

I39.   J. D. C. Little, "The use of storage water in a hydroelectric system," *J.O.R.S.A.* (1955).

XI. Theory of Games of Strategy

1. WORKS*

J1.   K. J. Arrow, "Social choice and individual values," *Cowles Comm. Res. Econ.* No. 12. Wiley, New York, 1951.

* The reader should also consult the bibliographies in J4, 7, 11, 15, 16, and 19.

J2.   K. J. Arrow, *Mathematical Models in Social Sciences*. Stanford Univ. Press, Stanford, California, 1951.

J3.   C. Berge, *Théorie générale des jeux à n personnes*. Gauthier-Villars, Paris, 1957.

J4.   D. Blackwell and M. A. Girshick, *Theory of Games and Statistical Decisions*. Wiley, New York, 1954.

J5.   T. Bonnessen and W. Fenchel, *Theorie der convexen Körper*. Springer Verlag, Berlin, 1934; or Chelsea, New York, 1948.

J6.   Burger, *Einführung in die Theorie der Spiels*. De Gruyter, 1960.

J7.   H. Chernoff and L. E. Moses, *Elementary Decision Theory*. Wiley, New York, or Chapman and Hall, London, 1959.

J8.   J. R. Gaenne and R. L. Sisson, *Dynamic Management and Decision Games*. Wiley, New York, 1959.

J9.   G. T. Guilbaud, "Stratégies et décisions économiques," *C.N.R.S.* (1954).

J10.   Herbemont (d'), "Aspects de la théorie statistique des décisions," *Informations Scientifiques, Bull.* (1961).

J11.   S. P. Jacot, "Stratégie et concurrence," *Thèse*, University of Lyon, 1961.

J12.   A. Kaufmann, R. Faure and A. Le Garff, *Les jeux d'entreprises*. Presses Univ. de France (Collection *Que sais-je?*), 1960.

J13.   T. C. Koopmans, *Activity Analysis of Production and Allocations*. Wiley, New York, 1951.

J14.   H. W. Kuhn and A. W. Tucker, "Contribution to the theory of games," *Annals of Mathematics Studies*, Vols. I and II. Princeton Univ. Press, Princeton, New Jersey, 1950 and 1953.

J15.   R. D. Luce and H. Raiffa, *Games and Decisions*. Wiley, New York, 1957.

J16.   J. C. C. MacKinsey, *An Introduction to Theory of Games*. McGraw-Hill, New York, 1952.

J17.   J. von Neumann and O. Morgenstern, *Theory of Games and Economic Behavior*. Princeton Univ. Press, Princeton, New Jersey, 1953.

J18.   M. Shubik, *Stratégie et structure des marchés*. Dunod, Paris, 1963 (*Strategy and Market Structure*. Wiley, New York, 1959).

J19.   R. M. Thrall, C. H. Coombs and R. L. Davis, *Decision Processes*. Wiley, New York; or Chapman and Hall, London, 1954.

J20.   S. Vajda, *Théorie des jeux et programmation linéaire*. Dunod, Paris, 1959 (*The Theory of Games and Linear Programming*. Methuen, London; or Wiley, New York, 1956).

J21.   A. Wald, *Statistical Decision Functions*. Wiley, New York, 1950.

J22.   L. Weiss, *Statistical Decision Theory*. McGraw-Hill, New York, 1961.

J23.   J. D. Williams, *La stratégie dans les actions humaines*. Dunod, Paris, 1956 (*The compleat Strategist*. McGraw-Hill, New York, 1954).

2. ARTICLES

J24.   M. Allais, "Le comportement de l'homme rationnel devant le risque: critique des postulats et axiomes de l'école américaine," *Econometrica* **21**, 503–546 (1953).

J25.   K. J. Arrow, "Alternative approaches to the theory of choice in risk-taking situations," *Econometrica* **19**, 404–437 (1951).

J26.   K. J. Arrow, Hurwicz's optimality criterion for decision-making under ignorance," Stanford Univ., Dept. of Economics and Statistics, Stanford, California, 1951.

J27.   T. Bayes, Facsimiles of two papers by Bayes prepared under the direction of W. E. Deming, The Graduate School Dept. of Agriculture, Washington D.C., Pt. 16, 52 pp.

J28. R. Bellman, "Decision making in the face of incertainty, *N.R.L.Q.* **1**, 230–232 and 327–332 (1954).

J29. D. Blackwell and M. A. Girshick, "Bayes and minimax solutions of sequential decision problems," *Econometrica* **17**, 213–243 (1949).

J30. G. B. Dantzig, "A proof of the equivalence of the programming problem and the game problem," *Dans J.* **13**, 330–338 (1951).

J31. G. B. Dantzig, "Constructive proof of the minimax theorem," *Pacific J. Math.* **6**, 25–33 (1956).

J32. R. Dorfman, "Application of the simplex method to a game theory problem," *Dans. J.* **13**, 348–358 (1951).

J33. D. Gale, "Convex polyhedral cones and linear inequalities," *Dans. J.* **13**, 287–297 (1951).

J34. D. Gale, H. W. Kuhn, and A. W. Tucker, "Reduction of games matrices," *Dans J.* **14**, 89–99 (1950).

J35. D. Gale, H. W. Kuhn, and A. W. Tucker, "Linear programming and the theory of Games," *Dans. J.* **13**, 317–329 (1951).

J36. D. Gale and S. Sherman, "Solutions of finite two-persons games," *Dans J.* **14**, 37–50 (1950).

J37. D. Gale and F. M. Stewart, "Infinite game with perfect information," *Dans J.* **14**, 245–266 (1953).

J38. M. Gerstenhaber, "Theory of convexe polyhedral cones," *Dans J.* **13**, 298–316 (1951).

J39. I. N. Herstein and J. Milnor, "An axiomatic approach to measurable utility," *Econometrica* **21**, 291–297 (1953).

J40. L. Hurwicz, "A criterion for decision-making under incertainty," Cowles Com. Discus. Paper, *Statistics*, No. 355.

J41. S. Karlin, "Continuous games," *Proc. Natl. Acad. Sci. USA* **37**, 220–223 (1951).

J42. H. W. Kuhn, "Lectures on the theory of games. Logistic prof." Office of Naval Res., Princeton Univ. Press, Princeton, New Jersey, 1953.

J43. R. D. Luce, "A definition of stability for *n*-persons game theory," *Math. models of Human behavior*, Dunlap and Assoc., Stanford, California, 1955.

J44. J. C. C. MacKinsey, "Isomorphism of games and strategic equivalence," *Dans J.* **14**, 117–130 (1950).

J45. J. Marschak and R. Radner, "Criteria for planning and incomplete information," Cowles Comm. Discus. Paper, *Economics*, No. 2018, 1951.

J46. J. Milnor, "Games against nature," *Dans J.* **19** (1954).

J47. J. F. Nash, "Equilibrium points in *n*-persons games," *Proc. Natl. Acad. Sci. USA* **36**, 38–49 (1950).

J48. L. J. Savage, "The theory of statistical decision," *J. Am. Stat. Ass.* pp. 238–248, 1947.

J49. M. Shubik, "Information, theories of competition, and the theory of games," *J. Political Econ.* **60**, 145–150 (1952).

J50. H. Weyl, "Elementary proof of a minimax theorem due to von Neumann," *Dans J.* **14**, 19–25 (1950).

J51. H. Weyl, "Elementary proof of a minimax theorem due to von Neumann," *Dans J.* **14**, 19–25 (1950).

J51. H. Weyl, "The elementary theory of convex polyhedra," *Dans J.* **14**, 3–18 (1950).

J52. "Colloques internationaux du Centre national de la Recherche Scientifique (C.N.R.S.)," *La décision, éd. C.N.R.S.*, Paris, 1961.

SUBJECT INDEX

In this index "n" is used after a page number (e.g., 104n) to indicate that the material cited appears in a footnote on that page.